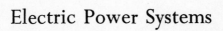

Electric Power Systems

# Electric Power Systems

## Second Edition

### B. M. WEEDY

*Senior Lecturer, Department of Electrical Engineering,
The University of Southampton, England*

**JOHN WILEY & SONS**

**LONDON   NEW YORK   SYDNEY   TORONTO**

Library of Congress Catalog Card No. 71–37109

ISBN 0 471 92445 8

*First edition*     *1967*

*Second printing 1970*

*Third printing   1971*

*Second edition   1972*

*Reprinted*     *1974*

Reprinted in litho by Ceuterick, Printers since 1804
Brusselse straat 153    B 3000 Louvain (Belgium)

# Preface to First Edition

In writing this book the author has been primarily concerned with the presentation of the basic essentials of power system operation and analysis to students in the final year of first degree courses at Universities and Colleges of Technology. The emphasis is on the consideration of the system as a whole rather than on the engineering details of its constituents, and the treatment presented is aimed at practical conditions and situations rather than theoretical nicety.

In recent years the contents of many undergraduate courses in electrical engineering have become more fundamental in nature with greater emphasis on electromagnetism, network analysis and control theory. Students with this background will be familiar with much of the work on network theory and the inductance, capacitance, and resistance of lines and cables, which has in the past occupied large parts of textbooks on power supply. In this book these matters have been largely omitted resulting in what is hoped is a concise account of the operation and analysis of electric power systems. It is the author's intention to present the power system as a system of interconnected elements which may be represented by models either mathematically or by equivalent electrical circuits. The simplest models will be used consistent with acceptable accuracy and it is hoped that this will result in the wood being seen as well as the trees. In an introductory text such as this no apology is made for the absence of sophisticated models of plant (synchronous machines in particular) and involved mathematical treatments as these are well catered for in more advanced texts to which reference is made.

The book is divided into four main parts, as follows:

(a) Introduction, including the establishment of equivalent circuits of the components of the system the performance of which, when interconnected, forms the main theme.

(b) Operation, the manner in which the system is operated and controlled to give secure and economic power supplies.

(c) Analysis, the calculation of voltage, power and reactive power in the system under normal and abnormal conditions. The use of computers is emphasized when dealing with large networks.

(d) Limitations of transmittable power owing to the stability of the synchronous machine, voltage stability of loads and the temperature rises of plant.

It is hoped that the final chapter will form a useful introduction to direct current transmission which promises to play a more and more important role in electricity supply.

The author would like to express his thanks to colleagues and friends for their helpful criticism and advice. To Mr. J. P. Perkins for reading the complete draft, to Mr. B. A. Carre on digital methods for load flow analysis and to Mr. A. M. Parker on direct current transmission. Finally thanks are due to past students who over several years have freely expressed their difficulties in this subject.

Southampton 1967                                          B. M. WEEDY

# Preface to Second Edition

The success of the first edition has encouraged the author to increase the scope of the text to give a more comprehensive coverage of the subject. In two new chapters insulation and protection are discussed. A chapter is now devoted exclusively to underground transmission which incorporates some of the material previously included under thermal limits. The existing chapters have been up-dated and revised to varying degrees and new material added.

Because of the considerable usage of the book in North America attention has been given to the various standards and practices there. It is hoped that in general the new text has a more international flavour.

The question of units is made difficult because of the potential non-European readership and the initial intention to use exclusively S.I. units has been modified. In a text on power systems this is not a major consideration and to try to meet both demands, a few units are given in both metric and existing British/American units.

The author is indebted to many teachers and engineers who have made valuable comments on the first edition and have pointed out errors In particular he would like to thank Dr. S. Sebo of Ohio State University, Columbus, Ohio, U.S.A., who has painstakingly read the new manuscript and made several valuable suggestions, especially on practice in the United States.

No apology is made for the inclusion in an introductory text of such futuristic material as ultra-high voltage lines and cryogenic cables. It is hoped that these topics will convince future readers that although electric power supply has existed for several decades it presents many technological challenges of vast scope and of tremendous importance to the future welfare of mankind.

Southampton 1971                                        B. M. WEEDY

# Contents

# Contents

# Symbols Used

$\mathbf{A, B, C, D}$ = Generalized circuit constants

$a\text{–}b\text{–}c$ = Phase rotation (alternatively $R\text{–}Y\text{–}B$)

$a$ = Operator $1/\underline{120°}$

$C$ = Capacitance (Farad)

$c_p$ = Specific heat (J/gm per °C)

$D$ = Diameter

$\mathbf{E}$ = e.m.f. generated

$F$ = Cost function (units of money per hour)

$f$ = Frequency (Hz)

$G$ = Rating of machine

$g$ = Thermal resistivity (°C cm/W)

$H$ = Inertia constant (seconds)

$h$ = Heat transfer coefficient (Watts/cm² per deg C)

$\mathbf{I}$ = Current (A)

$I_d$ = In phase current

$I_q$ = Quadrature current

$j$ = $1/\underline{90°}$ operator

$K$ = Stiffness coefficient of a system (MW/Hz)

$k$ = Thermal conductivity (W/(cm °C))

$L$ = Inductance (H)

ln = Natural logarithm

$M$ = Angular momentum (Joule-sec per rad or MJ-s per electrical degree)

$N$ = Rotational speed (revs/min, revs/sec, rads/sec)

$\mathbf{P}$ = Propagation constant ($\alpha + j\,\beta$)

$P$ = Power (W)

$\dfrac{\mathrm{d}p}{\mathrm{d}\delta}$ = Synchronizing power coefficient

p.f. = Power factor

$p$ = Iteration number

$Q$ = Reactive power (VAr)

$q$ = Loss dissipated as heat (Watts)

$R$ = Resistance (ohms) also thermal resistance (°C/W)

$R\text{–}Y\text{–}B$ = Phase rotation (British practice)

$\mathbf{S}$ = Complex power = $\mathbf{P} \pm j\,Q$

$s$ = Slip

S.C.R. = Short circuit ratio

$T$ = Absolute temperature (K)

$t$ = Time

$\Delta t$ = Interval of time

$t$ = Off-nominal transformer tap ratio

$U$ = Velocity

$V$ = Voltage; $\Delta V$ scalar voltage difference

$W$ = Volumetric flow of coolant ($m^3/s$)

$X'$ = Transient reactance of a synchronous machine

$X''$ = Subtransient reactance of a synchronous machine

$X_d$ = Direct axis synchronous reactance of a synchronous machine

$X_q$ = Quadrature axis reactance of a synchronous machine

$X_s$ = Synchronous reactance of a synchronous machine

$\mathbf{Y}$ = Admittance (p.u. or ohms)

$\mathbf{Z}$ = Impedance (p.u. or ohms)

$\mathbf{Z}_0$ = Characteristics or surge impedance (ohms)

$\alpha$ = Delay angle in rectifiers and inverters—d.c. transmission

$\alpha$ = Attenuation constant of line

$\alpha$ = Reflexion coefficient

$\beta$ = Phase-shift constant of line

$\beta$ = $(180 - \alpha)$—in inverters

$\beta$ = Refraction coefficient $(1 + \alpha)$

$\gamma$ = Commutation time in converters

$\delta$ = Load angle of synchronous machine or transmission angle across a system—electrical degrees

$\delta_0$ = Ionization angle of valve

$\varepsilon$ = Permittivity

$\eta$ = Viscosity (gm/(cm − sec))

$\theta$ = Temperature rise (deg C) above reference or ambient

$\lambda$ = Lagrange multiplier

$\rho$ = Electrical resistivity ($\Omega - m$)

$\rho$ = Density (gm/$cm^3$)

$\tau$ = Time constant

$\phi$ = Angle between voltage and current phasors (power factor angle)

$\omega$ = Angular frequency (pulsatance) rads/s

Subscripts 1, 2, and 0 = Refer to positive, negative and zero symmetrical components respectively

# I

# Introduction

## 1.1 Historical

Electricity supply on a commercial basis initially consisted of direct current generators feeding specialized loads such as street lighting, galleries and theatres. In January 1882 the Holborn Viaduct Generating Station in London commenced operation and this was one of the first power stations in the world that catered for private consumers generally, as opposed to the specialized loads. This scheme was comprised of a 60 kW generator driven by a horizontal steam engine; the voltage of generation was 110 V direct current. In the same year a similar station commenced operation in New York.

The first major alternating current station in Great Britain was at Deptford where power was generated by machines of 10,000 h.p. and transmitted at 10 kV to consumers in London. During this period the battle between the advocates of alternating current and direct current was at its most intense and acrimonious level. During this same period similar developments were taking place in the U.S.A. and elsewhere. Owing mainly to the invention of the transformer the advocates of alternating current prevailed and a steady development of local electricity generating stations commenced, each large town or load centre operating its own station.

In 1926 in Britain an Act of Parliament set up the Central Electricity Board with the object of interconnecting the best of the 500 stations then in operation with a high voltage network known as the *grid*. In 1948 the British supply industry was nationalized and two organizations set up: (a) the *area boards* which are mainly concerned with distribution and consumer service, and (b) the *generating boards* responsible for generation and the operation of the high voltage transmission network or grid. In most other countries except the U.S.A. there is also a large degree of central government control of the main power-supply system.

1

## 1.2 Characteristics Influencing Generation and Transmission

There are three main characteristics of electricity supply which, however obvious, have a profound effect on the manner in which it is engineered. They are as follows:

(a) Electricity, unlike gas and water, cannot be stored and the supplier has small control over the load at any time. The control engineers endeavour to keep the output from the generators equal to the connected load at the specified voltage and frequency; the difficulty of this task will be apparent from a study of the daily load curve in figure 1.1. It will be

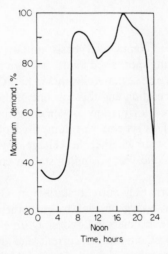

Figure 1.1   Typical winter daily load curve.

seen that the load consists basically of a steady component known as the base load plus peaks depending on the time of day, popular television programmes and other factors. The effect of an unusual television programme is shown in figure 1.2.

(b) There is a continuous increase in the demand for power which is roughly equivalent to a doubling in demand every ten years. This applies to most countries although in some under-developed countries it is even higher. A large and continuous process of adding to the system thus exists. Networks are evolved over the years rather than planned in a clear-cut manner and then left untouched.

(c) The distribution and nature of the *fuel* available. This aspect is of great interest as coal is mined in areas not necessarily the main load centres; hydro-electric power is usually remote from the large load centres. These two are the conventional sources of power in most countries and the problem of station siting and transmission distances is an involved exercise in economics. The greater use of oil and nuclear energy will tend to modify the existing pattern of supply.

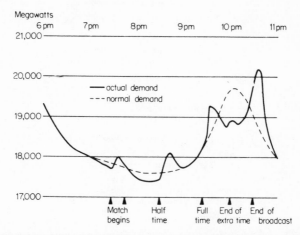

Figure 1.2 Effect of television programmes on demand—European Cup soccer match televised in Britain on May 29, 1968. Peaks caused by connexion of kettles, etc., in intervals and at end. (*Permission of Central Electricity Generating Board.*)

A further influence of a different nature which is becoming prominent is that of *amenity*. There are considerable pressures against the proliferation of overhead lines which occurs as the demand for electricity steadily increases. Although the raising of supply voltages increases substantially the power carried per overhead-line, the number of lines is rapidly increasing and it would seem likely that limited parts of the network will be undergrounded in spite of the much larger cost. These underground cables will be in addition to those normally used in built-up areas. For relatively small distances it is feasible to use high voltage alternating current with underground cables but for longer routes direct current at high voltages will have to be used for reasons which will be discussed later.

## 1.3 Representation of Power Systems

Modern electricity supply systems are invariably three-phase. The design of distribution networks is such that normal operation is reasonably close to balanced three-phase working, and often a study of the electrical conditions in one phase is sufficient to give a complete analysis. Equal loading on all three phases of a network is ensured by allotting as far as possible equal domestic loads to each phase of the low voltage distribution feeders; industrial loads usually take three-phase supplies.

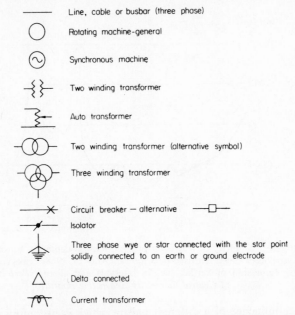

Figure 1.3    Symbols for representing the components of a three-phase
power system.

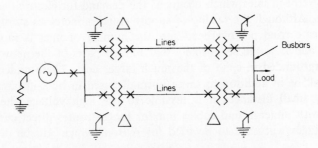

Figure 1.4    Line diagram of simple system.

A very useful and simple way of graphically representing a network is the schematic or line diagram in which three-phase circuits are represented by single lines. Certain conventions for representing items of plant are used and are shown in figure 1.3. A typical line or schematic diagram of a part of a power system is shown in figure 1.4. In this, the generator is star connected, with the star point connected to earth through a resistance. The nature of the connexion of the star point of rotating machines and transformers to earth is of vital importance when considering faults which produce electrical unbalance in the three phases. The generator feeds two three-phase lines (overhead or underground). The line voltage is increased from that at the generator terminals by transformers connected as shown. At the end of the lines the voltage is reduced for the secondary distribution of power. Two lines are provided to improve the *security* of the supply, i.e. if one line develops a fault and has to be switched out the remaining one still delivers power to the receiving end. It is not necessary in straightforward current and voltage calculations to indicate the presence of switches, etc., on the diagrams, but in some cases such as stability calculations the location of switches, current transformers and protection is very useful.

Although the use of jargon is avoided as much as possible certain terms will be constantly recurring which may be rather vague to the newcomer to the subject. A short list follows of such terms with explanations.

*Systems* This is used to describe the complete electrical network, generators, loads and prime movers.

*Load* This may be used in a number of ways; to indicate a device or collection of devices which consume electricity; to indicate the power required from a given supply circuit; the power or current being passed through a line or machine.

*Busbar* An electrical connexion of zero impedance joining several items such as lines, loads, etc. Often this takes the form of actual busbars of copper or aluminium.

*Earthing* (*Grounding*) The connexion of a conductor or frame of a device to the main body of the earth. This must be done in such a manner that the resistance between the item and the earth is below prescribed limits. This often entails the burying of large assemblies of conducting rods in the earth and the use of connectors of large cross-sectional area.

*Fault* This is a malfunctioning of the network usually due to the short-circuiting of two conductors or live conductors connecting to earth.

*Security of supply* Provision must be made to ensure continuity of supply to consumers even with certain items of plant out of action. Usually two circuits in parallel are used and a system is said to be secure when continuity is assured. This is obviously the item of first priority in design and operation.

## 1.4   Energy Conversion Employing Steam

The combustion of coal or oil in boilers produces steam at high temperatures and pressures which is passed to steam turbines. Oil has economic
advantages when it can be pumped from the refinery through pipelines
direct to the boilers of the generating station. The use of energy resulting
from nuclear fission is being progressively extended in electricity generation;
here also the basic energy is used to produce steam for turbines. The
axial flow type of turbine is in common use with several cylinders on the
same shaft.

The steam power-station operates on the Rankine cycle modified to
include superheating, feed-water heating and steam reheating. Increased
thermal efficiency results from the use of steam at the highest possible
pressure and temperature. Also for turbines to be economically constructed
the larger the size the less the capital cost. As a result turbogenerator sets
of 500 MW and over are now being used. With steam turbines of 100 MW
capacity and over the efficiency is increased by reheating the steam after
it has been partially expanded, by an external heater. The reheated steam
is then returned to the turbine where it is expanded through the final

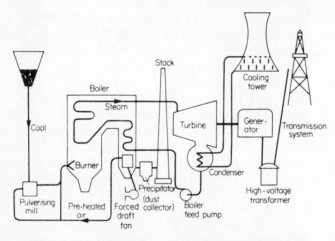

Figure 1.5   Schematic view of coal-fired generating station.

stages of blading. A schematic diagram of a coal-fired station is shown in
figure 1.5. In figure 1.6 the flow of energy in a modern steam station is
shown. Despite continual advances in the design of boilers and in the
development of improved materials the nature of the steam cycle is such
that efficiencies are comparatively low and vast quantities of heat are lost
in the condensate. However the great advances in design and materials in

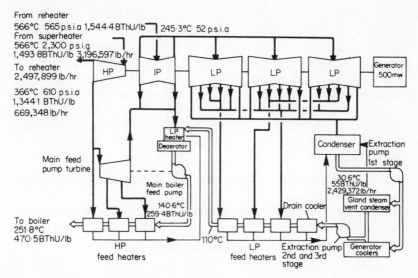

From reheater
566°C 565 p.s.i.a 1,544·4 BThU/lb  245·3°C 52 p.s.i.a
From superheater
566°C 2,300 p.s.i.g
1,493·8BThU/lb 3,196,597 lb/hr
To reheater
2,497,899 lb/hr
366°C 610 p.s.i.a
1,344·1 BThU/lb
669,348 lb/hr

HP    IP    LP    LP    LP    Generator 500mw

Main feed pump turbine
LP heater
Deaerator

Main boiler feed pump
140·6°C
259·4BThU/lb

Condenser  Extraction pump 1st stage

30·6°C
55BThU/lb
2,429,372lb/hr
Gland steam vent condenser

To boiler
251·8°C
470·5BThU/lb

HP feed heaters

110°C

Drain cooler

LP feed heaters

Extraction pump 2nd and 3rd stage

Generator coolers

Figure 1.6   Energy flow diagram for a 500 MW turbo-generator.
(*Permission of Electrical Review.*)

the last few years have increased the thermal efficiencies of steam stations to over 40 per cent. These low steam cycle efficiencies have resulted in much research effort into other means of producing electricity notably by fuel cells and magnetohydrodynamical methods.

### 1.5   Energy Conversion Using Water

Perhaps the oldest form of energy conversion is by the use of water power. In the hydro-electric station the energy is obtained free of cost. This attractive feature has always been somewhat offset by the very high capital cost of construction especially of the civil engineering works. Today, however, the capital cost per kilowatt of hydro-electric stations is becoming comparable with that of steam stations. Unfortunately the geographical conditions necessary for hydro-generation are not commonly found especially in Britain. In most highly developed countries hydro-electric resources are used to the utmost. There still exists great hydro-electric potential in many under-developed countries and this will doubtless be utilized as their load grows.

An alternative to the conventional use of water energy, pumped storage, enables water to be used in situations which would not be amenable to conventional schemes. The utilization of the energy in tidal flows in channels has long been the subject of speculation. The technical and economic difficulties are very great and few locations exist where such a scheme would be feasible. An installation using tidal flow has been constructed on the La Rance estuary in northern France where the tidal height range is 30 ft (9·2 m) and the tidal flow is estimated at 18,000 m³/sec.

Before discussing the types of turbine used, a brief comment on the general modes of operation of hydro-electric stations will be given. The vertical difference between the upper reservoir and the level of the turbines is known as the head. The water falling through this head gains kinetic energy which it then imparts to the turbine blades. There are three main types of installation as follows:

(a) *High Head* or *Stored*—the storage area or reservoir normally fills in over 400 hours;

(b) *Medium Head* or *Pondage*—storage fills in 200 to 400 hours;

(c) *Run of River*—storage fills in less than 2 hours and has 10–50 ft head.

A schematic diagram for type (c) is shown in figure 1.7.

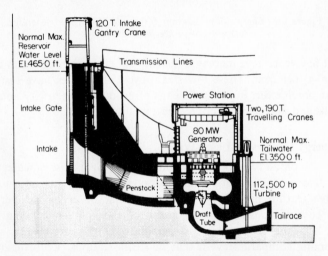

Figure 1.7   Hydro-electric scheme—Kainji, Nigeria. Section through the intake dam and power house. The scheme comprises an initial four 80 MW Kaplan turbine sets with the later installation of eight more sets. Running speed 115·4 rev/min. This is a large-flow low head scheme with penstocks 28 ft in diameter. (*Permission of Engineering.*)

Associated with these various heights or heads of water level above the turbines are particular types of turbine. These are:

(a) *Pelton* This is used for heads of 600–6000 ft (184–1840 m) and consists of a bucket wheel rotor with adjustable flow nozzles.

(b) *Francis* Used for heads of 120–1600 ft (37–490 m) and is of the mixed flow type.

(c) *Kaplan* Used for run of river and pondage stations with heads of up to 200 ft (61 m). This type has an axial flow rotor with variable-pitch blades.

Typical efficiency curves for each type of turbine are shown in figure 1.8. As the efficiency depends upon the head of water which is continually fluctuating, often water consumption in $m^3/kWh$ is used and is related to

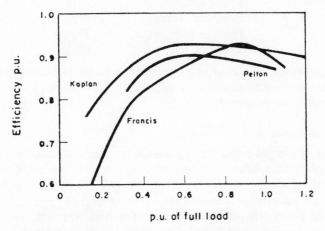

Figure 1.8  Typical efficiency curves of hydraulic turbines.

the head of water. Hydro-electric plant has the ability to start up quickly and the advantage that no losses are incurred when at a standstill. It has great advantages, therefore, for generation to meet peak loads at minimum cost, working in conjunction with thermal stations. By using remote control of the hydro sets, the time from the instruction to start-up to the actual connexion to the power network can be as short as two minutes and the rate of taking-up load of the machines in the order of 20 MW/min. As supplies of water at suitable heads are usually situated large distances from the load centres long transmission lines are needed.

*Pumped storage* A relatively recent method of obtaining the advantages of hydro plant where suitable water supplies are not available is by the use of pumped storage. This consists of an upper and a lower reservoir and turbine-generators which can be used as motor-pumps. The upper reservoir has sufficient storage usually for 4 to 6 hours of full load generation with a reserve of 1 to 2 hours.

The sequence of operation is as follows. During times of peak load on the network the turbines are driven by water from the upper reservoir in the normal manner. The generators then change to synchronous-motor action and, being supplied from the general power network, drive the turbine which is now acting as a pump. During the night when only base load stations are in operation, and electricity is being produced at its cheapest, the water in the lower reservoir is pumped back into the higher one ready for the next day's peak load. Typical operating efficiencies attained are:

| | |
|---|---|
| Motor and generator | 96% |
| Pump and turbine | 77% |
| Pipeline and tunnel | 97% |
| Transmission | 95% |

giving an overall efficiency of 67 per cent. A further advantage is that the synchronous machines can be easily used as synchronous compensators if required (see chapter 4).

### 1.6  Gas Turbines

The use of the gas turbine as a prime mover has certain advantages over steam plant, although with normal running it is less economical to operate. The main advantage lies in the ability to start and take up load quickly. Hence the gas turbine is coming into use as a method for dealing with the peaks of the system load. At the time of writing there are many installations, many generators being of 60 MW output. A further use for this type of machine is as a synchronous compensator to assist with maintaining voltage levels. Even on economic grounds it is probably advantageous to meet peak loads by starting up gas turbines from cold in the order of two minutes rather than running spare steam plant continuously.

The installation consists of a turbine, combustion chamber and a compressor driven by the turbine. The compressed air is delivered to the combustion chamber where continuous combustion of the injected fuel oil is maintained; the hot gases then drive the turbine.

### 1.7  Magnetohydrodynamic (M.H.D.) Generation

Whether the fuel used is coal, oil or nuclear, the result is the production of steam which then drives the turbine. Attempts are being made to generate electricity without the prime mover or rotating generator. In the magneto-hydrodynamic method gases at 2500°C are passed through a chamber in which a strong magnetic field has been created (figure 1.9). If the gas is hot enough it is electrically slightly conducting (it is seeded with potassium to improve the conductivity) and constitutes a conductor moving in the magnetic field. An e.m.f. is thus induced which can be collected at suitable electrodes. Great efforts are at present being made in several countries to construct an economical large M.H.D. generator. In its practical form

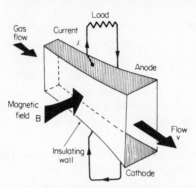

Figure 1.9 The principle of MDH power generation. (*Permission of English Electric Co. Ltd.*)

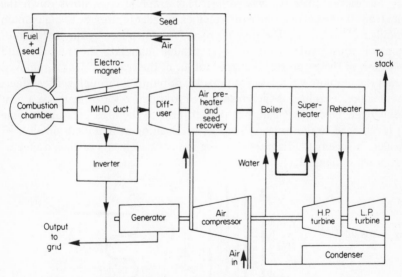

Figure 1.10 Block diagram of open cycles MHD power station. (*Permission of English Electric Co. Ltd.*)

it will be used in conjunction with traditional plant and figure 1.10 shows a schematic layout of a possible system incorporating normal steam operation plus M.H.D. generation. There are, however, great doubts as to the economic viability of this form of generation.

Other forms of direct generation include thermionic emission and fuel cells, both of which are being developed in attempts to improve their efficiency.

## 1.8 Nuclear Power

Nuclear fission takes place when a free neutron strikes the nucleus of a fissile material such as Uranium-235. On impact the nucleus splits into two particles, releasing energy which is manifested as heat. In this process some of the new neutrons released collide with other fissile nuclei which also split and a chain reaction is set up. The plant containing the fissile material is called a reactor or pile. The reactor produces heat which must be converted into electrical energy via a heat exchanger, turbine and generator.

There are a number of versions of the reactor in use with different coolants and types of fissile fuel. In Britain the Magnox reactor has been used in which natural uranium in the form of rods is enclosed in magnesium alloy cans. The fuel cans are placed in a structure or core of pure graphite made up of bricks called the *moderator*. This graphite core slows down the neutrons to the correct range of velocities in order to provide the maximum number of collisions. The fission process is controlled  by the insertion of control rods made of neutron-absorbing material; the number and position of these controls the heat output of the reactor. Heat is removed from the graphite via carbon monoxide gas pumped through vertical ducts in the core. This heat is then transferred to water to form steam via a heat exchanger. Once the steam has passed through the high pressure turbine it is returned to the heat exchanger for reheating as in a coal- or oil-fired boiler. A schematic diagram showing the basic elements of such a reactor is shown in figure 1.11.

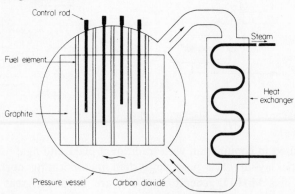

Figure 1.11    Schematic view, nuclear reactor- -British magnon type.

A more recent reactor similar to the Magnox is the *Advanced Gas-cooled Reactor* (*A.G.R.*). A reinforced-concrete steel-lined pressure vessel contains the reactor and heat exchanger. Enriched uranium dioxide fuel in pellet form, encased in stainless steel cans, is used; a number of cans form a cylindrical fuel element which is placed in a vertical channel in the core. Carbon dioxide gas at a higher pressure than in the Magnox type removes the heat. The control rods are made of boron steel. Spent fuel elements when removed from the core are stored in a special chamber for about a week and then dismantled and lowered into a pond of water where they remain until the level of radioactivity has decreased sufficiently for them to be removed from the station.

Nuclear generating stations require little fuel in terms of volume and weight and therefore may be sited independently of fuel supplies, although considerations of safety tend to keep them away from large centres of population. A disadvantage is their relatively high initial capital cost as against running costs. This requires that nuclear stations operate continuously as base load stations. With the development of the A.G.R. system the cost per unit of electricity by nuclear plant has been reduced considerably and the economics of such generation are now more favourable (see Chapter 2). It may be assumed, therefore, that in the 1970s an increasing percentage of generating plant will be nuclear. This may eventually influence transmission schemes if it becomes possible to locate nuclear plant near large urban concentrations.

In the U.S.A. pressurized water and boiling water reactors are used. In the *pressurized water type* the water is pumped through the reactor and acts as a coolant and moderator, the water being heated to 315°c. The

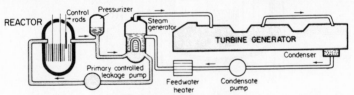

Figure 1.12    Schematic diagram of a pressurized water reactor.
(*Permission of Edison Electric Institute.*)

steam pressure is greater than the vapour pressure at this temperature and the water leaves the reactor at below boiling point. The fuel is in the form of pellets of uranium dioxide in bundles of stainless steel tubing. The *boiling water reactor* was developed later than the above type and is now used extensively. Inside the reactor heat is transferred to boiling water at a pressure of 690 $N/cm^2$. Schematic diagrams of these reactors are shown in figures 1.12 and 1.13.

All reactors that use uranium produce plutonium in the reaction. Most
of this is not utilized in the reactor. *Fast breeder reactors* which are being
developed at the present time breed new fuel in considerable quantities

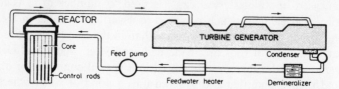

Figure 1.13   Schematic diagram of a boiling water reactor. (*Permission of Edison Electric Institute.*)

during the reaction as well as producing heat. In the liquid-metal fast-
breeder reactor, shown in figure 1.14, liquid sodium is the coolant which
leaves the reactor at 650°C at atmospheric pressure. The heat is then
transferred to a secondary sodium circuit which transfers it via a heat
exchanger to produce steam at 540°C. A similar development is the steam-
cooled fast-breeder reactor. In these the initial feed material is U-238
which breeds plutonium in the ratio 1 : 5. The breeder reactors show great

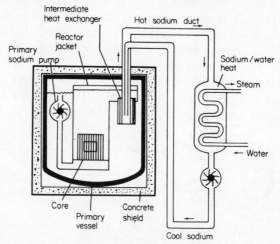

Figure 1.14   Schematic diagram of a liquid metal—fast breeder
reactor. (*Permission of Edison Electric Institute.*)

promise for the future. A considerable construction programme is under-
way in most countries. In the U.S.A. it is anticipated that the following
quantities of nuclear power generators will become available in the near
future: 1971 = 6000 MW; 1972 = 11,375 MW; 1973 = 13,433 MW; and
1974 = 9383 MW. (Source—Edison Electric Institute.)

### 1.9 Economical Operation of Generation Stations—Merit Order

With new and more efficient generating sets being brought into operation there exists a wide range of plant available for use. As previously mentioned, the load consists of a base plus a variable element depending on the time of day and other factors. Obviously the base load should be supplied by the more efficient plant which then runs 24 hours per day with the remaining load met by the less efficient stations. To facilitate this method of operation the generating sets are listed in order of cost per unit generated and the load is taken by each station in turn, subject to the necessary security of supply.

In addition to the machines supplying the load a certain proportion of available plant is held in reserve to meet sudden contingencies. A proportion of this reserve must be capable of being brought into operation in 5 minutes and hence some machines must be run at say 75 per cent full load to allow for this spare generating capacity.

### 1.10 Nature of Transmission and Distribution Systems

By transmission, is normally implied, the bulk transfer of power by high-voltage links between main load centres. Distribution on the other hand is mainly concerned with the conveyance of this power to consumers by means of lower voltage networks.

The machines usually generate voltage in the range 11–25kV which is increased by transformers to the main transmission voltage. At substations the connexions between the various components of the system, such as lines and transformers, are made and the switching of these components carried out. Large amounts of power are transmitted from the generating stations to the load-centre substations at 400 kV and 275 kV in Britain, and 345 kV and 500 kV in the U.S.A. The network formed by these very high-voltage lines is sometimes referred to as the *supergrid*. Most of the large and efficient stations feed through transformers directly into this network. This grid in turn feeds a subtransmission network operating at 132 kV in Britain and 115 kV in the U.S.A. Some of the older and less efficient stations feed into this system which in turn supplies networks which are concerned with distribution to consumers in a given area. In Britain these networks operate at 33 kV, 11 kV or 6·6 kV and supply the final consumer feeders at 415 V three-phase giving 240 V per phase. Other voltages exist in isolation in various places, e.g. the 66 kV London cable system. A typical part of a supply network is shown schematically in figure 1.15. The power system is thus made up of networks at various voltages. There exist in effect voltage tiers as represented in figure 1.16. Figure 1.17 shows the geographical disposition of the transmission network in England and Wales.

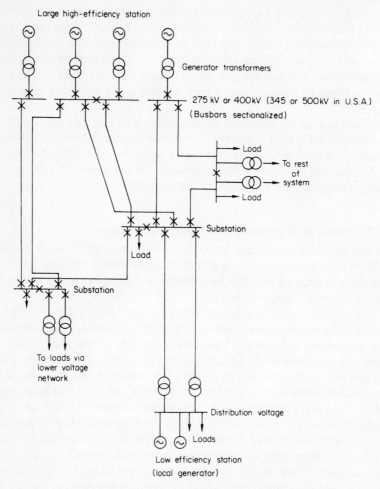

Figure 1.15   Part of a typical power system.

Summarizing, transmission networks deliver to wholesale outlets at 66 kV and above; subtransmission networks deliver to retail outlets at voltages from 11 kV to 132 kV, and distribution networks deliver to final step-down transformers at voltages from 2 to 33 kV.

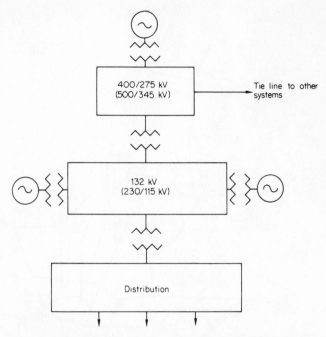

Figure 1.16 Schematic diagram of the constituent networks of a supply system. U.S.A. voltages in parentheses.

## 1.11 Reasons for Interconnexion

(a) Generating sets are becoming increasingly large and 2000 MW capacity stations are now in operation. Not only is the initial cost per kVA of stations of very large capacity lower than that of smaller ones but their efficiencies are substantially higher. Hence, regardless of geographical position, it is more economical to use these efficient stations to full capacity 24 hours a day and transmit energy considerable distances than to use less efficient more local stations. The main base load is therefore met by these large stations which must be interconnected so that they feed into the general system and not into a particular load.

(b) In order to meet sudden increases in load a certain amount of generating capacity known as the 'spinning reserve' is required. This consists of generators running at normal speed and ready to supply power instantaneously. If the machines are stationary a reasonable time is required (especially for steam turbo alternators) to run up to speed; this can approach 1 hour. It is more economical to have certain stations only serving this function than to have each station carrying its own spinning reserve.

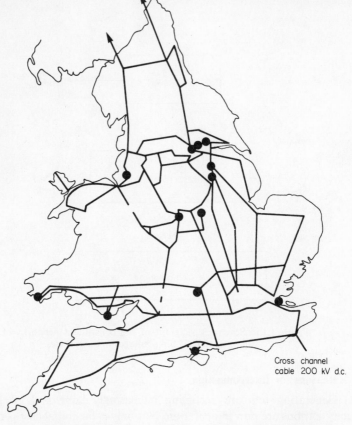

Figure 1.17    Geographical disposition of the British 400/275 kV
system. ● Power stations 2000 MW and above, —— 400/275 kV lines
– – – – 400 kV lines projected. (*Permission of the Electricity Council.*)

(c) The electricity supplies over the entire country are synchronized and
a common frequency exists.

(d) In an interconnected network consisting essentially of loops, continuity of supply is maintained, since substations can be fed from either
direction.

### 1.12   Statistics of Systems

Throughout the world the general form of power systems follows the same
pattern. Magnitudes of voltage vary from country to country, the differences originating mainly from geographical and historical reasons. Perhaps

one major difference in parameter is that of frequency, although today most of Europe operates on 50 Hz and North and South America on 60 Hz. In most countries electricity supply originated with many local and often privately owned utilities and this brought variations in voltage and general practice which are now disappearing as the systems become more integrated on a continental basis.

The general magnitudes of the systems in two countries will give an indication of the immense size and capital investment involved in power supply operations, the size of which approximately doubles every ten years.

*Britain.* It is forecast that the British system will have a simultaneous maximum demand of roughly 70,000 MW by the late 1970s, the major transmission voltage being 400 kV. The highest station thermal efficiency (1969) is 35 per cent and the load factor, 52·7 per cent. The main transmission circuits comprised the following lengths of overhead line in route-miles in 1969. 400 kV 1769, 275 kV 1475, 132 kV 5945, 66 kV 58; the corresponding lengths of underground cable were 6, 134, 765 and 183 at these respective voltages. The total number of towers is 52,213 and the number of substations 884.

For distribution systems the length (route-miles) of overhead line are, 33 kV 12,346, 11 kV 73,634, 6·6 kV to 11 kV 2150; the corresponding lengths of underground cable are, 7302, 41,926 and 12,169. The circuit mileage per 1000 consumers in the voltage range 33 kV to 650 V is 7·7 and below 650 V, 9. In 1969 the total energy generated was $187,064 \times 10^6$ kWH. and the total installed capacity 47,746 MW.

*North America.* In the *United States* systems are operated by both private companies and government agencies as opposed to most other countries in which power supply is basically controlled by the state. The individual systems are interconnected with tie lines so forming one massive synchronous network. The U.S. system is the largest in the world and the individual utilities are comparable with European countries. High electricity consumption and large distances have produced higher transmission voltages in North America than many other countries. The Canadian Hydro-Quebec and the American Electric Power companies already operate 765 kV lines. Research engineers are now considering the next voltage level above 765 kV which will be in the range from 1000 kV to 1500 kV.

In the U.S.A. total circuit mileages at the various voltage levels in 1968 were as follows: Below 69 kV — 136,000, 69 kV — 82,000, 115 kV — 77,000, 138 kV — 47,000, 230 kV — 37,000, 345 kV — 9000, 500 kV — 4000. The basic consumer voltage is 120 V single phase with preferred nominal distribution voltages below 69 kV of 46, 34·5, 23, 14·4, 13·2, 7·2,

4·8, and 2·4 kV. In December 1973 the peak demand for electricity is predicted to be 330,000 MW with a load factor of 63·1 per cent. Generation by means of nuclear power is expanding and in the scheduled expansion programme 38 per cent of the added generation will be nuclear. The reactors are of several different types with pressurized water and boiling-water reactors predominating.

The nature of the load in North America differs seasonally from one part of the country to another, and also load diversity occurs due to time zones. Generally the summer and winter loads are much more comparable than in Europe due to the extensive use of air-conditioning in the summer. The ratio (summer peak)/(peakload in the following December) is used to assess this characteristic, and in the early 1970s should be of the order of

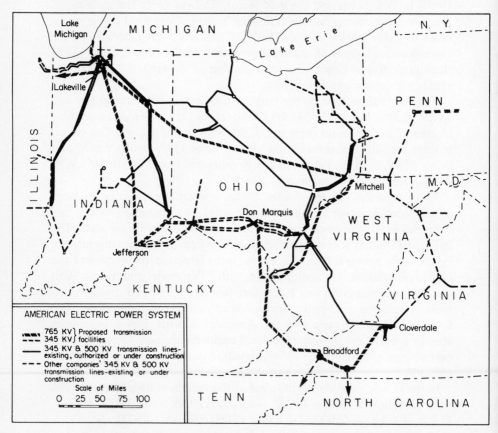

Figure 1.18    765 kV system of the American Electric Power Co.
(*Permission of the Institute of Electrical and Electronic Engineers.*)

1·06. Proposals to transmit excess generating capacity from one region to avoid the necessity for new generating plant in another are discussed in Appendix 6.

In figure 1.18 the 765 kV system of the American Electric Power Company is shown on a map of the area covered by this organization. It is expected that by 1980 the load will be 20,000 MW and by 1990, 34,000 MW. A number of factors influence the nature of the 765 kV network and these include system load, size·of generators, internal transmission

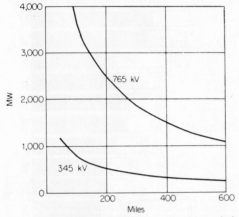

Figure 1.19 Relationships between power transmission capacity and circuit length for 765 kV and 345 kV circuits.

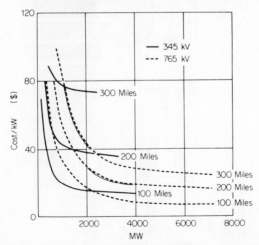

Figure 1.20 Relationship between transmission costs and power transmitted for 345 kV and 765 kV circuits of various lengths.

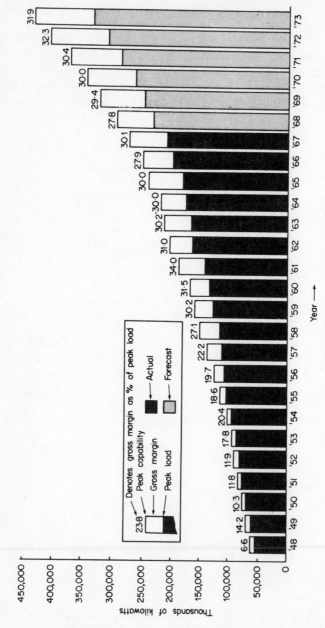

Figure 1.21 Peak production capabilities, peak load demands from 1948 to the present in the U.S.A. Values for December (note that in some areas the maximum demand occurs in the summer).

capability for handling power transfers and the strength of the interconnexions to outside systems. The system has 1150 miles of 765 kV line,
14,700 MVA of transformer capacity and 27 circuit breakers. There are
seven 765/345 kV, two 765/500 kV and two 765/138 kV, substations. In
addition there will be shunt reactors connected directly to the 765 kV line
terminals and of total capacity 4200 MVA. In figure 1.19 the relationship
between power transmittable over 765 kV and 345 kV circuits and length
of circuit is shown. The corresponding relationship between the cost of
transmission and power transmitted is shown in figure 1.20, showing the
cross-over powers for 345 kV and 765 kV for various distances.

Figure 1.21 shows the growth of total load in the United States from
1948 to the present day. Similar curves showing the increase in the electricity produced per head of population with time for four countries are
shown in figure 1.22. Sales of electricity (kilowatt-hours) in the U.S.A. by

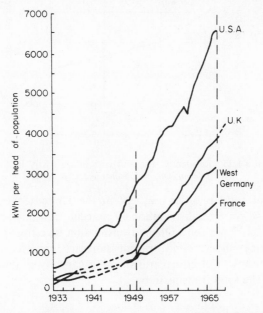

Figure 1.22  Comparative electricity production per head of population: broken lines denote provisional figures for years when information is not available. (*Permission of Institution of Electrical Engineers.*)

customer groupings in 1968 were as follows; industrial 51,8834 × 10⁶
(43 per cent) residential 367,692 × 10⁶ (31 per cent), commercial 265,151
(22 per cent), and others 50,644 × 10⁶ (4 per cent). This electricity is
mainly produced by several kinds of fuel and the total output for 1968 of

1100 billions of kilowatt-hours is made up as follows: coal — 683, fuel oil — 100, gas — 304, nuclear — 12·3. Of the total capacity of 290,365 MW in the U.S.A. in 1968, 83 per cent was provided by steam and the remaining 17 per cent by hydro generation. Investor owned companies amounted to 75·8 per cent of the total capacity the remainder being controlled by government agencies and cooperatives. It is predicted that the total capability of the investor-owned companies in 1980 will be of the order of 492,000 MW. In figure 1.23 the use and cost of electricity to residential consumers in the U.S.A. is summarized.

Total electric utility industry

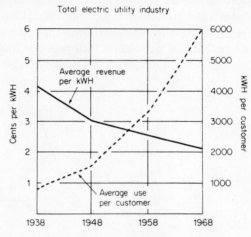

Figure 1.23   Residential use and cost of electricity in the U.S.A.
(*Permission of Edison Electric Institute.*)

In terms of total plant investment in the United States, generation accounts for 40 per cent, distribution 40 per cent, transmission 17 per cent, and offices, etc., 3 per cent. Of the transmission investment (of the order of $20 billion) underground transmission accounts for 10 per cent, although in terms of electrical transmission capacity it amounts to only 1 per cent of the total transmission.

## 1.13   Distribution Systems

Although in this book there is a bias towards transmission networks and the associated problems, the general area of distribution should not be considered as of secondary importance. Distribution networks differ from transmission networks in several ways quite apart from voltage magnitude. The number of branches and sources is much higher in distribution networks and the general structure or topology is different. A typical system consists of a step down (e.g. 66/11 kV) on-load tap-changing transformer at a bulk supply point feeding a number of cables which can vary in length from a few hundred metres to several kilometres. A series of step-down three-phase transformers, e.g. 11 kV/415 V in Britain or 4·16 kV/200 V in the U.S.A., are spaced along the cable and from these are supplied the consumer 3-phase, 4-wire networks which give 240 V or 120 V (in the U.S.A.) single-phase to houses and similar loads.

The structure of the network varies with location. In rural areas radial feeders are often used (usually overhead line) whereas in urban areas a well-defined low-voltage area or block is fed from the higher voltage network. For security reasons such areas are connected through fuses to neighbouring and similar areas which are fed from different feeders. In

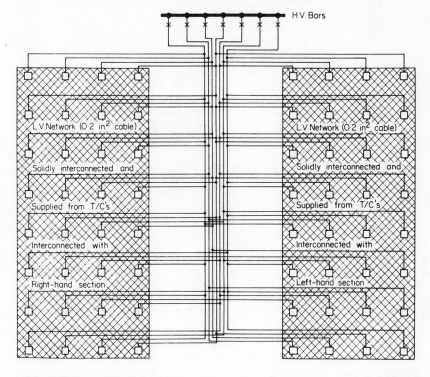

All main H.V. feeder cables 0·2 in$^2$ 3-core
All teed H.V. cables supplying transformer chambers 0·1 in$^2$ 3-core
All L.V. cables 0·2 in$^2$ 4-core
☐ 500 kVA Transformer chamber

Figure 1.24  Schematic diagram of interleaved interconnected low voltage distribution system. (*Permission of the Electricity Council.*)

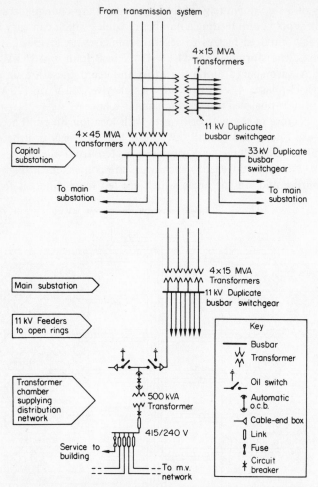

**Figure 1.25** Typical arrangement of supply to medium voltage (m.v.) network—British practice. (*Permission of Institution of Electrical Engineers.*)

such systems the network has essentially a topology of a loop nature. As most faults tend to be of a transient nature, auto-reclose circuit breakers are widely used. These open on a fault and then reclose after a short period. If the fault persists they reopen and the sequence continues three times. A part of a typical urban distribution network is shown in figure 1.24. This particular type is known as interleaved connexion and is popular in densely populated urban areas. Typical arrangements for supplying the distribution network are shown in figure 1.25. A typical system for rural areas is

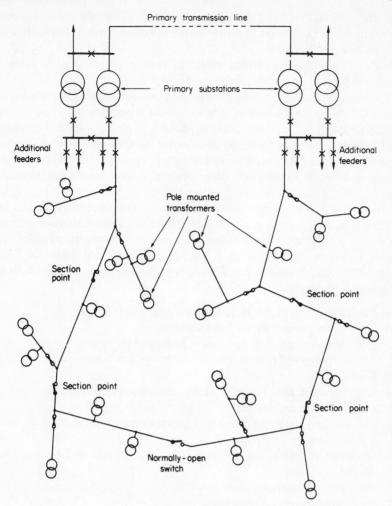

Figure 1.26 Typical rural distribution system. (*Permission of the Electricity Council.*)

shown in figure 1.26. Recent developments in single-phase high-voltage (12 kV) supplies to transformers are described in Chapter 8.

## 1.14 Use of Digital Computers

The most dramatic change in power-system analysis in the last decade has been the widespread use of computers. Their use has enabled very large systems to be analysed and controlled much more effectively and economically. The use of computers in power systems takes two main forms, off-line and on-line applications.

*Off-line* applications include research, routine calculation of system performance and data assimilation and retrieval.

*On-line* applications include date-logging and the monitoring of system state. It is an essential part of overall system control in the steady state including switching, safe interlocking, loading of plant, post-fault control, and load shedding. A recent application is its application to protective gear in which the high-speed measurement of system parameters is used to compare relevant quantities and so replace slower, more conventional, devices.

The use of a large central computer for economic and secure control of a large system at the moment presents difficulties, and an alternative is to subdivide it into smaller networks with a central coordinating facility. A typical sub-network could control, say, 100 generators and about 600 lines with 1200 circuit breakers. The functions of the on-line computer control would be as follows:

(a)  Prediction of demand at demand points.
(b)  Security of generation and transmission.
(c)  Economic operation of generating plant and the network system.
(d)  Maintaining system voltages and reactive power flows within specified limits.
(e)  Maintaining a file of current plant capability and constraints.
(f)  Costing proposed operations.
(g)  Despatching load magnitudes to generating units and ensuring that instructions are implemented.
(h)  Monitoring station performance relative to targets and costing and deviation.
(i)  Revising hot stand-by plant.
(j)  Revising network configuration.

In addition to the system aspects above, computers installed in generating stations control various processes such as the 'running up' from the cold state of turbine generators. These computers are completely local in their application.

### 1.15  Power System Equipment

The nature and parameters of the equipment comprising a power system will be discussed in detail in the following chapters. It is hoped, however, that a brief survey of certain aspects of this equipment at this stage might be helpful to newcomers to the subject.

### Turbine-generators

The ratings of steam-turbine driven round-rotor generators have increased dramatically over the last decade and sets of 1000 MW (1 GW) are now being installed. The advantages of such high output units lie in increased efficiency (on the steam side) and lower capital costs. The use of machines of 2 GW and above is forecast.

The most striking advances in machine design have been in the area of cooling and heat transfer. By the use of intense cooling methods the physical dimensions of generators have increased very much less than the corresponding electrical outputs. This is of great importance because of the restrictions in size and weight of plant which must be transported from the manufacturer's works to the generating station by road or rail. Large generators today are cooled by hydrogen or water pumped through the centre of the conductors, the stator winding is usually water-cooled and the rotor winding hydrogen-cooled. Machines are now appearing with water-cooled rotors as well as stators.

Typical statistics of a large generator are as follows:

500 MW, 588 MVA: winding cooling—rotor, water; stator, water; rotor diameter 1·12 m; rotor length 6·2 m; total weight 0·63 kg/VA; rotor weight 0·1045 kg/VA.

A cross-sectional view of a 1200 MW steam turbine generator is shown in figure 1.27.

### Transformers

These usually employ a three-limbed magnetic core and the core and windings are immersed in a tank filled with insulating oil. Channels between the windings are provided to allow the circulation of oil for cooling purposes. Size and weight are very important as with generators, and advances in more intense cooling have been achieved. An advantage with transformers is that at the very high ratings three separate single-phase units may be used, so alleviating the transportation problem.

### Overhead Lines and Underground Cables

The physical form of overhead lines is all too obvious to inhabitants of the developed parts of the world. With the consumption of electricity increasing (the rate of increase shows no sign of decreasing) the power to

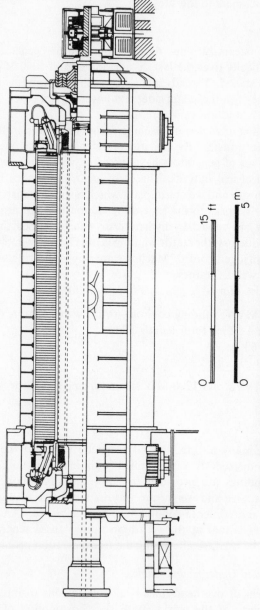

Figure 1.27   Sectional view of a 1000 MVA, 1800 rev/min. 60 Hz. generator. (*Permission of Institution of Electrical Engineers.*)

be transmitted increases and for this to be achieved economically transmission voltages must rise. Active consideration is now being given to line-to-line voltages of 1000 kV and above and the resulting size of structure (see Chapter 10) is going to pose serious problems of landscape conservation and amenity.

The relatively high cost of underground cables is a deterrent to their use. However, with artificial cooling techniques now available (see Chapter 8) ratings of single three-phase cable circuits can approach overhead line ratings. A drawback of such schemes is the need for cooling stations above ground along the route. Much research is directed at the present time to

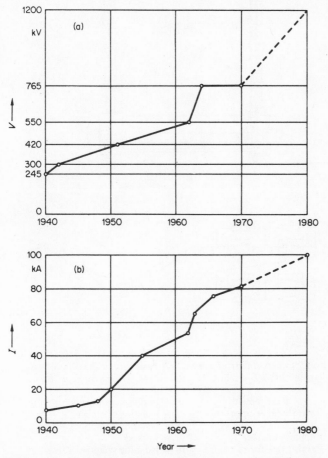

Figure 1.28 Relation between transmission voltage and short-circuit current (European systems). (*Permission of Brown Boveri Co.*)

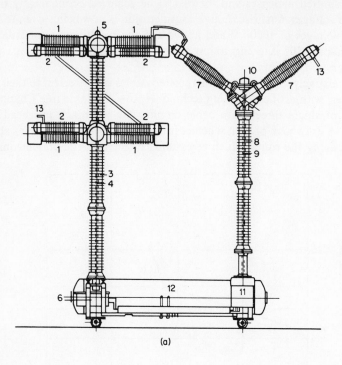

(a)

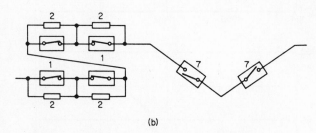

(b)

Figure 1.29   General arrangement and basic circuit diagram of one pole of a 245 kV airblast circuit-breaker. (*Permission of Brown Boveri Co.*)

a: General arrangement
b: Circuit (breaker open)

1 = Impulse chamber
2 = Parallel resistor to 1
3 = Impulse chamber column
4 = Tie rod
5 = Impulse chamber valve

6 = Impulse chamber control unit
7 = Isolating chamber
8 = Isolating chamber column
9 = Tie rod
10 = Isolating chamber valve
11 = Isolator control unit
12 = Air receiver
13 = Terminals

the development of superconducting cables through which very large amounts of power may be transmitted.

*Switchgear*

The switches by means of which the system is controlled are known as circuit breakers. At high voltages these may have several breaks (contacts) per phase and are very large and complicated structures (see Chapter 11). The very act of opening a circuit, which at domestic voltages is taken for granted, at very high voltages becomes a major engineering problem. At the higher transmission voltages the fault currents to be interrupted are very high, and the relation between their magnitude and the system (European) voltage is illustrated in figure 1.28a and b. A diagram of a 245 kV air-blast circuit breaker is shown in figure 1.29.

*Protection*

When the system malfunctions in any way, e.g. a short-circuit to earth on a line or transformer, rapid automatic means are required to remove the faulty equipment. At the same time the remainder of the system must be left operating intact. System protection embraces the various measuring and relaying schemes used to achieve this aim. In a large highly inter-connected supply network efficient protection is of the utmost importance, and its malfunction can bring about a wholesale electrical collapse of the supply system.

Much of the analysis discussed in this text has as its end-product the installation of the correct protection schemes. Protection is intimately connected with switchgear, the former representing the 'nervous system' and the latter the resulting physical action. At the present time in the British 400 kV system it is aimed to clear completely a fault in 140 msec, but with the increasing size and complexity a value of 80 msec is being actively considered in the future. This will allow, say, two cycles of 50 Hz for the protection to detect the fault and two cycles for the circuit breakers to open, a very stringent requirement.

**References**

BOOKS

1. *Electrical Transmission and Distribution Reference Book*, Westinghouse Electric Corp., East Pittsburgh, Pennsylvania.
2. Waddicor, H., *Principles of Electric Power Transmission*, Chapman and Hall, London, 5th Ed., 1964.
3. Guthrie Brown, F. (Ed.), *Hydro-Electric Engineering Practice*, Blackie, Glasgow, Vols. 1, 2 and 3, 1958.

4. Carr, T. H., *Electric Power Stations*, Chapman and Hall, London, Vols. I and II, 1944.
5. Sporn, P., *Research in Electric Power*, Pergamon, London, 1966.
6. Edison Electric Institute, *Nuclear Power*, Publication No. 67–68.
7. Zerban, A. H. and E. P. Nye, *Power Plants*, International Textbook Company, Scrantan, 2nd Ed., 1956.
8. Angrist, S. W., *Direct Energy Conversion*, Allyn and Bacon, New Jersey, 1965.
9. Sutton, G. W., *Direct Energy Conversion*, McGraw-Hill, New York, 1966.
10. Bary, C. W., *Operational Economics of Electrical Utilities*, Columbia University Press, New York, 1963.
11. Zoborszky J., and J. W. Rittenhouse, *Electric Power Transmission*, Rensselaer Bookstore, Troy, New York, 1969.
12. Hore, R. A., *Advanced Studies in Electrical Power System Design*, Chapman and Hall, London, 1966.


PAPERS

13. Edison Electric Institute, EHV Transmission Line Reference Book, 1968.
14. Connor, R. A. W., and R. A. Parkins, 'Operational Statistics in the Management of Large Distribution Systems', *Proc. I.E.E.*, **113**, 1823, (1966).
15. *The Economics of the Reliability of Supply*, IEE Conference Publication 34, 1967.
16. Milne, A. G., and J. H. Maltby, 'An Integrated System of Metropolitan Electricity Supply', *Proc. I.E.E.*, **114**, 745, (1967).
17. Papers describing the American Electric Power 765 kV System, *I.E.E.E.—PAS*, **88**, p. 1313, (1969).
18. Booth, E. S., *et al.*, 'The 400 kV grid system for England and Wales', *Proc. I.E.E.*, Part A (1962).
19. 'M.H.D. Electrical Power Generation', *Intern. Symp. I.E.E.* (1964).
20. Farmer, E. D., *et al.*, 'The development of automatic digital control of a power system from the laboratory to the field installation', *Proc. I.E.E.*, **116**, 436, (1969).

# 2

# Components of a Power System

## 2.1 Introduction

In this chapter the essential characteristics of the components of a power system will be discussed. It is essential that these be fully understood before the study of large systems of interconnexion of these components is undertaken. In all cases the simplest representation employing an equivalent circuit will be used not only to make the principles clearer but also because these simple models are used in practice. For more sophisticated treatments, especially of synchronous machines, the reader is referred to the more advanced texts given at the end of the chapter.

It is assumed that the reader has completed introductory courses in circuit theory and machines (or electromechanical energy conversion). A knowledge of basic control theory will be needed for the section on the dynamic response of machines although this may be omitted without essential loss of continuity in the text.

Loads are considered as components even though their exact composition and characteristics are not known with complete certainty. When designing a supply system or extending an existing one the prediction of the loads to be expected is required, statistical methods being used. This is the aspect of power supply known with least precision.

Before considering the components some basic ideas will be explained, notably the per unit system of notation which is used extensively in power network analysis. A brief revision of relevant network analysis is given in Appendix 2. In Appendix 1 is a discussion on the nature and sign of reactive power ($Q$) which all readers are advised to study at this stage.

It is stressed that most of the equivalent circuits used are single-phase and employ phase to neutral values. This assumes that the loads are balanced three-phase which is reasonable for normal steady-state operation. When unbalance exists between the phases, full treatment of all phases is required and special techniques for dealing with this are given in chapter 6.

35

## 2.2  The Per Unit System

In the analysis of power networks instead of using actual values of quantities it is usual to express them as fractions of reference quantities, such as rated or full-load values. These fractions are called per unit (denoted by p.u.) and the p.u. value of any quantity is defined as:

$$\frac{\text{the actual value (in any unit)}}{\text{the base or reference value in the same unit}}.$$

Some authorities express the p.u. value as a percentage. Although the use of p.u. values may at first sight seem a rather indirect method of expression there are in fact great advantages; they are as follows.

(a) The apparatus considered may vary widely in size; losses and volt drops will also vary considerably. For apparatus of the same general type the p.u. volt drops and losses are in the same order regardless of size.

(b) As will be seen later, the use of $\sqrt{3}$'s in three-phase calculations is reduced.

(c) By the choice of appropriate voltage bases the solution of networks containing several transformers is facilitated.

(d) P.u. values lend themselves more readily to automatic computation.

*Resistance*

$$R_{\text{p.u.}} = \frac{R(\Omega)}{\text{base } R(\Omega)}$$

$$= \frac{R(\Omega)}{\text{base voltage } (V_b)/\text{base current } (I_b)}.$$

$$= \frac{R(\Omega) \cdot I_b}{V_b}$$

$$= \frac{\text{voltage drop across } R \text{ at base or rated current}}{\text{base or rated voltage}}$$

Also,

$$R_{\text{p.u.}} = \frac{R(\Omega) I_b^2}{V_b I_b}$$

$$= \frac{\text{power loss at base current}}{\text{base power or volt-amperes}}$$

$\therefore$    the power loss (p.u.) at base or rated current $= R_{\text{p.u.}}$

Power loss (p.u.) at $I_{\text{p.u.}}$ current $= R_{\text{p.u.}} \cdot I_{\text{p.u.}}^2$

Similarly, p.u. Impedance $=$ Impedance in ohms$\left/ \dfrac{\text{base voltage}}{\text{base current}}\right.$

*Example 2.1* A d.c. series machine rated at 200 V, 100 A has an armature resistance of 0·1 Ω and field resistance of 0·15 Ω. The friction and windage loss is 1500 W. Calculate the efficiency when operating as a generator.

*Solution* Total series

$$R_{p.u.} = \frac{0·25}{200/100} = 0·125 \text{ p.u.}$$

where

$$V_{base} = 200 \text{ V, and } I_{base} = 100 \text{ A}$$

Friction and windage loss

$$= \frac{1500}{200 \times 100} = 0·075 \text{ p.u.}$$

At the rated load, the series-resistance loss

$$= 1^2 \times 0·125 \text{ p.u., and the total loss}$$
$$= 0·125 + 0·075 = 0·2 \text{ p.u.}$$

As the output = 1 p.u., the efficiency $= \dfrac{1}{1+0·2} = 0·83$ p.u.

*Three-phase circuits* A p.u. phase voltage has the same numerical value as the corresponding p.u. line voltage. With a line voltage of 100 kV and a rated line voltage of 132 kV, the p.u. value is 0·76. The equivalent phase voltages are $100/\sqrt{3}$ kV and $132/\sqrt{3}$ kV and hence the p.u. value is again 0·76. The actual values of $R$, $X_L$ and $X_c$ for lines, cables and other apparatus are phase values. When working with ohmic values it is less confusing to use the phase values of all quantities. In the p.u. system three-phase values of voltage, current and power can be used without undue anxiety about the result being a factor of $\sqrt{3}$ incorrect.

$$I_{base} = \frac{\text{base } VA}{\sqrt{(3)}V_{base}}$$

base $Z = \dfrac{V_{base}/\sqrt{3}}{I_{base}}$ assuming a star connected system.

$$= \frac{\dfrac{V_b}{\sqrt{3}} \cdot \dfrac{V_b}{\sqrt{3}}}{\dfrac{I_b \cdot V_b}{\sqrt{3}}} = \frac{V_b^2}{\sqrt{(3)}V_b I_b}$$

$$= \frac{(\text{base line voltage})^2}{\text{base } VA} \tag{2.1}$$

It should be noted that the same value for $Z_{base}$ is obtained using purely phase values.

Hence,

$$Z_{p.u.} = \frac{Z(\Omega) \times \text{base } VA}{(\text{base voltage})^2},$$

i.e. is directly proportional to the volt-ampere base and inversely to (base voltage)$^2$.

Hence, $Z_{p.u.}$(new bases)

$$= Z_{p.u.} \text{ (original base)} \times \left(\frac{\text{base } V_{old}}{\text{base } V_{new}}\right)^2 \times \left(\frac{\text{base } VA_{new}}{\text{base } VA_{old}}\right) \qquad (2.2)$$

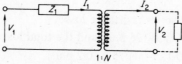

Figure 2.1   Equivalent circuit of single phase transformer.

*Transformers*  Consider a single-phase transformer in which the total series impedance of the two windings referred to the primary is $Z_1$ (figure 2.1). Then the p.u. impedance $= I_1 Z_1 / V_1$ where $I_1$ and $V_1$ are the rated or base values. The total impedance referred to the secondary

$$= Z_1 N^2 \text{ and this in p.u. notation}$$

$$= Z_1 N^2 \left(\frac{I_2}{V_2}\right)$$

$$= Z_1 N^2 \frac{I_1}{N} \frac{1}{V_1 N}$$

$$= \frac{Z_1 I_1}{V_1}$$

Hence the p.u. impedance of a transformer is independent of the winding considered.

In a circuit with several transformers care has to be taken regarding the different voltage levels. Consider the network in figure 2.2; in this two single-phase transformers supply a 10 kVA resistance load, the load voltage being maintained at 200 V. Hence the load resistance is $(200^2/10,000)$, i.e. 4 $\Omega$. In each of the circuits A, B and C a different voltage exists, so that each circuit will have its own base voltage, i.e. 100 V in A, 400 V in B, 200 V in C. Although it is not essential for rated voltages to be used as bases, it is essential that the *voltage bases used be related by the turns ratios of the transformers*. If this is not so the whole p.u. framework breaks down. The same volt-ampere base is used for all the circuits as $V_1 I_1 = V_2 I_2$ on each side of a transformer and is taken in this case as 10 kVA.

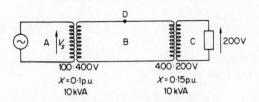

Figure 2.2 Network with two transformers—per-unit approach.

The base impedance in C

$$= \frac{200^2}{10000} = 4\Omega$$

The load resistance (p.u.) in C

$$= \frac{4}{4} = 1 \text{ p.u.}$$

In B the base impedance

$$= \frac{400^2}{10000} = 16\Omega$$

and the load resistance referred to B

$$= 4 \times 2^2 = 16\Omega$$

Hence the p.u. load referred to B

$$= 1 \text{ p.u.}$$

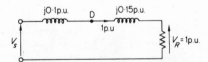

Figure 2.3 Equivalent circuit, with per-unit values, of network in figure 2.2.

Similarly the p.u. load resistance referred to A is also 1 p.u. Hence if the voltage bases are related by the turns ratios the load p.u. value is the same for all circuits. An equivalent circuit may be used as shown in figure 2.3. Let the volt-ampere base be 10 kVA; the voltage across the load ($V_R$) is

1 p.u. (as the base voltage in C is 200 V, if the load voltage had been maintained at 100 V, $V_R$ would be 0·5 p.u.)
The base current

$$= \frac{(VA)_b}{V_b} = \frac{10000}{200} = 50 \text{ A in C}$$

The corresponding currents in the other circuits are 25 A in B, and 100 A in A.
The actual load current $= 50/50 = 1$ p.u. (in phase with $V_R$ the reference phasor). Hence the supply voltage $V_s$

$$= 1(j0·1 + j0·15) + 1 \text{ p.u.}$$

$\therefore$ $\qquad\qquad V_s = 1·03$ p.u.

$$= 1·03 \times 100 = 103 \text{ V.}$$

The voltage at point D in figure 2.2

$$= 1 + j0·15 \times 1 = 1 + j0·15$$

$$= 1·012 \text{ p.u. modulus}$$

$$= 1·012 \times 400 = 404·8 \text{ V.}$$

It is a useful exercise to repeat this example using ohms, volts and amperes.

*Example 2.2*   In figure 2.4 the schematic diagram of a radial transmission system is shown. The ratings and reactances of the various components are shown along with the nominal transformer line-voltages. A load of 50 MW at 0·8 p.f. lagging is taken from the 33 kV substation which is to be maintained at 30 kV. It is required to calculate the terminal voltage of the synchronous machine. The line and transformers may be represented by series reactances.

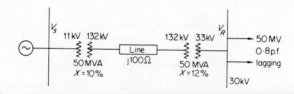

Figure 2.4   Line diagram of system for Example 2.2.

*Solution*   It will be noted that the line reactance is given in ohms; this is usual practice. The voltage bases of the various circuits are decided by the nominal transformer voltages, i.e. 11, 132 and 33 kV. A base of 100 MVA will be used for all circuits. The reactances (resistance is neglected) are expressed on the appropriate voltage and MVA bases.

Base impedance for the line

$$= \frac{132^2 \times 10^6}{100 \times 10^6}$$

$$= 174\Omega$$

Hence the p.u. reactance

$$= \frac{j100}{174} = j0\cdot575$$

P.U. reactance of the sending-end transformer

$$= \frac{100}{50} \times j0\cdot1 = j0\cdot2$$

P.U. reactance of the receiving-end transformer

$$= j0\cdot12 \times \frac{100}{50} = j0\cdot24$$

Load current

$$= \frac{50 \times 10^6}{\sqrt{(3)} \times 30 \times 10^3 \times 0\cdot8}$$

$$= 1200 \text{ A}$$

Base current for 33 kV, 100 MVA

$$= \frac{100 \times 10^6}{\sqrt{(3)} \times 33 \times 10^3}$$

$$= 1750 \text{ A}$$

Hence p.u. load current

$$= \frac{1200}{1750} = 0\cdot685$$

Voltage of load busbar

$$= \frac{30}{33} = 0\cdot91 \text{ p.u.}$$

The equivalent circuit is shown in figure 2.5.

Also,

$$\mathbf{V}_s = 0\cdot685(0\cdot8 - j0\cdot6)(j0\cdot2 + j0\cdot575 + j0\cdot24) + (0\cdot91 + j0)$$

$$= 1\cdot33 + j0\cdot555 \text{ p.u.}$$

$$\therefore \qquad V_s = 1\cdot44 \text{ p.u. or } 1\cdot44 \times 11 \text{ kV} = 15\cdot84 \text{ kV.}$$

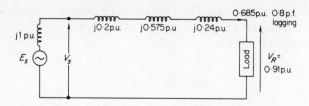

Figure 2.5 Equivalent circuit for Example 2.2.

## 2.3 Power, Reactive Power, Voltage Relationships in Simple Circuits

In this section equations for the power and reactive power transferred in terms of the voltages at nodes and the self and mutual impedances will be established. These equations will be found very useful in power system analysis. The simple circuit in figure 2.6 will be analysed in detail and

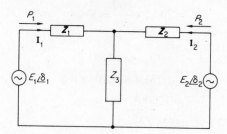

Figure 2.6 Network with two voltage sources.

the results extended to the general case of a network with n nodes. The power from voltage source $E_1$

$$= P_1 = \text{Real part of } \mathbf{E_1 I_1^*} = E_1 I_1 \cos \phi_1$$

($\mathbf{I_1^*}$ represents the conjugate of $\mathbf{I_1}$)

Similarly,

$$P_2 = \text{Real part of } \mathbf{E_2 I_2^*} = E_2 I_2 \cos \phi_2$$

$$Q_1 = \text{Imaginary part of } \mathbf{E_1 I_1^*} \text{ and so on}$$

Also,

$$\mathbf{I_1} = \mathbf{E_1 Y_{11}} - \mathbf{E_2 Y_{12}}$$

$$\mathbf{I_2} = \mathbf{E_2 Y_{22}} - \mathbf{E_1 Y_{21}}$$

$Y_{11}$ and $Y_{12}$ are the short-circuit driving point and transfer admittances respectively (refer to Appendix 2).

Hence

$$P_1 = \text{Real part } E_1 e^{j\delta_1}(E_1 e^{j\delta_1} Y_{11} e^{-j\theta_{11}} - E_2 e^{j\delta_2} Y_{12} e^{-j\theta_{12}})^*$$

$$= E_1 e^{j\delta_1}[E_1 Y_{11}\{\cos(\delta_1-\theta_{11})+j\sin(\delta_1-\theta_{11})\}$$
$$- E_2 Y_{12}\{\cos(\delta_2-\theta_{12})+j\sin(\delta_2-\theta_{12})\}]^*$$

Expanding the cosine and sine terms and substituting

$$\delta_{12}=\delta_1-\delta_2, \qquad \alpha_{11}=90-\theta_{11} \quad \text{and} \quad \alpha_{12}=90-\theta_{12},$$

$$P_1 = E_1^2 Y_{11}\sin\alpha_{11}+E_1 E_2 Y_{12}\sin(\delta_{12}-\alpha_{12})$$

Similarly,

$$P_2 = E_2^2 Y_{22}\sin\alpha_{22}+E_1 E_2 Y_{21}\sin(\delta_{21}-\alpha_{21}$$

$$= E_2^2 Y_{22}\sin\alpha_{22}-E_1 E_2 Y_{21}\sin(\delta_{12}+\alpha_{12})$$

as

$$\delta_{21}=-\delta_{12}$$

Also,

$$Q_1 = E_1^2 Y_{11}\cos\alpha_{11}-E_1 E_2 Y_{12}\cos(\delta_{12}-\alpha_{12})$$

and

$$Q_2 = E_2^2 Y_{22}\cos\alpha_{22}-E_1 E_2 Y_{21}\cos(\delta_{21}-\alpha_{21}).$$

For a network with $n$ sources,

$$P_m = E_m^2 Y_{mm}\sin\alpha_{mm}+E_m E_2 Y_{m2}\sin(\delta_{m2}-\alpha_{m2})+E_m E_3 Y_{m3}\sin(\delta_{m3}-\alpha_{m3})$$
$$+\dots \qquad (2.3)$$

When no resistance is present, $\alpha_{11}$, $\alpha_{12}$, etc. $= 0$ and

$$P_m = E_m E_2 Y_{m1}\sin\delta_{m2}+E_m E_3 Y_{m3}\sin\delta_{m3}+\dots \qquad (2.4)$$

*Calculation of sending and received voltages in terms of power and reactive power*   The determination of the voltages and currents in a network can obviously be achieved by means of complex notation but in power systems usually power $(P)$ and reactive power $(Q)$ are specified and often the resistance of lines is negligible compared with reactance. For example, if $R = 0\cdot1X$, the error in neglecting $R$ is $0\cdot49$ per cent and even if $R = 0\cdot4X$ the error is $7\cdot7$ per cent.

A simple transmission link is shown in figure 2.7a. It is required to establish the equations for $E$, $V$ and $\delta$. From figure 2.7b,

$$E^2 = (V+\Delta V)^2+\delta V^2$$

$$= (V+RI\cos\phi+XI\sin\phi)^2+(XI\cos\phi-RI\sin\phi)^2$$

$$\therefore \qquad E^2 = \left(V+\frac{RP}{V}+\frac{XQ}{V}\right)^2+\left(\frac{XP}{V}-\frac{RQ}{V}\right)^2 \qquad (2.5)$$

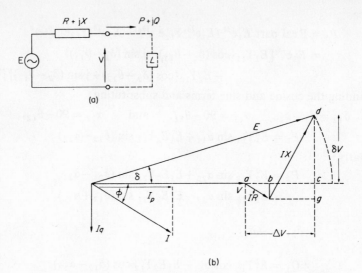

(a)

(b)

Figure 2.7   Phasor diagram for transmission of power through a
series impedance.

Hence,

$$\Delta V = \frac{RP + XQ}{V} \qquad (2.6)$$

and

$$\delta V = \frac{XP - RQ}{V} \qquad (2.7)$$

If

$$\delta V \ll V + \Delta V$$

$$E^2 = \left(V + \frac{RP + XQ}{V}\right)^2$$

and

$$E - V = \frac{RP + XQ}{V} = \Delta V$$

Hence the arithmetic difference between the voltages is approximately
given by

$$\frac{RP + XQ}{V}$$

If
$$R = 0,$$
$$E - V = \frac{XQ}{V}$$

and from equation 2.5 this is valid if
$$PX \ll V^2 + QX$$

The angle of transmission $\delta$ is obtained from $\sin^{-1} (\delta V/E)$.

Equations 2.6 and 2.7 will be used wherever possible owing to their great simplicity.

Consider a 275 kV line ($R$ and $X$ per mile = 0·055 and 0·52 $\Omega$ respectively), obviously $R \ll X$.

Assume a load of 600 MW, 300 MVAr (600 MVA is the thermal rating for a $2 \times 0·4$ in² line) and taking a base of 100 MVA, the base impedance is
$$\frac{275^2 \times 10^6}{100 \times 10^6} = 755 \, \Omega$$

For line of length 100 miles, $X = \frac{52}{755} = 0·069$ p.u.

and let the received voltage be 275 kV = 1 p.u.
$$PX = 6 \times 0·069 = 0·414$$
and
$$V^2 + QX = 1·207.$$

Hence the use of equation 2.6 will involve some inaccuracy.

Using equation 2.5 with $R$ neglected,
$$E^2 = \left(1 + \frac{0·069 \times 3}{1}\right)^2 + \left(\frac{0·069 \times 6}{V}\right)^2 = 1·207^2 + 0·414^2$$
$$E = 1·278 \text{ p.u.}$$

by equation 2.6,
$$E - V = \Delta V = \frac{0·069 \times 3}{1} = 0·207$$

$$\therefore \qquad E = 1·207 \text{ p.u.}$$

i.e. an error of 5·5 per cent.

With a 50 mile length of this line at the same load,
$$PX = 6 \times 0·0345 \quad \text{and} \quad V^2 + QX = (1 + 3 \times 0·0345)$$
$$= 0·207 \qquad\qquad\qquad = 1·103$$

The accurate formula gives
$$E = 1·125 \text{ p.u.}$$

and the approximate one gives
$$E = 1·104 \text{ p.u.}$$

i.e. an error of 1·9 per cent.

Hence a considerable length of double circuit line at rated load may be treated with the approximate formula. For a 275 kV, $2 \times 0.175$ in$^2$ conductor, line the rated power is 4·3 p.u. and thus a longer length could be considered than in the above case. With 132 kV lines on a 100 MVA base the reactance of a 100 mile 0·4 in$^2$ line is $100 \times 0.65/174$ or 0·37 p.u. and the rated power 1·5 p.u. ($Q = 0.75$ p.u. for same power factor)

$$PX = 1.5 \times 0.37 \quad \text{and} \quad V^2 + XQ = 1 + 0.277$$
$$= 0.55 \qquad\qquad\qquad = 1.28$$

Hence an error in the order of 5 per cent could be expected.

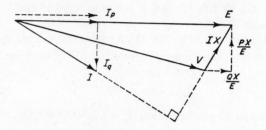

Figure 2.8   Phasor diagram when $E$ is specified.

If $E$ is specified and $V$ is required, the phasor diagram in figure 2.8 is used. From this,

$$V = \sqrt{\left[\left(E - \frac{QX}{E}\right)^2 + \left(\frac{PX}{E}\right)^2\right]} \qquad (2.8)$$

when $R = 0$.

If

$$\frac{PX}{E} \ll \frac{E^2 - QX}{E},$$

then

$$V = E - \frac{QX}{E} \qquad (2.9)$$

## SYNCHRONOUS MACHINES

### 2.4  Introduction

In this chapter those characteristics of the synchronous machine pertinent to power systems will be discussed. It is assumed that the reader has a basic knowledge of synchronous-machine theory. The two forms of rotor construction produce characteristics that influence the operation of the system to varying extents. In the round or cylindrical rotor the field winding is situated in slots cut axially along the rotor length; the diameter is relatively small (in the order of 90 cm) and the machine is suitable for operation at high speeds driven by a steam turbine. Hence it is known as the turbo-alternator. In the salient pole rotor the poles project as shown in figure 2.9, and low speed operation driven by hydraulic turbines is usual. The frequency of the generated e.m.f. and speed are related by the expression, $f = (np/60)$, where $n$ is the speed in rev/min and $p$ the number of pole pairs; a hydro-generator thus needs many poles to generate at normal frequencies.

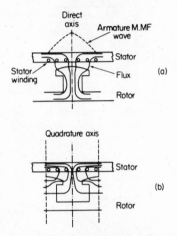

Figure 2.9a   Rotor pole and associated stator conductors—direct axis. Stator current power-factor such that M.M.F. wave is in the position shown.
Figure 2.9b   Rotor interpolar gap and associated stator conductors —quadrature axis.

The three-phase currents in the stator winding or armature generate a rotating magnetic field which is stationary relative to the rotor and its field. The effect of this armature M.M.F. wave on the rotor field is referred to as the armature reaction. The direct axis of the rotor pole system is shown in figure 2.9a, the reluctance of the flux path is a minimum and the flux linkage of each phase a maximum. The flux linkage per ampere is known as the direct-axis synchronous inductance with a corresponding reactance, $X_d$. The armature voltage induced by the change in flux linkages is in quadrature with the flux.

The axis through the interpolar gap (figure 2.9b) is called the quadrature axis and here magnetic reluctance is at its maximum. The value of reactance

$X_q$ corresponding to this position is therefore less than $X_d$. In a turbo-alternator rotor $X_q$ is almost equal to $X_d$ and for steady state operation they may be considered to be equal which considerably simplifies analysis. With hydro-generators, $X_q < X_d$ and saliency is said to exist; in analysis, fluxes, voltages and currents are resolved into their components along the direct and quadrature axes as will be explained later. The emphasis in this section will be on conditions where $X_q = X_d$, i.e. when only one value of reactance, current and voltage need be considered. This will simplify calculations and make the physical concepts clearer.

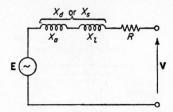

Figure 2.10   Equivalent circuit of a round-rotor synchronous generator (saliency ignored).

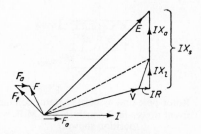

Figure 2.11   Phasor diagram corresponding to the equivalent circuit of a round-rotor machine.

In power system analysis the object is to use an equivalent-circuit model which exhibits the external characteristics of the generator with sufficient accuracy. For a round rotor machine the circuit in figure 2.10 is such a model. The effects of armature reaction and leakage flux are simulated by two reactances in series, the combined reactance being called the synchronous reactance $(X_s)$. The resistance of each phase of the armature winding $(R)$ is often negligible compared with $X_s$. Figure 2.11 shows the phasor diagram corresponding to the equivalent circuit. Phasors representing the M.M.F.'s set up by the field and armature currents are also shown in this diagram. The resultant air-gap M.M.F., $\mathbf{F} = \mathbf{F}_f + \mathbf{F}_a$; $\mathbf{F}_a$ is in phase with the armature current $I$ producing it.

*Effect of saturation on $X_s$—the short-circuit ratio*

The open-circuit characteristic is the graph of generated voltage against field current with the machine on open circuit and running at synchronous speed. The short-circuit characteristic is the graph of stator current against field current with the terminals short-circuited; curves for a modern machine are shown in figure 2.12. $X_s$ is equal to the open-circuit voltage

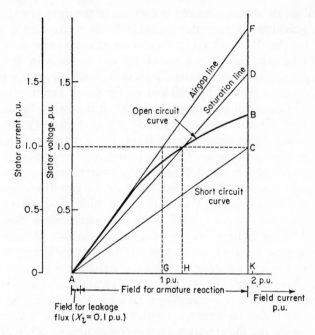

Figure 2.12  Open and short circuit characteristics of a synchronous machine. Unsaturated value of $X_s$ = FK/CK. With operation near nominal voltage, saturation line is used to give linear characteristic with saturation.

produced by the same field current that produces rated current on short-circuit, divided by this rated armature current. This value of $X_s$ is constant only over the linear part of the open-circuit characteristic (the air-gap line) and ignores saturation. The actual value of $X_s$ at full-load current will obviously be less than this value and several methods exist to allow for the effects of saturation [5, 6].

The short-circuit ratio (S.C.R.) of a generator is defined as the ratio between the field current required to give nominal open-circuit voltage and that required to circulate full-load current in the armature when short-circuited. In figure 2.12 the short-circuit ratio is $AH/AK$, i.e. 0·63. To allow for saturation it is common practice to assume that the synchronous reactance is $1/\text{S.C.R.}$ which for this machine is 1·58 p.u. Economy demands the design of machines of low S.C.R. and a value of 0·55 is common for modern machines.

## 2.5 Equivalent Circuit Under Balanced Short-Circuit Conditions

A typical set of oscillograms of the currents in the three armature phases when a synchronous generator is suddenly short-circuited is shown in figure 2.13a. In all three traces a direct-current component is evident and this is to be expected from a knowledge of transients in $R$–$L$ circuits. The magnitude of direct current present depends upon the instant at which the short-circuit is applied and on the power factor of the circuit. As there are three voltages mutually at 120° it is possible for only one to have a zero direct-current component. Often to clarify the physical conditions the direct component is ignored and a trace of short-circuit current as shown in figure 2.13b is considered. Immediately after the application of the short-circuit the armature current endeavours to create an armature reaction M.M.F., but the main air-gap flux cannot change to a new value immediately as it is linked with low-resistance circuits consisting of, (a) the rotor winding which is effectively a closed circuit, and (b) the damper bars, i.e. a winding which consists of short-circuited turns of copper strip set in the poles to dampen oscillatory tendencies. As the flux remains unchanged initially, the stator currents are large and can only flow through the medium of the creation of opposing currents in the rotor and damper windings by what is essentially transformer action. Owing to the higher resistance, the current induced in the damper winding decays rapidly and the armature current commences to fall. After this the currents in the rotor winding and body decay, the armature reaction M.M.F. is gradually established, and the generated e.m.f. and stator current fall until the steady-state condition on short-circuit is reached. Here the full armature reaction effect is operational and the machine represented by the synchronous reactance $X_s$. These effects are shown in figure 2.13 with the high initial current due to the damper winding and then the gradual reduction until the full armature reaction is established. A detailed analysis is given in reference 9.

To represent the initial short-circuit conditions two new models must be introduced. If in figure 2.13b the envelopes to the 50 Hz waves are traced, a discontinuity appears. Whereas the natural envelope continues to point 'a' the actual trace finishes at point 'b'; the reasons for this have been mentioned above. To account for both of these conditions, two new

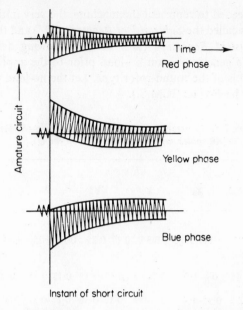

Figure 2.13a  Oscillograms of the currents in the three phases of a
generator when a sudden short-circuit is applied.

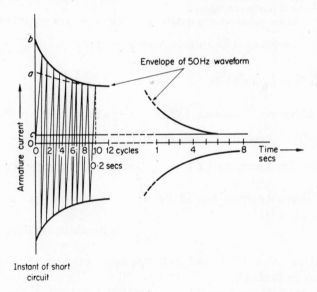

Figure 2.13b  Trace of short-circuit current when direct-current
component is removed.

reactances are needed to represent the machine, the very initial conditions requiring what is called the *Subtransient Reactance* ($X''$) and the subsequent period the *Transient Reactance* ($X'$). In the following definitions it is assumed that the generator is on no-load prior to the application of the short-circuit and is of the round-rotor type. Let the no-load phase voltage of the generator be $E$ volts (R.M.S.).

Table 2.1   Constants of synchronous machines—60 Hz
*(All values expressed as per unit on rating)*

| Type of machine | $X_s$ (or $X_d$) | $X_q$ | $X'$ | $X''$ | $X_2$ | $X_0$ | $r_a$ |
|---|---|---|---|---|---|---|---|
| Turbo-alternator | 1·2–2·0 | 1–1·5 | 0·2–0·35 | 0·17–0·25 | 0·17–0·25 | 0·04–0·14 | 0·003–0·008 |
| Salient pole (Hydro-electric) | 0·6–1·45 | 0·4–1·0 | 0·2–0·5 | 0·13–0·35 | 0·13–0·35 | 0·02–0·2 | 0·003–0·015 |
| Synchronous compensator | 1·5–2·2 | 0·95–1·4 | 0·3–0·6 | 0·18–0·38 | 0·17–0·37 | 0·03–0·15 | 0·004–0·01 |

$X_2$ = negative sequence reactance
$X_0$ = zero sequence reactance
$X'$ and $X''$ are the direct axis quantities for both direct and quadrature value (see reference 1)
$r_a$ = a.c. resistance of the armature winding per phase.

Then from figure 2.13b,

the Subtransient Reactance $(X'') = \dfrac{E}{0b/\sqrt{2}}$ where $0b/\sqrt{2}$ is the R.M.S. value of the subtransient current $(I'')$;

the Transient Reactance $(X') = \dfrac{E}{0a/\sqrt{2}}$ where $0a/\sqrt{2}$ is the R.M.S. value of the transient current, and finally

$$\frac{E}{0c/\sqrt{2}} = \text{Synchronous Reactance } X_s.$$

Typical values of $X''$, $X_s$ and $X_s$ for various types and sizes of machines are given in Table 2.1.

If the machine is previously on load the voltage applied to the equivalent reactance previously $E$ is now modified due to the initial load volt-drop.

Consider figure 2.14. Initially the load current is $I_L$ and the terminal voltage is $V$. The voltage behind the transient reactance $X'$ is

$$\mathbf{E}' = I_L(Z_L + jX')$$

$$= V + jI_L X'$$

and hence the transient current on short-circuit $= \mathbf{E}'/jX'$.

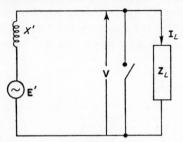

Figure 2.14 Modification of equivalent circuit to allow for initial load current.

## 2.6 Synchronous Generators in Parallel

Consider two machines A and B the voltages of which have been adjusted to equal values by means of the field regulators and the speeds of which are slightly different. In figure 2.15b the phase voltages are $E_{RA}$, etc. and the speed of machine A, $\omega_A$ rads/sec and of B, $\omega_B$ rads/sec.

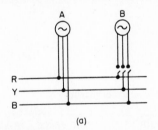

Figure 2.15a Generators in parallel.

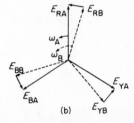

Figure 2.15b Corresponding phasor diagrams.

If voltage phasors of A are considered stationary those of B rotate at a relative velocity ($\omega_B - \omega_A$) and hence there are resultant voltages across the switch of ($E_{RA} - E_{RB}$), etc., which reduce to zero during each relative revolution. If the switch is closed at an instant of zero voltage the machines are connected (synchronized) without the flow of large currents due to the resultant voltages across the armatures. When the two machines are in synchronism they have a common terminal-voltage, speed, and frequency. Any tendency for one machine to accelerate relative to the other immediately results in a retarding or synchronizing torque being set up due to the current circulated.

Two machines operating in parallel are represented by the equivalent circuits shown in figure 2.16a with $E_A = E_B$ and on no external load. If A tries to gain speed the phasor diagram in figure 2.16b is obtained and $I = E_R/Z_A + Z_B$. The circulating current $\mathbf{I}$ lags $\mathbf{E}_R$ by angle $\tan^{-1}(X/R)$ and as in most machines $X \gg R$ this angle approaches $90°$. This current is a generating current for A and a motoring current for B, hence A is generating power and tending to slow down and B is receiving power from A

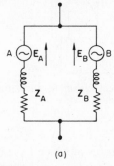

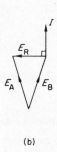

(a)                                                    (b)

Figure 2.16a   Two generators in           Figure 2.16b   Machine A in
parallel—equivalent circuit.                phase advance of machine B.

(c)

Figure 2.16c   Machine B in phase advance of machine A.

and speeding up. A and B therefore remain at the same speed, 'in step', or in synchronism. Figure 2.16c shows the state of affairs when B tries to gain speed on A. The quality of a machine to return to its original operating state after a momentary disturbance is measured by the *synchronizing power* and *torque*. It is interesting to note that as the impedance of the machines is largely inductive the restoring powers and torques are large; if the system were largely resistive it would be difficult for synchronism to be maintained.

Normally more than two generators operate in parallel and the operation of one machine connected in parallel with many others is of great interest. If the remaining machines in parallel are of such capacity that the presence of the generator under study causes no difference to the voltage and frequency of the others they are said to comprise an *Infinite Busbar System*, i.e. an infinite system of generation. In practice a perfect infinite busbar is never fully realized but if, for example, a 60 MW generator is removed from a 30,000 MW system the difference in voltage and frequency caused will be very slight.

### 2.7 The Operation of a Generator on Infinite Busbars

In this section in order to simplify the ideas as much as possible the resistance of the generator will be neglected; in practice this assumption is usually reasonable. Figure 2.17a shows the schematic diagram of a machine connected to an infinite busbar along with the corresponding phasor diagram. If losses are neglected the power output from the turbine is equal to the power output from the generator. The angle $\delta$ between the

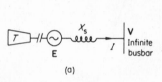
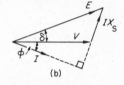

(a)    (b)

Figure 2.17a  Synchronous machine    Figure 2.17b  Corresponding
connected to an infinite busbar.    phasor diagram.

$E$ and $V$ phasors is known as the load angle and is dependent on the power input from the turbine shaft. With an isolated machine supplying its own load the latter dictates the power required and hence the load angle; when connected to an infinite-busbar system however, the load delivered by the machine is no longer directly dependent on the connected load. By changing the turbine output and hence $\delta$ the generator can be made to take any load the operator desires subject to economic and technical limits.

From the phasor diagram in figure 2.17b, the power delivered to the infinite busbar $= VI \cos \phi$ per phase but,

$$\frac{E}{\sin (90+\phi)} = \frac{IX_s}{\sin \delta}$$

hence

$$I \cos \phi = \frac{E}{X_s} \sin \delta$$

$\therefore$

$$\text{power delivered} = \frac{VE}{X_s} \sin \delta \qquad (2.10)$$

This expression is of extreme importance as it governs to a large extent the operation of a power system. It could have been obtained directly from equation 2.3, putting $\alpha = 0$.

Equation 2.10 is shown plotted in figure 2.18. The maximum power is obtained at $\delta = 90°$. If $\delta$ becomes larger than $90°$ due to an attempt to obtain more than $P_{max}$, increase in $\delta$ results in less power output and the machine becomes unstable and loses synchronism. Loss of synchronism results in the interchange of current surges between the generator and network as the poles of the machine pull into synchronism and then out again.

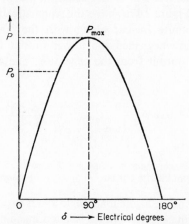

Figure 2.18 Power—angle curve of a synchronous machine.
Resistance and saliency neglected.

If the power output of the generator is increased by small increments with the no-load voltage kept constant, the limit of stability occurs at $\delta = 90°$ and is known as the *Steady-State Stability Limit*. There is another limit of stability due to a sudden large change in conditions such as caused by a fault, known as the *Transient Stability Limit*, and it is possible for the rotor to oscillate beyond $90°$ a number of times. If these oscillations diminish, the machine is stable. The load angle $\delta$ has a physical significance; it is the angle between like radial marks on the end of the rotor shaft of the machine and on an imaginary rotor representing the system. The marks are in identical physical positions when the machine is on no-load. The synchronizing power coefficient $= dP/d\delta$ W/rad and the synchronizing torque coefficient $= (1/\omega_S)/(dP/d\delta)$.

In figure 2.19a the phasor diagram for the limiting steady-state condition is shown. It should be noted that in this condition current is always leading. The following figures, 2.19b, c and d show the phasor diagrams for various

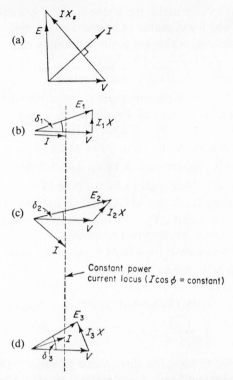

Figure 2.19 (a) Phasor diagram for generator at limit of steady-state stability, (c) and (d), Phasor diagrams for generator delivering constant power to the infinite busbar system but with different excitations. As $V$ is constant the in-phase component of $I$ must be constant. As $EV/X \sin \delta$ is constant as $E$ changes $\delta$ must change and $\delta_3 > \delta_1 > \delta_2$.

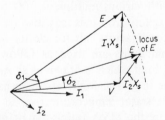

Figure 2.20 Operation at variable power and constant excitation.

operational conditions. Another interesting operating condition is variable power and constant excitation. This is shown in figure 2.20. In this case as $V$ and $E$ are constant when the power from the turbine is increased $\delta$ must increase and the power factor changes.

It is convenient to summarize the above types of operation in a single diagram or chart which will enable an operator to see immediately whether the machine is operating within the limits of stability and rating.

*The performance chart of a synchronous generator*

Consider figure 2.21a, the phasor diagram for a round rotor machine ignoring resistance. The locus of constant $IX_s$, $I$ and hence MVA, is a circle and the locus of constant $E$ a circle. Hence,

$0'S$ is proportional to $VI$ or MVA

$ps$ is proportional to $VI \sin \phi$ or MVAr

$sq$ is proportional to $VI \cos \phi$ or MW

To obtain the scaling factor for MVA, MVAr and MW the fact that at zero excitation, $E = 0$ and $IX_s = V$, is used, from which $I$ is $V/X_s$ at 90° leading to 00', corresponding to VAr/phase.

The construction of a chart for a 60 MW machine follows (figure 2.21b).

*Machine data*  60 MW, 0·8 p.f., 75 MVA

11·8 kV, S.C.R. 0·63, 3000 rev/min

Maximum exciter current 500 A

$$X_s = \frac{1}{0·63} \text{ p.u.} = 2·94 \, \Omega/\text{phase.}$$

The chart will refer to complete three-phase values of MW and MVAr. When the excitation and hence $E$ are reduced to zero, the current leads $V$ by 90° and is equal to $(V/X_s)$, i.e. $11800/\sqrt{(3)} \times 2·94$. The leading vars corresponding to this $= 11800^2/2·94 = 47$ MVAr.

With centre 0 a number of semi-circles are drawn of radii equal to various MVA loadings, the most important being the 75 MVA circle. Arcs with 0' as centre are drawn with various multiples of 00' (or $V$) as radii to give the loci for constant excitations. Lines may also be drawn from 0 corresponding to various power factors but for clarity only 0·8 p.f. lagging is shown. The operational limits are fixed as follows. The rated turbine output gives a 60 MW limit which is drawn as shown, i.e. line efg, which meets the 75 MVA line in g. The MVA arc governs the thermal loading of the machine, i.e. the stator temperature rise, so that over portion gh the output is decided by the MVA rating. At point h the rotor heating becomes more decisive and the arc hj is decided by the maximum excitation current allowable, in this case assumed to be 2·5 p.u. The remaining limit is that governed by loss of synchronism at leading power factors. The theoretical limit is the line perpendicular to 00' at 0' (i.e. $\delta = 90°$) but in practice a safety margin is introduced to allow a further increase in load of either 10 or 20 per cent before instability. In figure 2.21 a 10 per cent margin is used and is represented by ecd: it is constructed in the following manner. Considering point 'a' on the theoretical limit on the $E = 1$ p.u.

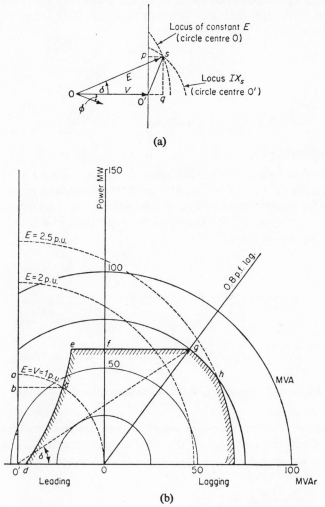

(a)

(b)

Figure 2.21  Performance chart of a synchronous generator.

arc, the power 0'a is reduced by 10 per cent of the rated power (i.e. by 6 MW) to 0'b; the operating point must, however, still be on the same $E$ arc and b is projected to c which is a point on the new limiting curve. This is repeated for several excitations giving finally the curve ecd.

The complete operating limit is shown shaded and the operator should normally work within the area bounded by this line.

As an example of the use of the chart, the full-load operating point g (60 MW, 0·8 p.f. lagging) will require an excitation $E$ of 2·3 p.u. and the measured load angle $\delta$ is 33°. This can be checked by using, power = $VE/X_s \sin \delta$.

i.e.

$$60 \times 10^6 = \frac{11800^2 \times 2\cdot3}{2\cdot94} \sin \delta,$$

from which

$$\delta = 33\cdot3°.$$

## 2.8   Salient-Pole Generators

So far the direct-axis reactance has been assumed equal to the quadrature-axis value. In the case of turbo-alternators this assumption is reasonably accurate for steady-state conditions but for salient-pole machines a further axis between the main poles known as the quadrature axis is utilized for accurate representation.

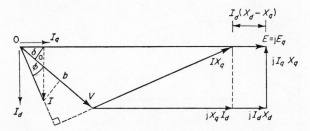

Figure 2.22   Phasor diagram of synchronous generator—full two-axis representation.

To account for the armature reaction there are two reactances $X_{ad}$ for the direct axis and $X_{aq}$ for the quadrature axis and $X_{ad} > X_{aq}$. Let $X_d = X_l + X_{ad}$ and $X_q = X_l + X_{aq}$, where $X_l$ = leakage reactance; then from the phasor diagram figure 2.22

$$\mathbf{E} = \mathbf{V} + j\mathbf{I}_d X_d + j\mathbf{I}_q X_q$$

$$= \mathbf{V} + j\mathbf{I}_q X_q + j\mathbf{I}_d(X_d - X_q) \quad \text{or} \quad \mathbf{E}_{qd} + j\mathbf{I}_d(X_d - X_q) \quad (2.11)$$

As $j\mathbf{I}_d(X_d - X_q)$ is in phase with $E$ knowing $V$, $I$ and $\phi$, $IX_q$ can be drawn and hence the direction of $E$ determined and consequently $\mathbf{I}_d$ and $\mathbf{I}_q$. Knowing $\mathbf{I}_d$ and $\mathbf{I}_q$ the remaining phasors can be drawn to complete the diagram. Values of $X_d$ and $X_q$ can be found by test.

*Output power*

The output power $= VI\cos\phi$.

From figure 2.22,

$$I\cos\phi = ob = oa + ab$$

$$= I_q\cos\delta + I_d\sin\delta.$$

If the armature resistance is negligible,

$$V\sin\delta = I_qX_q$$

$$V\cos\delta = E - I_dX_d$$

$$\therefore \qquad P = V\left[\frac{V\sin\delta}{X_q}\cos\delta + \frac{E - V\cos\delta}{X_d}\sin\delta\right]$$

$$= \frac{VE}{X_d}\sin\delta + \frac{V^2(X_d - X_q)}{2X_dX_q}\sin 2\delta \qquad (2.12)$$

and

$$\frac{dP}{d\delta} = \frac{VE}{X_d}\cos\delta + \frac{V^2(X_d - X_q)}{X_dX_q}\cos 2\delta \qquad (2.13)$$

$$= \text{Synchronizing Power Coefficient.}$$

The plot of $P$ against $\delta$ is shown in figure 2.23, from which it is evident that the steady-state stability limit occurs at a lower $\delta$ than $90°$.

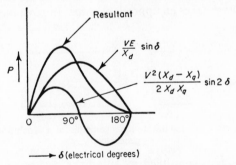

Figure 2.23  Power-angle curve—salient pole machine.

Equations 2.12 refer to operation in the steady state. Under transient conditions the direct-axis reactance becomes $X_d'$ or $X_d''$ as previously discussed, in the quadrature axis it is usual to assume $X_q = X_q'$. In the transient state the excitation voltage $E = jE_q$ still lies along the quadrature axis and the voltage 'behind' the transient reactance $E_1 = E_{q1}$ also lies on this axis. The transient power output equation is

$$P = \frac{E_{q1}V}{X'_d} \sin \delta + V^2 \frac{X_q - X'_d}{2X'_d X_q} \sin 2\delta \qquad (2.14)$$

In steady-state stability calculations the effects of saliency may usually be ignored, especially compared with the effects of saturation.

Consider a salient-pole machine with the following parameters $X_d = 1.2$ p.u., $X_q = 0.9$ p.u., excitation voltage $E = 1.5$ p.u. The maximum value of power output to an infinite busbar of 1 p.u. voltage is,

$$P = \frac{EV}{X_d} \sin \delta + V^2 \frac{X_d - X_q}{2X_d X_q} \sin 2\delta$$

$$= 1.25 \sin \delta + 0.14 \sin 2\delta.$$

Obtaining $P_{max}$ graphically or by other means, $P_{max} = 1.25$ p.u. When this exercise is repeated with lower values of $E$ the error introduced by ignoring saliency is seen to increase. Provided that $E > V$, saliency may generally be ignored in steady-state studies.

If the machine is connected to an infinite busbar through a link of reactance $X_L$ then $_T X_d = X_d + X_L$ and $_T X_q = X_q + X_L$, where the prefix $T$ refers to the total effective reactance of the system.

## 2.9 Automatic Voltage Regulators

In the previous pages the synchronous generator has been considered to operate with a fixed excitation and any changes in excitation have been assumed carried out manually. Any adjustment of the terminal voltage $V$ after a load change takes an appreciable time. In most modern generators

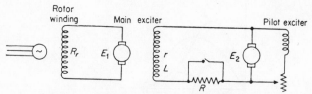

Figure 2.24   Excitation arrangements for a synchronous generator.

the output voltage is controlled by automatic devices so as to remain at a constant prearranged value. In this section the modes of operation of *Automatic Voltage Regulators* (A.V.R.) are discussed and their effect on the operation of synchronous machines.

The excitation arrangements for the rotor of a synchronous generator is shown in figure 2.24. The main exciter, a direct current shunt generator, generates a voltage $E_1$ which is supplied direct to the rotor winding through sliprings. The field of the main exciter is supplied from a further exciter which is self excited. Both the direct-current machine armatures are driven from the main rotor shaft. In series with the main exciter field winding is a resistor $R$ which can be short-circuited as dictated by a voltage sensing mechanism actuated by the terminal voltage of the synchronous machine. An aspect of major importance is the speed of response of the regulator, i.e. the time elapsing between the voltage deviation and the

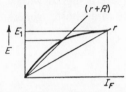

Figure 2.25    Main exciter self excited—effect of voltage regulator.

return to the prescribed value. Considering the circuit of figure 2.24 if $R$ is short-circuited the final excitation current is $E_2/r$ and assuming that the machine is unsaturated, the growth curve is exponential and of time constant $\tau = L/r$. If the main exciter had been self excited then the voltage $E_1$ is given by the magnetization curve as shown in figure 2.25.

The speed of response of the exciter is obtained in the following manner. Let the machine be on open circuit and adjusted to give the normal full-load voltage $E_1$. When the resistor $R$ is short-circuited the voltage rises as shown by the curve in figure 2.26. For a time interval of 0·5 seconds a

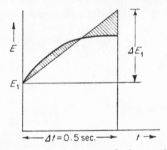

Figure 2.26    Response speed of an exciter.

line is drawn such that equal areas above and below the curve are obtained. The ratio $\Delta E_1/\Delta t$ is the nominal speed of response and $\Delta E_1/\Delta t/E_1$ is the relative speed of response. For large machines $E_1$ is a few hundred volts and with $E_1 = 200$ V and $\Delta E/\Delta t = 200$ V/sec the relative response is 1 which is a medium value.

One method of obtaining a high response for the exciter is to use a design giving a high maximum voltage when $R$ is short-circuited; this voltage is known as the *ceiling voltage*. In figure 2.27 the optimum exciter characteristic (i.e. infinitely fast exciter) is line abc and $E_c$ is the ceiling voltage which is usually in the order of $2E_1$. The relative speed of response is $dg/ag/E_1$ and if $E_c = 2E_1$ this equals 4. This gives an upper limit to the response for large separately-excited machines. Special exciters exist for forcing the rotor field when system faults occur and in these ceiling voltages of four times $E_1$ are achieved with a rate of rise of 2000 V/sec. This is obtained by the use of a booster generator in series with the main exciter.

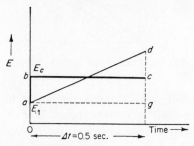

Figure 2.27   Optimum speed of response.

The above discussion has been limited to the exciter but the response at the synchronous machine terminals must also be studied. If a sudden change is made to the exciter field-circuit resulting in a sudden voltage being applied, then the e.m.f. generated by the exciter $e = E_1(1 - e^{-t/\tau_e})$ where $\tau_e$ is the time constant and $E_1$ the initial voltage. If the rotor iron is not saturated, then $e_r = R_r i + L_r \dfrac{di}{dt}$ where $R_r$ and $L_r$ refer to the rotor resistance and inductance.
But

$$e_r = E_1(1 - e^{-t/\tau_e}) = R_r i + L_r \frac{di}{dt};$$

hence the rotor current and combined time-constant may be determined.

*Types of automatic voltage regulator*

The detailed study of regulators is a specialist field and it is not intended here to discuss types in any detail but rather to indicate their general effect. There are two broad divisions of automatic regulators both of which set out to control the output voltage of the synchronous generator by controlling the exciter field. In general the deviation of the terminal voltage from a prescribed value is passed to control circuits and thus the field current varied and it is in the manner and speed in achieving this that the division occurs.

The first and older type can be broadly classed as *electromechanical*. A well-known variety of this is the carbon-pile regulator. In this a voltage proportional to the deviation voltage operates a solenoid assembly to

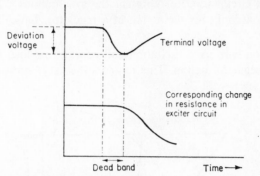

Figure 2.28  The effect of a dead band in a regulator.

vary the pressure exerted on a carbon-pile resistor in the exciter field, thus varying its resistance. Another type depends upon the conversion of the deviation voltage into a torque by means of a 'torque motor'; according to the angular position of the shaft of the motor, certain resistors in a resistor chain are cut out of circuit and hence the exciter field-current

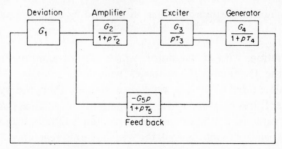

Figure 2.29 Block diagram of a continuously acting closed
loop automatic voltage regulator.

changes. There are various other types including the popular Vibrating Reed regulator. All these types suffer from the disadvantages of being relatively slow acting and possessing dead-bands, i.e. a certain deviation must occur before the mechanism operates; this is illustrated in figure 2.28.

The other main group of regulators is known as continuously acting and these are faster than the above and have no dead bands. A general block diagram of a typical control system is given in figure 2.29. The

amplifier can take a variety of forms including magnetic amplifiers and
rotary amplifiers such as Amplidyne, Metadyne and Magnavolt (these
are special types of direct current generators).

*Automatic voltage regulators and generator characteristics*
The equivalent circuit used to represent the synchronous generator can
be modified to account for the action of a regulator. Basically there are
three conditions to consider.

(i) Operation with fixed excitation and constant no-load voltage ($E$),
    i.e. no regulator action. This requires the usual equivalent circuit
    of $E$ in series with $Z_s$.

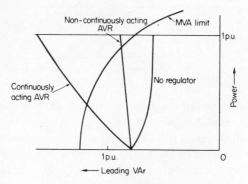

Figure 2.30   Performance chart as modified by the use of automatic voltage regulators.

(ii) Operation with a regulator which is not continuously acting,
     i.e. the terminal voltage varies with load changes. This can be
     simulated by $E'$ and a reactance smaller than the synchronous
     value. It has been suggested by experience in practice that a reason-
     able value would be the transient reactance although some authori-
     ties suggest taking half of the synchronous reactance. This mode
     will apply to most modern regulators.

(iii) Terminal voltage constant. This requires a very fast-acting regulator
      and the nearest approach to it exists in the forced-excitation re-
      gulators used on generators supplying very long lines; in these the
      exciter voltage is increased very rapidly in the order of 2000 V/sec.

Each of the above representations will give significantly different values
of maximum power output. The degree to which this happens depends on
the speed of the A.V.R. and the effect on the operation chart of the
synchronous generator is shown in figure 2.30 which indicates clearly

the increase in operating range obtainable. It should be noted however that operation in these improved leading power-factor regions may be limited by the heating of the stator winding. The actual power-angle curve may be obtained by a step-by-step process by using the gradually increasing values of $E$ in $EV/X \sin \delta$.

When a generator has passed through the steady-state limiting angle of $\delta = 90°$ with a fast acting A.V.R. it is possible for synchronism to be retained. The A.V.R. in forcing up the voltage, increases the power output of the machine so that instead of the power falling after $\delta = 90°$ it is maintained and $dP/d\delta$ is still positive.

*Example 2.3*   Determine the limiting powers for the system shown in figure 2.31 for the three types of voltage regulation. All values are on 100 MVA, 132 kV bases. It may be assumed that the lines and transformers are each represented by a single series reactance.

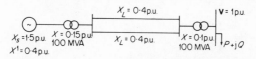

Figure 2.31   Line diagram of system for Example 2.3. Normal operating load $P = 0.8$ p.u. $Q = 0.5$ p.u.

*Solution*   (a) No control, constant excitation voltage.

$$X = 1.5 + 0.15 + 0.1 + \frac{0.4}{2} = 1.95 \text{ p.u.}$$

From figure 2.7,

$$E = \sqrt{\left[\left(1 + \frac{0.5 \times 1.95}{1}\right)^2 + \left(\frac{0.8 \times 1.95}{1}\right)^2\right]}$$

$$= 2.52 \text{ p.u.}$$

∴     $$P_{max} = \frac{EV}{X} = \frac{2.52 \times 1}{1.95} = 1.29 \text{ p.u.}$$

(b) Non-continuously acting A.V.R.

$$\cdot X = 0.4 + 0.15 + 0.1 + 0.2 = 0.85 \text{ p.u.}$$

$$E' = \sqrt{\left[\left(1 + \frac{0.5 \times 0.85}{1}\right)^2 + \left(\frac{0.8 \times 0.85}{1}\right)^2\right]}$$

$$= 1.59 \text{ p.u.}$$

$$\therefore \qquad P_{max} = \frac{1\cdot59 \times 1}{0\cdot85} = 1\cdot87 \text{ p.u.}$$

(c)

$$X = 0\cdot15 + 0\cdot1 + 0\cdot2 = 0\cdot45 \text{ p.u.}$$

Terminal voltage of generator $V_T$ is constant.

$$V = \sqrt{\left[\left(1 + \frac{0\cdot5 \times 0\cdot45}{1}\right)^2 + \left(\frac{0\cdot8 \times 0\cdot45}{1}\right)^2\right]}$$
$$= 1\cdot28 \text{ p.u.}$$

$$P_{max} = \frac{1\cdot28 \times 1}{0\cdot45} = 2\cdot85 \text{ p.u.}$$

It is interesting to compare the power limit when one of the two lines is disconnected; the transient-reactance representation will be used.

$$E' = \sqrt{\left(1 + \frac{0\cdot5 \times 1\cdot05}{1}\right)^2 + \left(\frac{0\cdot8 \times 1\cdot05}{1}\right)^2}$$
$$= 1\cdot74 \text{ p.u.}$$

$$P_{max} = \frac{1\cdot74 \times 1}{1\cdot05} = 1\cdot65 \text{ p.u.}$$

A recent innovation is the use of solid-state rectifiers to supply the d.c. to the rotor field. These are mounted on the main shaft and rotate with it and are supplied by slip rings. The associated A.V.R. is electronic in nature.

## LINES, CABLES AND TRANSFORMERS

### 2.10  Overhead Lines—Types and Parameters

Overhead lines are suspended from insulators which are themselves supported by towers or poles. The span between two towers is dependent upon the allowable sag in the line and for steel towers with very high voltage lines the span is normally 1200–1500 ft (370–460 m). Typical supporting structures are shown in figures 2.32 and 2.33. There are two main types of tower:

(a) those for straight runs in which the stress due to the weight of the line alone has to be withstood;
(b) those for changes in route, called deviation towers; these withstand the resultant forces set up when the line changes direction.

When specifying towers and lines, ice and wind loadings are taken into account as well as extra forces due to a break in the lines on one side of a tower. For lower voltages and distribution circuits wooden or reinforced concrete poles are used with conductors supported in horizontal formations.

The live conductors are insulated from the towers by insulators which take two basic forms, the *pin type* and the *suspension type*. The pin type

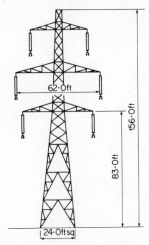

Figure 2.32 400 kV double-circuit overhead line tower. Two conductors per phase (bundle conductors). (*Permission of Institution of Electrical Engineers.*)

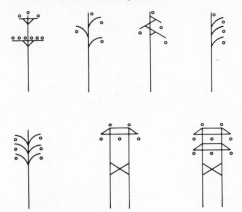

Figure 2.33 Typical pole-type structures.

insulator is shown in figure 2.34 and it is seen that the conductor is supported on the top of the insulator. This type is used for lines up to 33 kV. The two or three porcelain 'sheds' or 'petticoats' provide an adequate

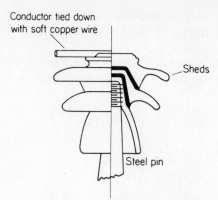

Figure 2.34   Pin-type insulators.

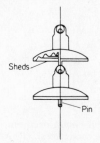

Figure 2.35   Suspension-type insulators.

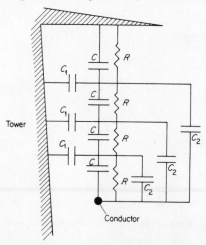

Figure 2.36   Equivalent circuit of a string of four suspension insulators. $C$ = self capacitance of disc, $C_1$ = capacitance disc to earth,
$C_2$ = capacitance disc to line, $R$ = leakage resistance.

leakage path from the conductor to earth and are shaped to follow the equipotentials of the electric field set up by the conductor–tower system. Suspension insulators (figure 2.35) consist of a string of interlinking separate discs made of glass or porcelain. A string may consist of many discs depending upon the line voltage; for 400 kV lines, 19 discs of overall length 12 ft 7 in (3·84 m) are used. The conductor is held at the bottom of the string which is suspended from the tower. Owing to the capacitances existing between the discs, conductor and tower, the distribution of voltage along the insulator string is not uniform, the discs nearer the conductor being the more highly stressed. Methods of calculating this voltage distribution are available but of dubious value owing to the shunting effect of the leakage resistance (see figure 2.36). This resistance depends on the presence of soot and dirt on the insulator surfaces and is considerably modified by rain and fog.

*Parameters*

The parameters of interest for circuit analysis are inductance, capacitance, resistance and leakage resistance. The derivation of formulae for the calculation of these quantities is given in reference 3. It is intended here to merely quote these formulae and discuss their application.

The inductance of a single-phase, two-wire line

$$= \frac{\mu_0}{4\pi}\left[1 + 4\ln\left(\frac{d-r}{r}\right)\right] H/m \tag{2.15}$$

where $d$ is the distance between the centres and $r$ is the radius, of the conductors. When performing load flow and balanced-fault analysis on three-phase systems it is usual to consider one phase only with the appropriate angular adjustments made for the other two phases. Phase voltages therefore are used and the inductances and capacitances are the equivalent phase or line-to-neutral values. For a three-phase line with equilateral spacing (figure 2.37) the inductance and capacitance with respect to the hypothetical neutral conductor are used and this inductance can be shown

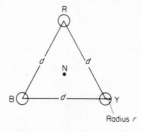

Figure 2.37   Overhead line with equilateral spacing of the conductors.

to be half the loop inductance of the single-phase line, i.e. the inductance of one conductor.

The line-neutral inductance for equilateral spacing

$$= \frac{\mu_0}{8\pi}\left[1 + 4\ln\left(\frac{d-r}{r}\right)\right]H/m \qquad (2.16)$$

The capacitance of a single-phase line

$$C = \frac{\pi\epsilon_0}{\ln(d-r/r)}F/m \qquad (2.17)$$

With three-phase conductors spaced equilaterally the capacitance of each line to the hypothetical neutral is double that for the two-wire circuit, i.e.

$$\frac{2\pi\epsilon_0}{\ln(d/r)}F/m \ (d \gg r) \qquad (2.18)$$

In practice the conductors are rarely spaced in the equilateral formation, and it can be shown that the average value of inductance or capacitance for any formation of conductors can be obtained by the representation of the system by one of equivalent equilateral spacing. The equivalent spacing $d_{eq}$ between conductors is given by,

$$d_{eq} = \sqrt[3]{(d_{12} \cdot d_{23} \cdot d_{31})}$$

Often two, three-phase circuits are electrically in parallel; if physically remote from each other the reactances of the lines are identical. When the two circuits are situated on the same towers, however, the magnetic interaction between them should be taken into account. The use of bundle conductors, that is more than one conductor per insulator, reduces the reactance; it also reduces conductor-surface voltage gradients and hence corona loss and radio interference. Unsymmetrical conductor spacing

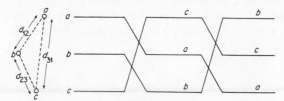

Figure 2.38   Transposition of conductors.

results in different inductances for each phase which causes an unbalanced voltage drop even when the load currents are balanced. The residual or resultant voltage or current induces unwanted voltages into neighbouring communication lines. This can be overcome by the interchange of conductor positions at regular intervals along the route, a practice known as *Transposition*, see figure 2.33. In practice lines are rarely transposed at

Table 2.2(a)   Overhead line constants at 50 Hz
(*per phase, per mile*)

| Voltage No. and area of conductors (in²) | | 132 kV 1× 0·175 | 1× 0·4 | 275 kV 2× 0·175 | 2× 0·4 | 400 kV 2× 0·4 | 4× 0·4 |
|---|---|---|---|---|---|---|---|
| Resistance ($R$) | Ohms | 0·25 | 0·11 | 0·125 | 0·055 | 0·055 | 0·027 |
| Reactance ($X_L$) | Ohms | 0·66 | 0·65 | 0·54 | 0·52 | 0·52 | 0·435 |
| Susceptance ($1/X_c$) | Micro-Mhos | 4·72 | 4·72 | 5·92 | 5·92 | 5·92 | 6·58 |
| Charging current ($I_c$) | Amps | 0·36 | 0·36 | 0·94 | 0·94 | 1·36 | 1·52 |
| Surge impedance | Ohms | 373 | 371 | 302 | 296 | 296 | 258 |
| Natural load | MW | 47 | 47 | 250 | 255 | 540 | 620 |
| $X_L/R$ Ratio | — | 2·6 | 5·9 | 4·3 | 9·5 | 9·5 | 15·8 |
| Thermal rating: | | | | | | | |
| Cold weather (below 40°F) | MVA | 125 | 180 | 525 | 760 | 1100 | 2200 |
| Normal (40°–65°F) | MVA | 100 | 150 | 430 | 620 | 900 | 1800 |
| Hot weather (above 65°F) | MVA | 80 | 115 | 330 | 480 | 790 | 1580 |

regular intervals and transposition is carried out where physically convenient, for example, at substations. In many cases the degree of unbalance existing without transposition is small and may be neglected in calculations.
*Resistance* Overhead line conductors usually comprise a stranded steel core (for mechanical strength) surrounded by aluminium wires which form the conductor. The resistance at power frequencies whether stranded or solid is higher than the direct-current resistance due to the skin effect; also the effect of the temperature coefficient of resistance is significant.

Leakage resistance is usually negligible for most calculation purposes and is very difficult to assess because of its dependence on the weather. This resistance represents the combined effect of all the various paths to earth from the line. The main path is that presented by the surfaces of the line insulators, the resistance of which depends on the condition of these surfaces. This varies considerably according to location; in industrial areas there will be layers of dirt and soot whilst in coastal districts deposits of salt occur. The bushings of circuit breakers and transformers form other leakage paths. The leakage losses on 132 kV lines vary between 0·3 to 1 kW per mile. Table 2.2(a) gives the parameters for various overhead-line circuits for the line voltages operative in Britain, Table 2.2(b) gives values for international lines and Table 2.2(c) relates to the United States.

Table 2.2(b)  Typical characteristics of bundled-conductor E.H.V. lines

| Number of subconductors in bundle | Country | Line voltage and number of circuits (in parentheses) kV | Diameter of subconductors mm | Radius of circle on which subconductors are arranged mm | Resistance of bundle Ω/km | Inductive reactance at 50 Hz Ω/km | Susceptance at 50 Hz μmho/km |
|---|---|---|---|---|---|---|---|
| 1 | Japan | 275 (2) | 27·9 | — | 0·0744 | 0·511 | 3·01 |
|  | Canada | 300 (2) | 35·0 | — | 0·0451 | 0·492 | 2·33 |
|  | Australia | 330 (1) | 45·0 | — | 0·0367 | 0·422 | 2·78 |
|  | USSR | 330 (1) | 30·2 | — | 0·065 | 0·404 | 2·82 |
|  | Italy | 380 (1) | 50·0 | — | 0·0294 | 0·398 | 2·84 |
|  | Average (one conductor) |  |  |  | 0·048 | 0·442 | 2·75 |
| 2 | Japan | 275 (2) | 25·3 | 200 | 0·0444 | 0·374 | 4·35 |
|  | Canada | 360 (1) | 28·1 | 229 | 0·04 | 0·314 | 3·57 |
|  | Australia | 330 (1) | 31·8 | 190·5 | 0·0451 | 0·341 | 3·34 |
|  | USA | 345 (1) | 30·4 | 228·6 | 0·0315 | 0·325 | 3·55 |
|  | Italy | 380 (1) | 31·5 | 200 | 0·0285 | 0·315 | 3·57 |
|  | USSR | 330 (1) | 28·0 | 200 | 0·04 | 0·321 | 3·43 |
|  | Average (two conductors) |  |  |  | 0·0400 | 0·323 | 3·58 |
| 3 | Sweden | 380 (1) | 31·68 | 260 | 0·018 | 0·29 | 3·85 |
|  | USSR | 525 (1) | 30·2 | 230 | 0·0212 | 0·294 | 3·88 |
| 4 | Germany | 380 (2) | 21·7 | 202 | 0·0316 | 0·260 | 4·3 |
|  | Germany | 500 (1) | 22·86 | 323 | 0·0285 | 0·272 | 4·0 |
|  | Canada | 735 (1) | 35·04 | 323 | 0·0121 | 0·279 | 3·9 |

Table 2.2(c)   Overhead line parameters—U.S.A.
(Permission Edison Electric Institute)

| Line voltage (kV) | 345 | 345 | 500 | 500 | 735 | 735 |
|---|---|---|---|---|---|---|
| Conductors per phase at 18 in. spacing | 1 | 2 | 2 | 4 | 3 | 4 |
| Conductor code | Expanded | Curlew | Chuker | Parakeet | Expanded | Pheasant |
| Conductor diameter (inches) | 1·750 | 1·246 | 1·602 | 0·914 | 1·750 | 1·382 |
| Phase spacing (ft) | 28 | 28 | 38 | 38 | 56 | 56 |
| GMD (ft) | 35·3 | 35·3 | 47·9 | 47·9 | 70·6 | 70·6 |
| 60 Hz inductive reactance Ω/mile $\{X_A$ | 0·3336 | 0·1677 | 0·1529 | 0·0584 | 0·0784 | 0·0456 |
| $X_D$ | 0·4325 | 0·4325 | 0·4694 | 0·4694 | 0·5166 | 0·5166 |
| $X_A+X_D$ | 0·7661 | 0·6002 | 0·6223 | 0·5278 | 0·5950 | 0·5622 |
| 60 Hz capacitive reactance MΩ-miles $\{X'_A$ | 0·0777 | 0·0379 | 0·0341 | 0·0126 | 0·0179 | 0·0096 |
| $X'_D$ | 0·1057 | 0·1057 | 0·1147 | 0·1147 | 0·1263 | 0·1263 |
| $X'_A+X'_D$ | 0·1834 | 0·1436 | 0·1488 | 0·1273 | 0·1442 | 0·1359 |
| Zo (Ω) | 374·8 | 293·6 | 304·3 | 259·2 | 292·9 | 276·4 |
| Natural loading MVA | 318 | 405 | 822 | 965 | 1844 | 1955 |
| Conductor d.c. resistance at 25°c (Ω/mile) | 0·0644 | 0·0871 | 0·0510 | 0·162 | 0·0644 | 0·0709 |
| Conductor a.c. resistance (60 Hz) at 50°c (Ω/mile) | 0·0728 | 0·0979 | 0·0599 | 0·179 | 0·0728 | 0·0805 |

$X_A$ = Component of inductive reactance due to flux within a 1 ft radius.
$X_D$ = Component due to other phases.
Total reactance per phase = $X_A+X_D$

$$X_A = 0\cdot2794 \log_{10} \left( \frac{1}{[N(GMR)\,(A)^{N-1}]^{1/N}} \right) \Omega/\text{mile}$$

$X_D = 0\cdot2794 \log_{10} (GMD) \ \Omega/\text{mile}$
$X'_A$ and $X'_D$ are similarly defined for capacitive reactance and

$$X'_A = 0\cdot0683 \log_{10} \left( \frac{1}{[Nr(A)^{N-1}]^{1/N}} \right) M\Omega\text{-miles}; \quad X'_D = 0\cdot0683 \log_{10} (GMD) \ M\Omega\text{-miles}.$$

$GMR$ = Geometric mean radius in feet.
$GMD$ = Geometric mean diameter in feet.
$N$ = Number of conductors per phase.
$A = S/2 \sin(\pi/N); N>1.$
$A = 0; N = 1.$
$S$ = Bundle spacing in feet.
$r$ = Conductor radius in feet.

### 2.11    Representation of Lines

The manner in which lines and cables are represented depends very much on their length and the accuracy required. There are three broad classifications of length—short, medium and long. The actual line or cable is

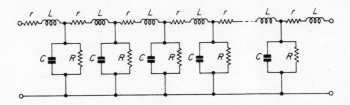

Figure 2.39    Distributed constant representation of a line: $L =$ Inductance line to neutral per unit length, $r =$ a.c. resistance per unit length, $C =$ capacitance line to neutral per unit length, $R =$ leakage resistance per unit length.

a distributed-constant circuit, i.e. it has resistance, inductance, capacitance and leakage resistance distributed evenly along its length as shown in

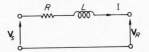

Figure 2.40    Equivalent circuit of a short line—representation under balanced three-phase conditions.

figure 2.34. Except for long lines the total resistance, inductance, capacitance and leakage resistance of the line are concentrated to give a lumped-constant circuit.

*The short line (up to 50 miles—80 km.)*

The equivalent circuit is shown in figure 2.40 and it will be noticed that both shunt capacitance and leakage resistance have been neglected. The four-terminal network constants are

$$\mathbf{A} = 1, \quad \mathbf{B} = \mathbf{Z}, \quad \mathbf{C} = 0 \quad \text{and} \quad \mathbf{D} = 1$$

The drop in voltage along a line is important and the *Regulation* is defined as,

$$\frac{\text{received voltage on no load} - \text{received voltage on load}}{\text{received voltage on load } (V_R)}$$

It should be noted that if $\mathbf{I}$ is leading $V_r$ in phase, i.e. a capacitive load, then $V_R > V_S$, as shown in figure 2.41.

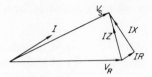

Figure 2.41   Phasor diagram for short line on leading load.

*Medium length lines (up to 150 miles—240 km.)*

Owing to the increased length the shunt capacitance is now included to form either a $\pi$ or a T network. The circuits are shown in figure 2.42.

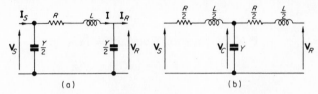

Figure 2.42a   Medium length line—$\pi$ equivalent circuit.
b   Medium length line—T equivalent circuit.

Of these two versions the $\pi$ representation tends to be in more general use but there is little difference in accuracy between the two. For the $\pi$ network;

$$\mathbf{V}_S = \mathbf{V}_R + \mathbf{IZ}; \quad \mathbf{I} = \mathbf{I}_R + \mathbf{V}_R\frac{\mathbf{Y}}{2}; \quad \mathbf{I}_S = \mathbf{I} + \mathbf{V}_S\frac{\mathbf{Y}}{2},$$

from which $\mathbf{V}_S$ and $\mathbf{I}_S$ are obtained in terms of $\mathbf{V}_R$ and $\mathbf{I}_R$ giving the following constants;

$$\mathbf{A} = \mathbf{D} = 1 + \frac{\mathbf{ZY}}{2}, \quad \mathbf{B} = \mathbf{Z} \quad \text{and} \quad \mathbf{C} = \left(1 + \frac{\mathbf{ZY}}{4}\right)\mathbf{Y}$$

Similarly for the T network;

$$\mathbf{V}_S = \mathbf{V}_C + \frac{\mathbf{ZI}_S}{2}; \quad \mathbf{V}_C = \mathbf{V}_R + \frac{\mathbf{ZI}_R}{2}; \quad \mathbf{I}_S = \mathbf{I}_R + \mathbf{V}_C\mathbf{Y},$$

giving

$$\mathbf{A} = \mathbf{D} = 1 + \frac{\mathbf{ZY}}{2}, \quad \mathbf{B} = \left(1 + \frac{\mathbf{ZY}}{4}\right)\mathbf{Z} \quad \text{and} \quad \mathbf{C} = \mathbf{Y}.$$

*The long line (above 150 miles)*

Here the treatment assumes distributed parameters. The changes in voltage and current over an elemental length $\Delta x$ of the line, $x$ metres from

the sending end are determined and conditions for the whole line obtained
by integration.

Let  $R$  = resistance/unit length        $z$  = impedance/unit length
    $L$  = inductance/unit length        $y$  = shunt admittance/unit length
    $G$  = leakage/unit length          $Z$  = total series impedance of line
    $C$  = capacitance/unit length      $Y$  = total shunt admittance of line

The voltage and current $x$ metres from the sending end are given by

$$\left. \begin{array}{l} \mathbf{V}_x = \mathbf{V}_s \cosh \mathbf{P}x - \mathbf{I}_s\mathbf{Z}_0 \sinh \mathbf{P}x \\[3mm] \mathbf{I}_x = \mathbf{I}_s \cosh \mathbf{P}x - \dfrac{\mathbf{V}_s}{\mathbf{Z}_0} \sinh \mathbf{P}x \end{array} \right\} \qquad (2.19)$$

where

$$\mathbf{P} = \text{Propagation Constant} = (\alpha + j\beta)$$

$$= \sqrt{[(R+j\omega L)(G+j\omega C)]} = \sqrt{(\mathbf{z} \cdot \mathbf{y})}$$

and

$$\mathbf{Z}_0 = \text{Characteristic Impedance}$$

$$= \sqrt{\frac{R+j\omega L}{G+j\omega C}} \qquad (2.20)$$

$\mathbf{Z}_0$ is the input impedance of an infinite length of the line; hence if any
line is terminated in $\mathbf{Z}_0$ its input impedance is also $\mathbf{Z}_0$.

The Propagation Constant $\mathbf{P}$ represents the changes occurring in the
transmitted wave as it progresses along the line. $\alpha$ measures the attenuation
and $\beta$ the angular phase-shift. With a loss-free line, $\mathbf{P} = j\omega\sqrt{(LC)}$ and
$\beta = \omega\sqrt{(LC)}$. With a velocity of propagation of $3 \times 10^5$ km/sec the wavelength of the transmitted voltage and current at 50 c/s is 6000 km. Thus
lines are much shorter than the wavelength of the transmitted energy.

Usually conditions at the load are required, when $x = l$ in equations
2.19.

$$\therefore \qquad\qquad \left. \begin{array}{l} \mathbf{V}_r = \mathbf{V}_s \cosh \mathbf{P}l - \mathbf{I}_s\mathbf{Z}_0 \sinh \mathbf{P}l \\[3mm] \mathbf{I}_r = \mathbf{I}_s \cosh \mathbf{P}l - \dfrac{\mathbf{V}_s}{\mathbf{Z}_0} \sinh \mathbf{P}l \end{array} \right\} \qquad (2.21)$$

and

Alternatively,

$$\left. \begin{array}{l} \mathbf{V}_s = \mathbf{V}_r \cosh \mathbf{P}l + \mathbf{I}_R\mathbf{Z}_0 \sinh \mathbf{P}l \\[3mm] \mathbf{I}_s = \dfrac{\mathbf{V}_r}{\mathbf{Z}_0} \sinh \mathbf{P}l + \mathbf{I}_R \cosh \mathbf{P}l \end{array} \right\} \qquad (2.22)$$

and

The parameters of the equivalent four-terminal network are thus,

$$A = D = \cosh \sqrt{(ZY)}$$

$$B = \sqrt{\frac{Z}{Y}} \sinh \sqrt{(ZY)}$$

and

$$C = \sqrt{\frac{Y}{Z}} \sinh \sqrt{(ZY)}$$

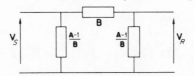

Figure 2.43 Equivalent circuit to accurately represent the terminal conditions of a long line.

The easiest way to handle the hyperbolic functions is to use the appropriate series.

$$A = \cosh \sqrt{(ZY)} = 1 + \frac{YZ}{2} + \frac{Y^2 Z^2}{24} + \frac{Y^3 Z^3}{720}$$

$$B = Z\left(1 + \frac{YZ}{6} + \frac{Y^2 Z^2}{120} + \frac{Y^3 Z^3}{5040}\right)$$

$$C = Y\left(1 + \frac{YZ}{6} + \frac{Y^2 Z^2}{120} + \frac{Y^3 Z^3}{5040}\right)$$

Usually not more than three terms are required and for (overhead) lines less than 500 km in length the following expressions for the constants hold approximately,

$$A = D = 1 + \frac{ZY}{2}, \quad B = Z\left(1 + \frac{ZY}{6}\right), \quad C = Y\left(1 + \frac{ZY}{6}\right).$$

An exact equivalent circuit for the long line can be expressed in the form of the $\pi$ section shown in figure 2.43. The application of simple circuit laws will show that this circuit yields the correct four terminal network equations. If only the first term of the expansions is used, $B = Z$ and $\frac{A-1}{B} = \frac{Y}{2}$, i.e. the medium-length $\pi$ representation. Figure 2.43 is exact only for conditions at the ends of the line; if intermediate points are to be investigated the full equations must be used.

*Example 2.4*   The conductors of a one mile long, 3·3 kV, overhead line are in horizontal formation with 30 in between centres. The effective diameter of the conductors is 0·138 in. The resistance per mile of the conductors is 0·66 Ω. Calculate the line to neutral inductance of the line If the sending-end voltage is 3·3 kV and the load 1 MW at 0·8 p.f. lagging, calculate the voltage at the load busbar, the power loss in the line, and the efficiency of transmission.

*Solution*   The equivalent equilateral spacing is given by

$$d_e = \sqrt[3]{(d_{12}d_{23}d_{31})}$$

In this case

$$d_e = \sqrt[3]{(30 \times 30 \times 60)} = 30\sqrt[3]{2} = 37\cdot8 \text{ in.}$$

The inductance (line to neutral) =

$$= \frac{\mu_0}{2 \times 4\pi}\left[1 + 4\ln\left(\frac{d-r}{r}\right)\right]\text{H/m}$$

$$= \frac{4\pi}{10^7 \times 2 \times 4\pi}\left[1 + 4\ln\left(\frac{37\cdot8 - 0\cdot069}{0\cdot069}\right)\right]\text{H/m}$$

$$= \frac{1}{2 \times 10^7}\left(\frac{26\cdot21}{1}\right)\text{H/m}$$

Total inductance

$$= 1\cdot0 \times 5280 \times 0\cdot3048 \times \frac{13\cdot105}{10^7} = 0\cdot00212 \text{ H}$$

Inductive Reactance

$$= 2\pi f L = 2\pi \times 50 \times 0\cdot00212 = 0\cdot66 \ \Omega$$

The line is obviously in the category of short and will be treated accordingly.

The load can be expressed as,

$$P = 1 \text{ MW}, \quad Q = 0\cdot75 \text{ MVAr.}$$

The load can be represented by an equivalent shunt impedance in which case $P$ and $Q \propto V_R^2$. If $P$ and $Q$ are constant regardless of $V_R$ then an iterative procedure must be used. The latter will be used here. As the load voltage is unknown to obtain the current the nominal voltage of 3·3 kV will be assumed. The current is then,

$$\frac{10^6}{\sqrt{(3)} \times 3300 \times 0\cdot8} = 218\underline{/-\cos^{-1} 0\cdot8} \text{ A}$$

$$V_s = V_R + IZ$$

$$\frac{3300}{\sqrt{3}} = V_R + (218 \times 0 \cdot 8 - j218 \times 0 \cdot 6)(0 \cdot 66 + j0 \cdot 66)$$

$$= V_R + 115 \cdot 0 + 86 \cdot 7 + j(115 \cdot 0 - 86 \cdot 7)$$

$$1900 = V_R + 201 \cdot 7 + j28 \cdot 3$$

$$V_R = 1698 \cdot 3 - j28 \cdot 3$$

$$V_R = 1698 \text{ V with negligible phase shift.}$$

A quicker method would be to use,

$$\Delta V = \frac{RP + XQ}{V_R}$$

$$= \frac{\frac{1}{3} \times 0 \cdot 66 \times 10^6 + \frac{1}{3} \times 0 \cdot 66 \times 0 \cdot 75 \times 10^6}{1900} = \frac{385}{1 \cdot 9} = 202 \text{ V}$$

$$\therefore \qquad V_R = 1900 - 202 = 1698 \text{ V.}$$

As, however, the load is 1 MW at the receiving end $I$ should be recalculated and the above process repeated. As the approximate method is more convenient, it will be used.

$$\Delta V' = \frac{385}{1 \cdot 698} = 227 \text{ V}$$

$$V'_R = 1900 - 227 = 1673 \text{ V}$$

Iterating again,

$$\Delta V'' = \frac{385}{1 \cdot 673} = 230 \text{ V}$$

$$V''_R = 1670 \text{ V}$$

Again,

$$\Delta V''' = \frac{385}{1670} = 230.$$

The final value of $V_R$ is therefore 1670 V.

$V_R$ can be obtained analytically without recourse to iterative procedures, but this is more lengthy in calculation. The type of load in this question in which $P$ and $Q$ remain constant regardless of the voltage is called *stiff*.

$$\text{The line loss} = I^2R = 218^2 \times 0 \cdot 66 \text{ per phase}$$
$$= 94 \text{ kW for three phases.}$$

Efficiency of transmission

$$= \frac{\text{output}}{\text{output} + \text{losses}} = \frac{1000}{1000 + 94} \times 100 \text{ per cent}$$

$$= 91 \cdot 4 \text{ per cent.}$$

*Example 2.5*   A 275 kV, three-phase transmission line of length 300 miles is rated at 840 A. The values of resistance, inductance and susceptance per phase per mile are 0·125 Ω, 1·7 mH and 5·92 μmhos respectively. The receiving-end voltage is 275 kV when full load is transmitted at 0·85 power factor lagging. Calculate the sending-end voltage.

Both the nominal $\pi$ and long-line methods will be used to solve this problem.

*Solution*   (a) The equivalent $\pi$ section is shown in figure 2.42

$$\mathbf{Z} = R + jX$$

where

$$R = 300 \times 0 \cdot 125 = 37 \cdot 5 \, \Omega$$

$$X = 300 \times 2\pi \times 50 \times \frac{1 \cdot 7}{1000} = 160 \, \Omega.$$

The leakage resistance is infinite.

$$\mathbf{Y} = j\frac{5 \cdot 92 \times 300}{10^6} \text{ mhos}$$

$$= j1776 \times 10^{-6} \text{ mhos}.$$

Take the receiving-end voltage as reference phasor.
Load current

$$\mathbf{I}_R = 840 \text{ A}, \, 0 \cdot 85 \text{ p.f. lagging}$$

$$= (720 - j436) \text{ A}$$

$$\mathbf{I}_L = (720 - j436) + \left( \frac{275 \times 10^3}{\sqrt{3}} \times \frac{j1776}{2 \times 10^6} \right)$$

$$= 720 - j436 + j141$$

$$= 720 - j295 \text{ A}$$

$$\mathbf{I}_L \mathbf{Z}_1 = (720 - j295)(37 \cdot 5 + j160)$$

$$= 74300 + j104100$$

$$\therefore \mathbf{V}_S = 159000 + 74300 + j104100$$

$$\mathbf{V}_S = 256 \text{ kV.}$$

Line voltage at sending-end

$$= \sqrt{(3)} \times 256 \text{ kV} = 443 \text{ kV.}$$

From the long line equations

$$\mathbf{A} = \mathbf{D} = \cosh \sqrt{(\mathbf{ZY})}$$

$$B = \sqrt{\frac{Z}{Y}} \sinh \sqrt{(ZY)}$$

$$C = \sqrt{\frac{Y}{Z}} \sinh \sqrt{(ZY)}$$

Using the appropriate series,

$$A = D = 1 + \frac{YZ}{2} + \frac{Y^2 Z^2}{24}$$

$$YZ = (j1776 \times 10^{-6})(37 \cdot 5 + j160) = -0 \cdot 284 + j0 \cdot 0666$$

$$(Y^2 Z^2) = 0 \cdot 0807 - 0 \cdot 00444 - j0 \cdot 0375 = 0 \cdot 08514 - j0 \cdot 0378$$

$$A = D = 1 + -0 \cdot 142 + j0 \cdot 0333 + 0 \cdot 0035 - j0 \cdot 001575$$

$$= 1 - 0 \cdot 1385 + j0 \cdot 031725 = 0 \cdot 8615 + j0 \cdot 0317$$

$$B = Z\left(1 + \frac{YZ}{6} + \frac{Y^2 Z^2}{120}\right)$$

$$= (37 \cdot 5 + j160)(1 + -0 \cdot 0463 + j0 \cdot 0108)$$

$$= 33 \cdot 97 + j152 \cdot 9$$

$$C = Y\left(1 + \frac{YZ}{6} + \frac{Y^2 Z^2}{120}\right)$$

$$= j\frac{1776}{10^6}(0 \cdot 954 + j0 \cdot 0108)$$

$$= (-19 \cdot 2 + j1693)10^{-6}$$

$$V_S = AV_R + BI_R$$

$$I_S = CV_R + DI_R$$

Hence

$$V_S = (0 \cdot 8165 + j0 \cdot 0317)(159000) + (33 \cdot 97 + j152 \cdot 9)(720 - j436)$$

$$= 221,100 + j100250$$

$$V_S = 243 \text{ kV (phase).}$$

It is seen that the use of the nominal $\pi$ model gives a value for $V_S$ which is about 5 per cent high. This error is high and the long-line solution should be used. It is evident, however, that little error would be incurred by the use of only two terms of the series instead of three, i.e. $(1 + YZ/2)$ and $(1 + YZ/6)$ thus reducing the work involved. This line however is very long and such lengths are not usual except in hydro-electric schemes in the U.S.A., Sweden and the U.S.S.R. In Britain the longest lines used are in the order of 100 miles in length and the use of the nominal $\pi$ will yield reasonable results.

The drop in voltage along the line is (243–159), i.e. 84 kV giving a regulation of 34·6 per cent. This is a very large volt-drop and could not be tolerated in practice especially as the range of tap-changing transformers is usually only up to 15 per cent. In practice such a line would normally incorporate series capacitors to reduce the series reactance and the load-current power factor would be increased from 0·9 lag to near unity by the use of shunt capacitors or synchronous compensators at the receiving end.

### The natural load

The characteristic impedance $Z_0$ is also known as the surge impedance. When a line is terminated in its characteristic impedance the power delivered is known as the natural load. For a loss-free line under natural-load conditions the reactive power absorbed by the line is equal to the reactive power generated, i.e.

$$\frac{V^2}{X_C} = I^2 X_L$$

and

$$\frac{V}{I} = Z_0 = \sqrt{(X_L X_C)} = \sqrt{\frac{L}{C}}$$

At this load $V$ and $I$ are in phase all along the line and optimum transmission conditions obtained; in practice however the load impedances are seldom in the order of $Z_0$. Values of $Z_0$ for various line voltages are as follows, values of the corresponding natural loads are shown in brackets:

132 kV, 150 $\Omega$ (50 MW); 275 kV, 315 $\Omega$ (240 MW);
380 kV, 295 $\Omega$ (490 MW).

The angle of the impedance varies between 0 and $-15°$. For underground cables $Z_0$ is roughly one tenth of the overhead-line value.

### 2.12  Parameters of Underground Cables

In cables with three conductors contained within the lead or aluminium sheath the electric field set up has components tangential to the layers of impregnated-paper insulation in which direction the dielectric strength is poor. At voltages over 11 kV therefore, each conductor is separately screened (Höchstadter Type) to ensure only radial stress through the paper. The capacitance of single conductor and individually screened three-conductor cables is readily calculated. For three-conductor unscreened cables resort must be made to empirical design data. Cross-sections of typical high voltage cables are shown in figures 2.44a and 2.44b. The capacitance ($C$) of single-core cables may be calculated from design data by the use of the formula

$$C = \frac{2\pi\epsilon_0\epsilon_r}{\ln (R/r)} \tag{2.23}$$

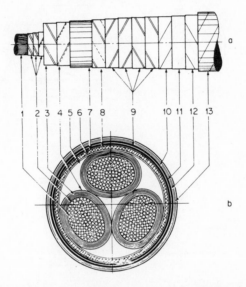

Figure 2.44. Typical gas-filled cables, 1 conductor, 2 conductor screen, 3 dielectric, 4 dielectric screen, 5 jute fillers, 6 binder, 7 lead sheath, 8 copper-wire woven fabric tape, 9 reinforcing tapes, 10 bituminized cotton tape, 11 rubber sheath, 12 carbon-impregnated tape, 13 hessian tapes, a single-core, 132 kV, b 3-core, 33 kV. (*Permission Institution of Electrical Engineers.*)

where $r$ and $R$ are the inner and outer radii of the dielectric and $\epsilon_r$ is the relative permittivity of the dielectric, usually in the order of 3·5. This also holds for three-core cables with each conductor separately screened.

The various capacitances present in three-core unscreened cables may be represented as shown in figure 2.45, in which the conductor to conductor capacitance is $C_1$ and the conductor to sheath capacitance is $C_0$. The equivalent circuit is shown in figure 2.45b and circuits obtained by star-delta or delta-star transformations in (c) and (d). The value of

$$C_3 = \frac{C_0}{3}$$

and similarly $C_4 = 3C_1$. The parallel combination of $C_4$ and $C_0$ gives the equivalent line to neutral value of the cable capacitance. $C_0$ and $C_1$ may be measured as follows. The three cores are short-circuited and the capacitance between them and the sheath measured; this measurement gives $3C_0$. Next the capacitance between two cores is measured, the third

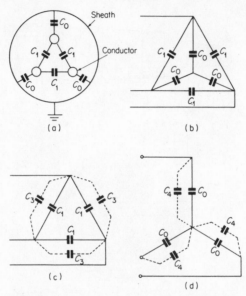

Figure 2.45   Capacitances in a three-core cable.

being connected to the sheath. It is easily shown that the measured value is now $C_1 + [(C_0 + C_1)/2)]$; hence $C_0$ and $C_1$ are obtained.

Owing to the symmetry of the cable the normal phase-sequence values of $C$ and $L$ are the same as the negative phase-sequence values (i.e. for reversed phase rotation). The series resistance and inductance are complicated by the magnetic interaction between the conductor and sheath (see Chapter 8). For a single-phase cable it can be shown that the conductor reactance is given by,

$$2\omega \left( 2 \ln \frac{d}{r} + \frac{1}{2} \right) 10^{-7} \Omega/\text{metre},$$

where          $d$ = spacing between conductor centres
               $r$ = radius of the solid conductor.

The values for three-phase circuits of single core or three screened cores, cables are given in references 1 and 2.

The effective resistance of the conductor is the direct-current resistance modified by the following factors: the skin effect in the conductor; the eddy currents induced by adjacent conductors (the proximity effect); the equivalent resistance to account for the $I^2R$ losses in the sheath. The determination of these effects is complicated and is discussed further in Chapter 8.

Table 2.3   Underground cable constants at 50 Hz
(*per mile*)

|  |  | 132 kV | | 275 kV | | 400 kV |
| --- | --- | --- | --- | --- | --- | --- |
| Size | sq. in. | 0·55 | 1·00 | 1·50 | 1·75 | 3·00 |
| Rating (soil $g$ = 120°c cm/W) | Amps | 550 | 870 | 1,100 | 1,100 | 1,600 |
|  | MVA | 125 | 200 | 525 | 525 | 1,100 |
| Resistance ($R$) at 85°c | Ohms | 0·104 | 0·056 | 0·040 | 0·033 | 0·021 |
| Reactance ($X_L$) | Ohms | 0·206 | 0·222 | 0·352 | 0·216 | 0·352 |
| Charging current ($I_c$) | Amps | 12·7 | 17·2 | 25·3 | 28·2 | 38·4 |
| $X_L/R$ Ratio |  | 2·0 | 4·0 | 8·8 | 6·6 | 16·8 |

The parameters of the cable having been determined the same equivalent circuits as for overhead lines are used, paying due regard to the selection of the correct model for the appropriate length of cable. Owing to the high capacitance of cables the charging current especially at high voltages is an important factor in deciding the permissible length to be used. Table 2.3 gives the charging currents for various cables and indicates the severe limitation on length for very high-voltage installations.

## 2.13   Transformers

The equivalent circuit of one phase of a transformer referred to the primary winding is shown in figure 2.46. The resistances and reactances can be found from the well-known open- and short-circuit tests. In the absence of complete information for each winding, the two arms of the T network can each be assumed to be half the total transformer impedance. Also

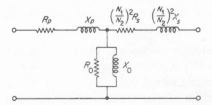

Figure 2.46   Equivalent circuit of a two winding transformer.

little accuracy is lost in transferring the shunt branch to the input terminals forming a cantilever circuit.

In power transformers the current taken by the shunt branch is usually a small percentage of the load current and this branch neglected.

*Phase shifts in three-phase transformers*

Consider the transformer shown in figure 2.47a. The red phases on both sides are taken as reference and the transformation ratio is $1:N$. The corresponding phasor diagrams are shown in figure 2.47b. Although no neutral point is available in the delta side the effective voltages from

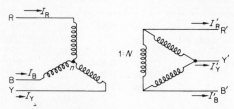

Figure 2.47a   Star-delta transformer with turns ratio $1:N$.

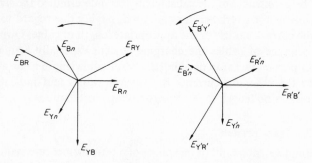

Figure 2.47b   Corresponding phasor diagrams ($N$ taken as 1 in diagrams).

line to earth are denoted by $\mathbf{E}_{R'n}$, $\mathbf{E}_{Y'n}$ and $\mathbf{E}_{B'n}$. Comparing the two phase diagrams, the following relationships are seen:

$\mathbf{E}_{R'n}$ = line-earth voltage on the delta side
     $= N\mathbf{E}_{Rn}\underline{/30°}$,

i.e. the positive sequence or normal balanced voltage of each phase is advanced through 30°. Similarly it can be shown that the positive sequence currents are advanced through 30°.

By a consideration of the negative phase-sequence phasor diagrams (these are phasors with reversed rotation, i.e. *R–B–Y*) it will readily be seen that the phase currents and voltages are shifted through $-30°$. When using the per unit system the transformer ratio does not directly appear in calculations. In star−star and delta−delta connected transformers there are no phase shifts; hence transformers having these connexions and those with star−delta should not be connected in parallel. To do so introduces a resultant voltage acting in the local circuit formed by the usually low, transformer impedances. In figure 2.48 is shown the general practice on the British network with regard to transformers with

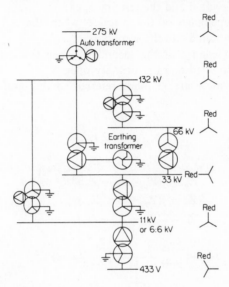

Figure 2.48   Typical phase shifts in a power system.

phase shifts. It is seen that the reference phasor direction is different at different voltage levels. The larger than 30° phase shifts are obtained by suitable re-arrangement of the winding connexions. Tertiary or third windings are provided to give sufficient fault current to operate protective gear and to provide a path for third-harmonic currents.

*Three-winding transformers*   Many transformers used in power systems have three windings per phase, the third winding being known as the tertiary. This type can be represented under balanced three-phase conditions by a single-phase equivalent circuit of three impedances star-connected as shown in figure 2.49. The values of the equivalent impedances $Z_p$, $Z_s$ and $Z_t$ may be obtained by test. It is assumed that the no-load currents are negligible.

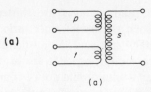

(a)

Figure 2.49a    Three winding transformer.

Let,    $Z_{ps}$ = impedance of the primary when the secondary is short-circuited and the tertiary open;

$Z_{pt}$ = impedance of the primary when the tertiary is short-circuited and the secondary open;

$Z_{st}$ = impedance of the secondary when the tertiary is short-circuited and the primary open.

The above impedances are in ohms referred to the same voltage base. Hence,

$$Z_{ps} = Z_p + Z_s$$
$$Z_{pt} = Z_p + Z_t$$
$$Z_{st} = Z_s + Z_t$$

$$\left.\begin{array}{l} Z_p = \tfrac{1}{2}(Z_{ps} + Z_{pt} - Z_{st}) \\ Z_s = \tfrac{1}{2}(Z_{ps} + Z_{st} - Z_{pt}) \\ Z_t = \tfrac{1}{2}(Z_{pt} + Z_{st} - Z_{ps}) \end{array}\right\} \qquad (2.24)$$

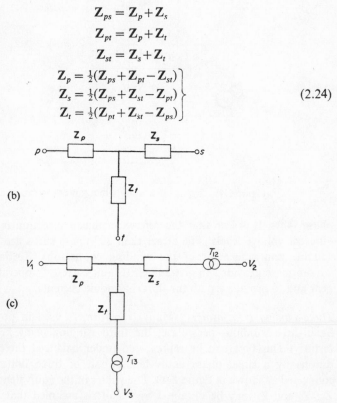

Figure 2.49b and c    Equivalent circuits.

It should be noted that the star point in figure 2.49b is purely fictitious and that the diagram is a single-phase equivalent circuit. In most large transformers the value of $Z_s$ is very small and can be negative. All impedances must be referred to common volt-ampere and voltage bases. The complete equivalent circuit is shown in figure 2.49c.

*Auto-transformers* The symmetrical auto-transformer may be treated in the same manner as two and three winding transformers. This type of transformer shows to best advantage when the transformation ratio is small and it is widely used for the interconnexion of the supply networks working at different voltages, e.g. 275 kV to 132 kV. The neutral point is solidly grounded, i.e. connected directly to earth without intervening resistance.

*Earthing (grounding) transformers* A means of providing an earthed point or neutral in a supply derived from a delta-connected transformer may be obtained by the use of a zig-zag transformer shown schematically in figure 2.48. By the interconnexion of two windings on each limb a node of zero potential is obtained.

*Harmonics* Due to the non-linearity of the magnetizing characteristic for transformers the current waveform is distorted and hence contains harmonics; these flow through the system impedances and set up harmonic voltages. In transformers with delta connected windings the third and ninth harmonics circulate round the delta and are less evident in the line currents. Another source of harmonics is a rectifier load.

On occasions the harmonic content can prove important due mainly to the possibility of resonance occurring in the system, e.g. resonance has occurred with fifth harmonics. Also the third-harmonic components are in phase in the three conductors of a three-phase line and if a return path is present these currents add and cause interference in neighbouring communication circuits.

When analysing systems with harmonics it is sufficient to use the normal values for series inductance and shunt capacitance; the effect on resistance is more difficult to assess: However it is usually only required to assess the presence of harmonics and the possibility of resonance.

*Tap-changing transformers* A method of controlling the voltages in a network lies in the use of transformers the turns ratio of which may be changed. In figure 2.50a a schematic diagram of an off-load tap changer is shown; this however requires the disconnexion of the transformer when the tap setting is to be changed. Most transformers now have on-load

changers the basic form of which is shown in figure 2.50b. In the position
shown the voltage is a maximum and the current divides equally in the
two halves of coil *R* resulting in zero resultant flux and minimum impedance. To reduce the voltage, $S_1$ opens and the total current passes
through the other half of the reactor. Selector switch *B* then moves to the

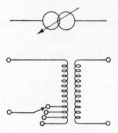

Figure 2.50a   Tap changing transformer.

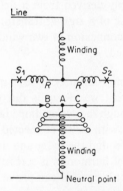

Figure 2.50b   On load tap changing transformer. $S_1$, $S_2$ transfer
switches; *R* centre-tapped reactor.

next contact and $S_1$ closes. A circulating current now flows in *R* superimposed on the load current. $S_2$ now opens and *C* moves to the next
tapping; $S_2$ then closes and the operation is complete. Six switch operations
are required for one change in tap position.

The voltage change between taps is often 1·25 per cent of the nominal
voltage. This small change is necessary to avoid large voltage disturbances
at consumer busbars. A schematic block diagram of the on-load tap
system is shown in figure 2.51. The line drop compensator (LDC) is used
to allow for the voltage drop along the feeder to the load point, so that
the actual load voltage is seen and corrected by the transformer. The total
range of tapping varies with the transformer usage, a typical figure for
generator transformers is +2 to −16 per cent in 18 steps.

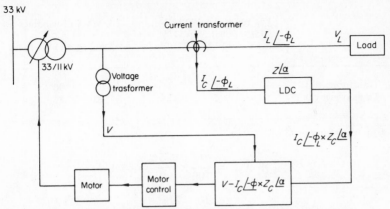

Figure 2.51 Schematic diagram of a control system for an on-load tap changing transformer incorporating *line drop compensation* (LDC).

*Typical parameters for transformers* The leakage reactances of two-winding transformers increase slightly with their rating for a given voltage, i.e. from 3·2 per cent at 20 kVA to 4·3 per cent at 500 kVA at 11 kV. For larger sizes, i.e. 20 MVA upwards, 10 per cent is a typical value at all voltages. For auto-transformers the impedances are usually less than for double wound. Parameters for large transformers are as follows: (a) 400/275 kV auto-transformer, 500 MVA, 12 per cent impedance, tap range +10 to −20 per cent; (b) 400/132 kV double wound, 240 MVA, 20 per cent impedance, tap range +5 to −15 per cent.

*Example 2.6* Two, three-phase, 275/66/11 kV transformers, are connected in parallel and the windings are employed as follows:

the 11 kV winding supplies a synchronous compensator (i.e. a synchronous motor running light),
the 66 kV winding supplies a network,
the 275 kV winding is connected to the primary transmission system.

The winding data are as shown below.
The line diagram is shown in figure 2.52a.

It is required to determine the division of power between the two 275/66 kV windings. Resistance may be neglected.

*Solution* To obtain the equivalent star circuits the reactances will be expressed on common bases of 90 MVA, 275 kV.

| | $X\%$ referred to 90 MVA | | | MVA of windings | | |
| --- | --- | --- | --- | --- | --- | --- |
| | 275/ 66 kV | 66/ 11 kV | 11/ 275 kV | 275 kV | 66 kV | 11 kV |
| Transformer (1) | 10·5 | 20 | 8 | 95 | 90 | 60 |
| Transformer (2) | 10·5 | 5 | 15 | 90 | 85 | 45 |

For transformer (1),

$$X_{275} + X_{66} = \frac{10 \cdot 5}{100} \times \frac{275^2}{90} = 88 \cdot 4 \ \Omega$$

$$X_{66} + X_{11} = \frac{20}{100} \times \frac{275^2}{90} = 168 \ \Omega$$

$$X_{11} + X_{275} = \frac{8}{100} \times \frac{275^2}{90} = 67 \cdot 2 \ \Omega$$

The corresponding quantities for transformer (2) are $88 \cdot 4 \ \Omega$, $42 \cdot 1 \ \Omega$ and $126 \cdot 3 \ \Omega$ respectively.

For transformer (1),

$$\begin{aligned} X_p(275 \text{ kV side}) &= \tfrac{1}{2}(X_{ps} + X_{pt} - X_{st}) \\ &= \tfrac{1}{2}(88 \cdot 4 + 67 \cdot 2 - 168) \\ &= -6 \cdot 2 \ \Omega \end{aligned}$$

Similarly,

$$X_S(66 \text{ kV side}) = 94 \cdot 6 \ \Omega$$

and

$$X_T(11 \text{ kV side}) = 73 \cdot 4 \ \Omega$$

For transformer (2),

$$X_p = 86 \cdot 75 \ \Omega, \quad X_S = 1 \cdot 65 \ \Omega \text{ and } X_T = 40 \ \Omega.$$

$X_p$ in (1) and $X_S$ in (2) will be neglected.

The equivalent circuit is shown in figure 2.52b which in turn reduces to figure 2.52c.

The power will divide according to the reactances of the two transformers, see chapter 5, hence,

$$P_{(1)} = \frac{94 \cdot 6 \times 80}{94 \cdot 6 + 86 \cdot 75} = 41 \cdot 7 \text{ MW}$$

and

$$P_{(2)} = \frac{86 \cdot 75 \times 80}{86 \cdot 75 + 94 \cdot 6} = 38 \cdot 3 \text{ MW}.$$

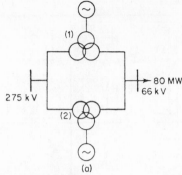

(a)

Figure 2.52a   Line diagram for system in Example 2.6.

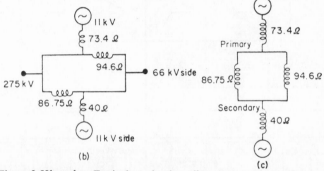

(b)

(c)

Figure 2.52b and c   Equivalent circuits: all reactance values refer
to bases of 275 kV, 90 MVA.

## 2.14   Connexion of Three-phase Transformers

In Section 2.13 the phase changes occuring in star-delta transformers are briefly discussed and sufficient treatment given for the following analysis to be understood. From the practical view, however, the connexion of such transformers requires further discussion to be fully appreciated. It is seen that for the connexion shown in figure 2.47a the output line voltages on the delta side are 30° in advance of the input line voltages to the star winding for normal, a – b – c, phase notation. Also the output line voltage is equal to the input line voltage multiplied by $(N/\sqrt{3})$ and impedances

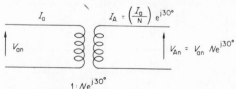

Figure 2.53   Equivalent single-phase circuit of a three-phase transformer with a phase shift from primary to secondary.

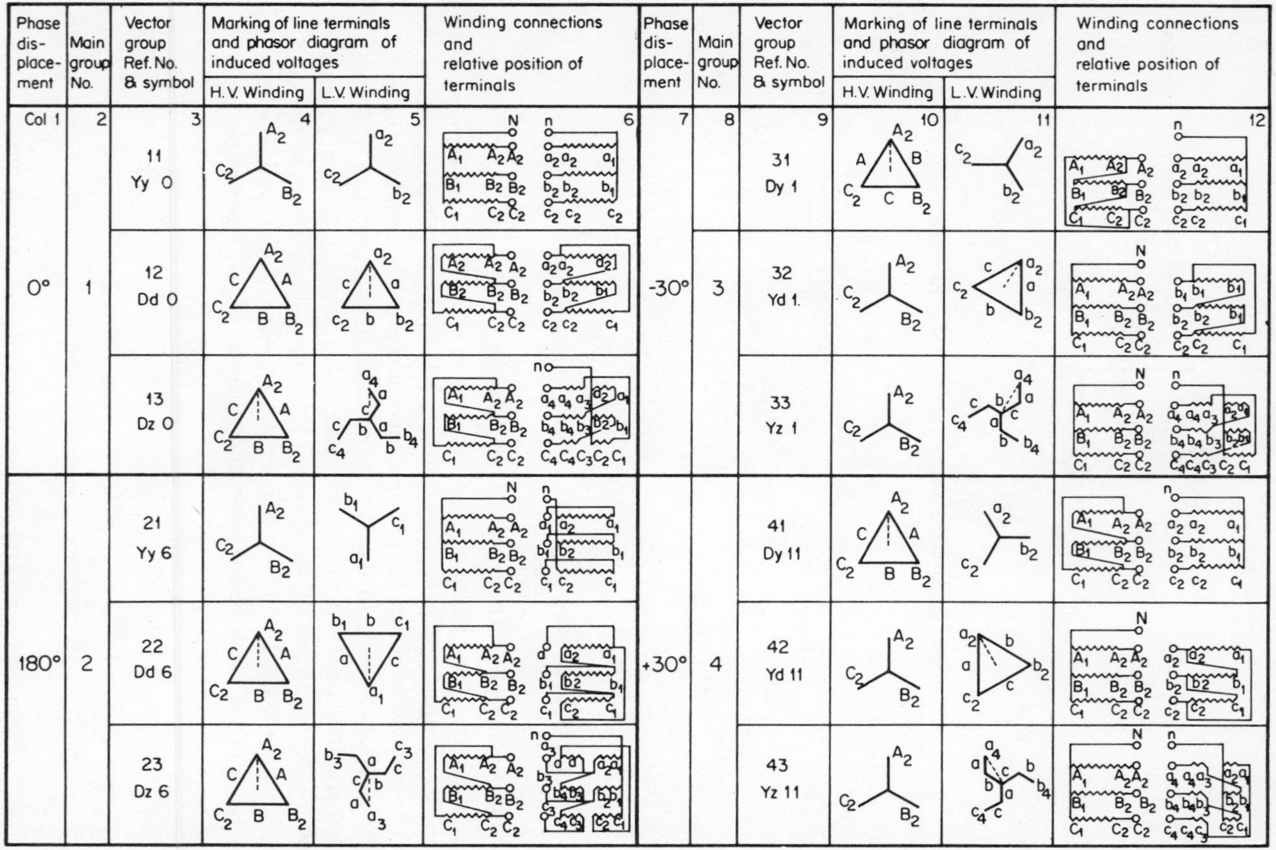

Figure 2.54 British Standard for connexions of three-phase transformers and resulting phase shifts.

transferred from one side to the other are modified by $(N/\sqrt{3})^2$ and vice versa. The true single-phase equivalent circuit is therefore as shown in figure 2.53.

However, by suitable rearrangement of the winding connexions, phase shifts of 30, 90, 150, 210, 270 or 330 electrical degrees may be obtained, and there is a need for standardization of nomenclature and procedure. Should a phase difference exist between the secondary output voltages of transformers connected to a common supply, dangerously large circulating currents will occur and the circuit protection should operate. For uniformity the transformer windings of figure 2.47a are relabelled and shown in figure 2.54 with capitals referring to the H.V. winding. It is seen that with the star-delta a shift of $-30°$ can also be produced by a rearrangement of the terminal connexions. British practice involves the use of a 'vector group reference' number to describe transformer connexions. The first number indicates the phase shift, e.g. 1 indicates 0°, 2—180°, 3—($-30°$), 4—30°; the next number indicates the interphase connexion of the secondary, e.g. 1—star, 2—delta, 3—interconnected star. Letters D and y indicate star (y) or delta (D) connexion, the H.V. winding coming first. Finally comes the phase displacement angle in the form of a clock-reference, the low voltage winding induced voltage being given by the hour hand and the H.V. voltage by the minute hand, e.g. 0° = 12 o'clock, 180° = 6 o'clock, +30° = 11 o'clock and $-30°$ = 1 o'clock; $A_2$ in the phasor diagrams is always at 12 o'clock. The application of this principle to the most widely used transformer connexions is shown in figure 2.54.

From an analytical viewpoint it may be desirable to have a phase shift of 90° thus facilitating the use of operator 'j'. In this type of work the exact arrangement is not important provided the system adopted is used consistently and the winding arrangement is seldom the cause for concern. However, when dealing with the connexion of actual equipment in the system the correct arrangement of the transformer winding is of vital importance.

LOADS

### 2.15 Introduction

Of the various parameters of electrical power systems the load required by the consumers is the most difficult to assess scientifically. The magnitude of the load varies second by second and with millions of consumers each taking energy as individually required the problem of its future assessment is a statistical one. A typical daily load curve is shown in figure 1.1. This curve is an indication of a nation's social habits. For example, in certain European countries the load remains higher later into the evening and night than in others. This difference in load peaks has been utilized by the connexion of England and France, each country having spare generating capacity at different times of the day. The development of off-peak electrical storage heating is an attempt to even out the daily load curve. There are schemes for the central control of such heating, and if these materialize the supply engineer will have some degree of control over the load.

The composition of loads can be broadly divided between industrial and domestic consumers. The proportion of industrial load increases from about one third at peak loads to one half at minimum loads. A very important difference between the two types of consumer is the high proportion of induction motors in the industrial load (about 60 per cent), whereas the domestic variety consists largely of lighting and heating. Loads influence the design and operation of power systems both economically and electrically.

*Economical aspects*

Quantities used in the measurement of loads are defined as follows:

*Maximum load*   The average load over the half hour of maximum output.

*Load factor*   The units of electricity exported by the generators in a given period divided by the product of the maximum load in this period and the length of the period in hours. The Load Factor should be high; if it is unity all the plant is being used over all of the period. It varies with the type of load, being poor for lighting (about 12 per cent) and high for industrial loads, e.g. 100 per cent for pumping stations.

*Diversity factor*   This is defined as the sum of individual maximum demands of the consumers, divided by the maximum load on the system. This factor measures the diversification of the load and is concerned with the installation of sufficient generating and transmission plant. If all the demands occurred simultaneously, i.e. unity Diversity Factor, many more generators would have to be installed. Fortunately, the factor is much higher than unity, especially for domestic loads.

A high Diversity Factor could be obtained with four consumers by compelling them to take load as shown in figure 2.55. Although compulsion obviously cannot be used, encouragement can be provided in the form of tariffs. An example is the two-part tariff in which the consumer has to pay an amount dependent on the maximum demand he requires, plus a charge for each unit of energy consumed. Sometimes the charge is based on kVA instead of power to penalize loads of low power-factor.

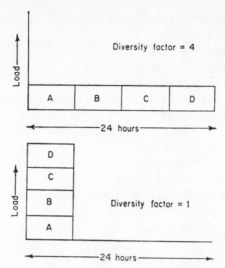

Figure 2.55 Two extremes of Diversity Factor in a system with four consumers.

The cost of generation and distribution is governed approximately by the expression $(a + b \times kW + c \times kW\ hr)$ per annum, where $a$ is a fixed charge for the central organization and similar activities and is independent of the power output; $b$ depends on the maximum demand on the system and hence on the interest and depreciation on the plant installed; $c$ depends on the units generated and hence on the fuel costs, wages of personnel directly .engaged in the production of electricity. The effect of load factor and diversity factor on these costs may be seen from the following example.

*Example 2.7*  Compare the overall cost per unit generated by a conventional coal station with that from a certain type of nuclear station assuming the load factor, $\left( \dfrac{\text{average load}}{\text{installed capacity}} \right)$ of the coal station to be 50 per cent and of the nuclear station 60 per cent. The following data applies:

> Capital cost (£/kW) 40 (Coal), 80 (Nuclear)
> Interest and depreciation per annum, 10 per cent (Coal), 12 per cent (Nuclear)
> Running cost per unit (p/kWh) 0·167 (Coal), 0·05 (Nuclear)
> (Note £1 = 100 p)

*Solution*  It is convenient to work on the basis of 1 kW output on an annual basis. The average loads are therefore 0·5 kW and 0·6 kW over the year, for coal and nuclear, respectively.

The capital costs on an annual basis require,

$$£\left(40 \times \frac{10}{100}\right) \text{ for coal, and } £\left(80 \times \frac{12}{100}\right) \text{ for nuclear}$$

The energy (kWh) generated per annum
$= 0.5 \times 365 \times 24 = 4380$ for the coal station
and $0.6 \times 365 \times 24 = 5250$ for the nuclear station
Cost of these units $= 0.167 \times 4380 = 730$ p (coal)
$\qquad\qquad\qquad\quad 0.05 \times 5250 = 263$ p (nuclear)
The total costs per annum are,
Coal, $400 + 730 = 1130$ p
Nuclear, $960 + 263 = 1223$ p
The corresponding costs per unit are:

$$\text{coal, } \frac{1130}{4380} = 0.258 \text{ p and}$$

$$\text{nuclear, } \frac{1223}{5250} = 0.233 \text{ p}$$

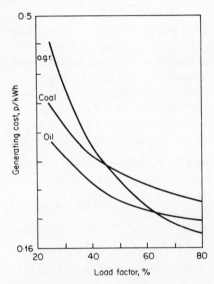

Figure 2.56 Estimated fuel and power costs of a British 2000 MW station, using different types of fuels. Capital costs are estimated to be £90/kW for an advanced gas-cooled reactor (a.g.r.), £45/kW for gas, £47/kW for oil and £55/kW for coal, based on an interest rate of 7.5% over 30 years for a conventional station and 20 years for a nuclear one (latest thinking is to base calculations on a 25-year life); initial fuel charges are included in the capital cost. (*Permission of Institution of Electrical Engineers* (1970).)

Hence, on the basis of the cost figures stated, this type of nuclear station is more economical than the coal-fired one provided it is used with a higher load factor.

The importance of the load factor is evident in this example, this is further illustrated in figure 2.56.

*Load Loss-factor*

The average losses in the system, e.g. transmission line losses, depend upon the load curves for any given period. The *Load Loss-Factor* is defined as the ratio of the actual losses over a period to the losses obtaining if the maximum load is maintained. A detailed study of load curves peculiar to Great Britain has shown that the load loss-factor is related to the load factor by the following expression,

Load Loss-Factor = 0·2 (Load Factor) + 0·8 (Load Factor)²

## 2.16 Voltage Characteristics of Loads

The variation of the power and reactive power taken by a load with various voltages is of importance when considering the manner in which such loads are to be represented in load flow and stability studies. Usually in

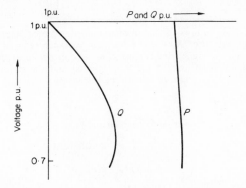

Figure 2.57 P–V and Q–V curves for a synchronous motor.

such studies the load on a substation has to be represented and is a composite load consisting of industrial and domestic consumers. A typical composition of a substation load is as follows:

| | |
|---|---|
| Induction Motors | 50–70% |
| Lighting and Heating | 20–25% |
| Synchronous Motors | 10% |
| (Transmission Losses | 10–12%) |

*Lighting*   This is independent of frequency and consumes no reactive power. The power consumed does not vary as the (Voltage)$^2$ but approximately as (Voltage)$^{1\cdot6}$.

*Heating*   This maintains constant resistance with voltage change, and hence the power varies with (Voltage)$^2$.

The above loads may be described as static.

*Synchronous Motors*   The power consumed is approximately constant. For a given excitation the vars change in a leading direction with voltage reduction. The *P–V*, *Q–V* generalized characteristics are shown in figure 2.57.

*Induction Motors*   The *P–V*, *Q–V* characteristics may be determined by the use of the simplified equivalent circuit shown in figure 2.58. It is assumed that the mechanical load on the shaft is constant.

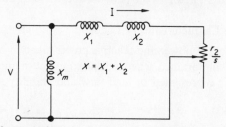

Figure 2.58   Equivalent circuit of an induction motor, $X_1$ = stator leakage reactance, $X_2$ = rotor leakage reactance referred to the stator, $X_m$ = magnetizing reactance, $r_2$ = rotor resistance, $s$ = slip p.u. Magnetizing losses have been ignored and the stator losses lumped in with the line-losses.

The electrical power,

$$P_{\text{electrical}} = P_{\text{mechanical}} = P$$

$$= 3I^2\frac{r_2}{s} = \text{constant}$$

$$\therefore \qquad\qquad s = \frac{3I^2r_2}{P}$$

The reactive power consumed

$$= \frac{3V^2}{X_m} + 3I^2(X_1 + X_2) \qquad\qquad (2.25)$$

Also from figure 2.58,

$$P = 3I^2\frac{r_2}{s} = \frac{3V^2}{(r_2/s)^2 + X^2}\cdot\frac{r_2}{s}$$

$$= \frac{3V^2r_2s}{r_2^2 + (sX)^2} \qquad\qquad (2.26)$$

The graphs of $V$–$P$ and $V$–$Q$ are shown in figure 2.59. The $P$ and $Q$ scales have been so arranged that $P$ and $Q$ are both equal to 1 p.u. at a voltage of 1 p.u. The effect of shaft load is also shown. Similar analysis

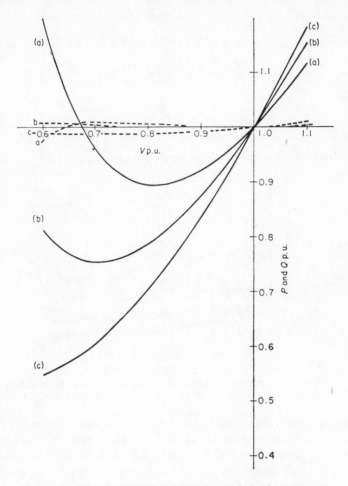

Figure 2.59  Graphs of P–V and Q–V for an induction motor—shaft torque constant. Curve, (a) full load, (b) 0·75 full load, (c) 0·5 full load, - - - - P, —— Q.

can be performed for the induction motor with Torque $\propto$ (Speed)[2] and Torque $\propto$ Speed.

The well-known power-slip curves for an induction motor are shown in figure 2.60. It is seen that for a given mechanical torque there is a critical voltage and a corresponding critical slip. If the voltage is reduced further the motor becomes unstable and stalls. This critical point occurs when

$$\frac{dP}{ds} = 0,$$

i.e. when

$$V^2 r_2 \cdot \frac{r_2^2 - (sX)^2}{(r_2^2 + (sX)^2)^2} = 0$$

so that

$$s = \frac{r_2}{X} \quad \text{and} \quad P_{max} = \frac{V^2}{2X}.$$

Alternatively for a given output power $P$, $V_{critical} = \sqrt{2PX}$.

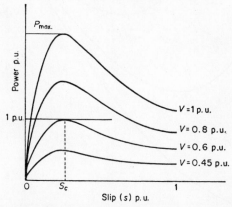

Figure 2.60   Power-slip curves for an induction motor. If voltage falls to 0·6 p.u. at full load $P = 1$, the condition is critical (slip $s_c$).

*Composite Loads*   Except in isolated cases, such as very large induction motor drives in refineries and steel mills, it is the composite load which is of interest. Tests to obtain $P$–$V$ and $Q$–$V$ characteristics of complete substation loads have been few (the consumers object!). The results of one such test on the Polish Network[19] are given in table 2.4. Instead of actual characteristics, values of $\partial P / \partial V$ and $\partial Q / \partial V$ at 1 p.u. voltage are given. As will be seen later these quantities are of considerable interest. Composite load characteristics can be obtained by summing the characteristics of the constituent loads.

In practice great difficulty is experienced in assessing the degree to which motors are loaded and this materially affects the shape of the characteristics. Loads which require the same power at reduced voltage are called stiff; loads, such as heating, in which the power falls off rapidly at lower voltages are termed soft.

In much analytical work loads are represented by constant impedances and hence it is assumed that $P \propto V^2$ and $Q \propto V^2$. If the load is $P+jQ$ p.u. the current at voltage $V$ is,

$$\frac{P-jQ}{V} \text{ p.u. (remembering } \mathbf{VI^*} = S\text{)}$$

Hence the impedance to represent this load

$$= \frac{\mathbf{V}}{\mathbf{I}} = \frac{V^2}{P-jQ} \text{ p.u.}$$

The $P$-$V$, $Q$-$V$ characteristics can be used if they are known, but this process is rather tedious. Some authorities suggest that the representation of the load by a constant-current consuming device gives a good approximation to practical conditions. If analogue or digital methods are used the

Table 2.4   Polish network test results
(*Bogucki, A. and Wojcik, M., Loads Static Characteristics*)

| $dP/dV$ at nominal voltage | Morning | Evening | Night |
|---|---|---|---|
| Large towns | 0·9–1·2 | 1·5–1·7 | 1·5–1·6 |
| Small towns | 0·6–0·7 | 1·4–1·6 | 1·4–1·6 |
| Villages | 0·5–0·6 | 1·5 | 1·5 |
| Street lighting | — | 1·55 | 1·55 |
| Mines | 0·6–0·76 | 0·75–0·89 | 0·65–0·78 |
| Ironworks | 0·6–0·7 | 0·6–0·75 | 0·6–0·7 |
| Textile industries | 0·5–0·6 | 0·6–0·7 | 0·6–0·65 |
| Large chemical industries | 0·6–0·7 | 0·6–0·75 | 0·6–0·7 |
| Machine industry | 0·5–0·55 | 0·6–0·65 | 0·6–0·65 |
| Junction points in the H.V. network | 0·6–0·7 | 0·8–1·1 | 0·8–1·0 |

| $dQ/dV$ at nominal voltage | Morning | Evening | Night |
|---|---|---|---|
| Large towns | 2·5–3·0 | 2·2–2·5 | 2·4–3·0 |
| Small towns | 2·6–3·5 | 2·0–2·5 | 2·5–3·0 |
| Villages | 2·6–3·5 | 2·0–2·5 | 2·4–3·5 |
| Industry (total) | 1·8–2·7 | 1·8–2·3 | 1·8–2·7 |
| Junction points in 6 kV network | 2·2–2·9 | 1·8–2·5 | 2·0–2·6 |
| Junction points in 110 kV network | 1·8–2·6 | 1·5–2·2 | 1·7–2·3 |

correct representation of loads presents no great difficulties. When extensive reduction of a network to a simpler form is necessary then the constant impedance representation is usually resorted to, although the load this represents seldom occurs in practice.

*Load-Frequency Characteristics*   Often voltage changes are accompanied by frequency changes. Again information regarding the characteristics of composite loads with frequency is scarce. With the small frequency changes allowed in practice the effects on the power and reactive power consumed are usually neglected in calculations.

## References

BOOKS

1. *Electrical Transmission and Distribution Reference Book*, Westinghouse Electric Corp., East Pittsburgh, Pennsylvania, 1964.
2. Waddicor, H., *Principles of Electric Power Transmission*, Chapman and Hall, London, 5th ed., 1964.
3. Stevenson, W. D. Jr., *Elements of Power System Analysis*, McGraw-Hill, New York, 2nd ed., 1962.
4. Cohn, N., *Control of Generation and Power Flow on Interconnected Systems*, Wiley, New York, 1961.
5. Dahl, O. G. C., *Electric Power Circuits*, Vols. 1 and 2, McGraw-Hill, New York, 1938.
6. Kimbark, E. W., *Power System Stability*, Vol. 3. *Synchronous Machines*, Wiley, New York, 1962.
7. Slemon, G. R., *Magnetoelectric Devices*, Wiley, New York, 1965.
8. Blume, L. F., *et. al.*, *Transformer Engineering*, Wiley, New York, 2nd ed., 1951.
9. Fitzgerald, A. E. and Kingsley, C. Jr., *Electric Machinery*, McGraw-Hill, New York, 2nd ed., 1961.
10. Langlois-Berthelot, R., *Transformers and Generators for Power Systems*, (*translated by* H. M. Clarke), MacDonald, London, 1960.
11. Zoborszky, J., and J. W. Rittenhouse, *Electric Power Transmission*, Rensselaer Bookstore, Troy, New York, 1969.
12. Edison Electric Institute, *EHV Transmission Line Reference Book*, 1968.
13. Adkins, B., *The General Theory of Electrical Machines*, Chapman and Hall, London, 1964.
14. Knable, A. H., *Electrical Power Systems Engineering, Problems and Solutions*, McGraw-Hill, New York, 1961.
15. Meisel, J., *Principles of Electromechanical Energy Conversion*, McGraw-Hill, New York, 1966.
16. Barnes, C. C., *Electric Cables*, Chapman and Hall, London, 1964.
17. Steven, R. E., *Electromechanics and Machines*, Chapman and Hall, London, 1970.

PAPERS

18. Park and Robinson, 'The Reactance of Synchronous Machines', *Trans. A.I.E.E.*, **47**, 1928.

19. Bogucki, A. and M. Wojcik, 'Static Characteristics of Loads', *Energetyka B.*, No. 3 (*Polish*), 1959.
20. Gross, E. T. B., and A. H. Weston, 'Transposition of high voltage overhead lines and elimination of electrostatic unbalance to ground', *A.I.E.E. Trans.*, **70**, 1837, (1951).
21. American Standards Association, *Terminal Markings and Corrections for Distribution and Power Transformers*, December, 1964.

## Problems

2.1. When two 4 pole, 50 Hz, synchronous generators are paralleled their phase displacement is 2° mechanical. The synchronous reactance of each machine is 10 Ω/phase and the common busbar voltage is 6·6 kV. Calculate the synchronizing torque. (690 lb. ft.)

2.2. A synchronous generator has a synchronous impedance of 10 Ω and a resistance of 1·2 Ω It is connected to an infinite busbar of 1·0 kV (phase). If the generated (no load) e.m.f. is 1·1 kV (phase) calculate the current and power factor for maximum output. (139 A, 0·7 leading.)

2.3. A 6·6 kV synchronous generator has negligible resistance and synchronous reactance of 4 Ω/phase. It is connected to an infinite busbar and gives 2000 A at unity power factor. If the excitation is increased by 25 per cent find the maximum power output and the corresponding power factor. State any assumptions made. (31·6 MW; 0·96 leading.)

2.4. A synchronous generator whose characteristic curves are given in figure 2.12 delivers full load at the following power factors, 0·8 lagging, 1·0, and 0·9 leading. Determine the percentage regulation at these loads.

2.5. A salient pole 75 MVA 11 kV synchronous generator is connected to an infinite busbar through a link of reactance 0·3 p.u. and has $X_d = 1·5$ p.u. and $X_q = 1$ p.u. and negligible resistance. Determine the power output for a load angle 30° if the excitation e.m.f. is 1·4 times the rated terminal voltage. Calculate the synchronizing coefficient in this operating condition. All p.u. values are on a 75 MVA base. ($P = 0·48$ p.u., $\partial P/\partial \delta = 0·78$ p.u.)

2.6. A synchronous generator of open circuit terminal voltage 1 p.u. is on no-load and then suddenly short-circuited; the trace of current against time is shown in figure 2.13b. In figure 2.13b the current oc $= 1·8$ p.u., oa $= 5·7$ p.u. and ob $= 8$ p.u. Calculate the values of $X_s$, $X'$ and $X''$. Resistance may be neglected. If the machine is delivering 1 p.u. current at 0·8 power factor lagging at the rated terminal voltage before the short circuit occurs sketch the new envelope of the 50 c/s current waveform. ($X_s = 2·36$ p.u., $X' = 0·3$ p.u., $X'' = 0·2$ p.u.)

2.7. Construct a performance chart for a 22 kV, 500 MVA, 0·9 p.f. generator having a short circuit ratio of 0·55.

2.8. A 275 kV 3-phase transmission line of length 60 miles is rated at 800 A. The values of resistance, inductance and capacitance per phase per mile are 0·125 Ω, 1·7 mH and 0·018 $\mu$F, respectively. The receiving end voltage is 275 kV when full load is transmitted at 0·9 power factor lagging. Calculate the sending-end voltage and current, and the transmission efficiency and compare with the answer obtained by the short line representation. Use the nominal $\pi$ and T methods of solution. The frequency is 50 Hz. ($V_s = 175,000$ V per phase.)

2.9. A 220 kV, 60 Hz three-phase transmission line is 200 miles long and has the following constants per phase per mile:

| Inductance | 2·1 mH |
| Capacitance | 0·14 $\mu$F |
| Resistance | 0·25 $\Omega$ |

Ignore leakage conductance.

If the line delivers a load of 300 A, 0·8 power-factor lagging, at a voltage of 220 kV, calculate the sending-end voltage. Determine the $\pi$ circuit which will represent the line. Calculate the error resulting from the use of the medium-length $\pi$ section. (Series $(11 + j116)$ $\Omega$; shunt $(0·44 + j0·72) \times 10^{-3}$ $\Omega^{-1}$.)

2.10. In a 3-core cable, the capacitance between the 3 cores short circuited together and the sheath is 0·54 $\mu$F/mile, and that between 2 cores connected together and the third core is 0·52 $\mu$F/mile.

Determine the kVA required to keep 10 miles of this cable charged when the supply is 33 KV, 3 phase, 50 Hz. (270 kVA.)

2.11. Calculate the A B C D constants for a 275 kV overhead line of length 300 miles. The parameters per mile are as follows:

| Resistance | 0·125 $\Omega$ |
| Reactance | 0·53 $\Omega$ |
| Admittance (shunt capacitive) | 5·92 $\times$ 10$^{-6}$ mhos |

The shunt conductance is 0.

Answer:

$$[A = 0·8615 + j\,0·0317, \; B = 34·0 + j\,153, \; C = (-19·2 + j\,1693)\,10^{-6}]$$

2.12. A 132 kV, 60 Hz transmission line has the following generalized constants:

$$A = 0·9696\underline{/0·49°}$$
$$B = 52·88\underline{/74·79°} \; \Omega$$
$$C = 0·001177\underline{/90·15°} \; \text{mhos}$$

If the receiving end-voltage is to be 132 kV when supplying a load of 125 MVA 0·9 p.f. lagging, calculate the sending end voltage and current.

Answer:                    (95·9 kV phase, 554 A)

2.13. Two identical transformers each have a nominal or no-load ratio of 33/11 kV and a reactance of 2 $\Omega$ referred to the 11 kV side; resistance may be neglected. The transformers operate in parallel and supply a load of 9 MVA, 0·8 p.f. lagging. Calculate the current taken by each transformer when they operate five tap steps apart (each step is 1·25 per cent of the nominal voltage). (307A, 194A for 3-ph. transformers)

2.14. An induction motor, the equivalent circuit of which is shown in figure 2.61 is connected to supply busbars which may be considered as possessing a voltage and frequency which is independent of the load. Determine the reactive power consumed for various busbar voltages and construct the $Q$-$V$ characteristic. Calculate the critical voltage at which the motor stalls and the critical slip assuming that the mechanical load is constant. ($S_{cr} = 0·2$, $V_c = 0·63$.)

2.15. In the system shown in figure 2.4, Example 2.2, the generator has a synchronous reactance of 1·0 p.u. The internal voltage (no-load) of the generator is held at 17 kV (line). Determine the voltage at the load busbar for the same load as before, i.e. 50 MW, 0·8 p.f. lagging. The load may be represented by a constant impedance. (0·68 p.u.)

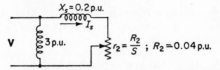

Figure 2.61    Equivalent circuit of 500 kW, 6·6 kV, induction motor in Problem 2.12. All p.u. values refer to rated voltage and power (P = 1 p.u. and V = 1 p.u.).

2.16. A 100 MVA round-rotor generator of synchronous reactance 1·5 p.u. supplies a sub-station (*L*) through a step-up transformer of reactance 0·1 p.u., two lines each of reactance 0·3 p.u. in parallel and a step-down transformer of reactance 0·1 p.u. The load taken at *L* is 100 MW at 0·85 lagging. *L* is connected

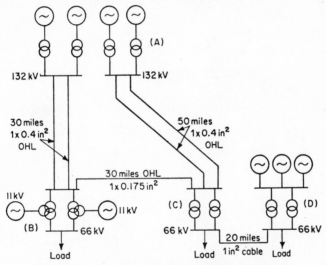

Figure 2.62    Line diagram of system in Problem 2.17

to a local generating station which is represented by an equivalent generator of 75 MVA and synchronous reactance of 2 p.u. All reactances are expressed on a base of 100 MVA.

Draw the equivalent single-phase network. If the voltage at *L* is to be 1 p.u. and the 75 MVA machine is to deliver 50 MW, 20 MVAr, calculate the internal voltages of the machines. ($E_1 = 2$ p.u., $E_2 = 1·72$ p.u., $\delta_{2L} = 35·45°$.)

2.17. The following data applies to the power system shown in figure 2.62;

Generating Station A. Four identical turboalternators each rated at 16 kV, 125 MVA and of synchronous reactance 1·5 p.u. Each machine supplies a 125 MVA, 0·1 p.u., transformer connected to a busbar sectioned as shown.

Substation B. Two identical 150 MVA, three-winding transformers each having the following reactances between windings: 132/66 kV windings 10 per cent, 66/11 kV windings 20 per cent: 132/11 kV windings 20 per cent; all on a 150 MVA base.

The secondaries supply a common load of 200 MW at 0·9 p.f. lagging. To each tertiary winding is connected a 30 MVA synchronous compensator of synchronous reactance 1·5 p.u.

Substation C. Two identical 150 MVA transformers each of 0·15 p.u. reactance supply a common load of 300 MW at 0·85 p.f. lagging.

Generating Station D. Three identical 11 kV, 75 MVA generators each of 1 p.u. synchronous reactance supply a common busbar which is connected to an outgoing 66 kV cable through two identical 100 MVA transformers. Load 50 MW 0·8 p.f. lagging.

Determine the equivalent circuit for balanced operation giving component values on a base of 100 MVA. Neglect resistance where appropriate and treat the loads as impedances. Reduce the network to the simplest form.

# Control of Power and Frequency

## 3.1 Introduction

Although to a certain extent the control of power and frequency is interrelated to the control of reactive power and voltage, it is hoped that by dealing with power and frequency separately from voltage control a better appreciation of the operation of power systems will be obtained. In a large interconnected system many generation stations large and small are synchronously connected and hence all have the same frequency. The following remarks refer mainly to networks in which the control of power is carried out by the decisions and actions of engineers as opposed to systems in which the control and allocation of load to machines is effected completely automatically. The latter are sometimes based on a continuous load-flow calculation by analogue or digital computers. The allocation of the required power among the generators has to be decided before the advent of the load which must therefore be predicted. An analysis is made of the loads experienced over the same period in previous years, account is also taken of the value of the load immediately previous to the period under study and of the weather forecast. The probable load to be expected, having been decided, is allocated to the various turbine-generators in the order of their efficiencies, as described in chapter 1.

A daily load cycle is shown in figure 1.1. It is seen that the rate of growth of the load is very high and varies between 2 MW/min/1000 MW of peak demand in the early morning to about 8 MW/min/1000 MW of peak demand between 7.00 and 8.00 hours. Hence the ability of machines to increase their output quickly from zero to full load is important. It is extremely unlikely that the output of the machines at any instant will exactly equal the load on the system. If the output is higher than the demand the machines will tend to increase in speed and the frequency will rise, and vice versa. Hence the frequency is not a constant quantity but continuously varies; these variations are normally small and have no noticeable effect on most consumers. The frequency is continuously monitored against standard time-sources and when long term tendencies to rise or fall are noticed the control engineers take appropriate action by regulating the generator outputs.

Should the total generation available be insufficient to meet the demand the frequency will fall. If the frequency falls by more than 1 Hz the reduced speed of power station pumps, fans etc. may reduce the station output and a serious situation arises. In this type of situation although the reduction of frequency will cause a reduction in power demand voltage must be reduced and if this is not sufficient loads will have to be disconnected and continue to be disconnected until the frequency rises to a reasonable level.

When a permanent increase in load occurs on the system the speed and frequency of all the interconnected generators fall, since the increased energy requirement is met from the kinetic energy of the machines. This causes an increase in steam admitted to the turbines due to the operation of the governors and hence a new load balance is obtained. Initially the boilers have a thermal reserve by means of which sudden changes can be supplied until the new firing rate has been established.

### 3.2 The Turbine Governor

In the following discussion of governor mechanisms both steam and water turbines will be considered together and points of difference indicated where required. A simplified schematic diagram of a traditional governor

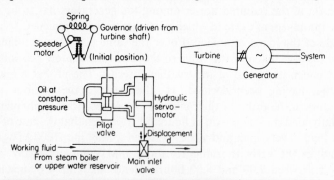

Figure 3.1   Governor control system employing the Watt governor as sensing device and a hydraulic servo-system to operate main supply valve. Speeder-motor gear determines initial setting of the governor position.

system is shown in figure 3.1. The sensing device, which is sensitive to change of speed, is the time-honoured Watt centrifugal governor. In this two weights move radially outwards as their speed of rotation increases and thus move a sleeve on the central spindle. This sleeve movement is transmitted by a lever mechanism to the pilot-valve piston and hence the servo-motor is operated. A dead-band is present in this mechanism, i.e. the speed must change by a certain amount before the valve commences to operate, due to friction and mechanical back-lash. The time taken for the main steam valve to move due to delays in the hydraulic pilot-valve and servo-motor systems is appreciable, 0·2 to 0·3 seconds. The governor characteristic for a large steam turbo-alternator is shown in figure 3.2 and it is seen that there is a 4 per cent drop in speed between no load and full load on the turbine.

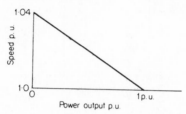

Figure 3.2   Idealized governor characteristic of a turbo-alternator.

An important feature of the governor system is the mechanism by means of which the governor sleeve and hence main-valve positions can be changed and adjusted quite apart from when actuated by the speed changes. This is accomplished by the speed changer or speeder motor, as it is sometimes termed. The effect of this adjustment is the production of a family of parallel characteristics, as shown in figure 3.3. Hence the power output of the generator at a given speed may be adjusted at will and this is of extreme importance when operating at optimum economy.

The torque of the turbine may be considered to be approximately proportional to the displacement of the main inlet valve $d$. Also the expression for the change in torque with speed may be expressed approximately by the equation:

$$T = T_0(1 - kN) \tag{3.1}$$

where $T_0$ is the torque at speed $N_0$ and $T$ the torque at speed $N$; $k$ is a constant for the governor system. As the torque depends on both the main-valve position and the speed, $T = f(d,N)$.

There is a time delay between the occurrence of a load change and the new operating conditions. This is due not only to the governor mechanism but also the fact that the new flow-rate of steam or water must accelerate or decelerate the rotor in order to attain the new speed. In figure 3.4 typical curves are shown for a turbogenerator which has a sudden decrease in the electrical power required, and hence the retarding torque on the turbine shaft is suddenly much smaller. In the ungoverned case the considerable time-lag between the load change and the attainment of the new

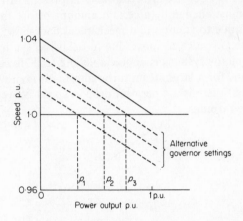

Figure 3.3   Effect of speeder gear on governor characteristics $P_1$, $P_2$ and $P_3$ = outputs at various settings but at same speed.

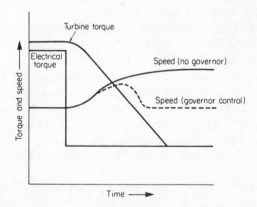

Figure 3.4   Graphs of turbine torque, electrical torque and speed against time when the load on a generator suddenly falls.

steady speed is obvious. With the regulated or governed machine, due to the dead band in the governor mechanism, the speed-time curve starts to rise, the valve then operates and the fluid supply is adjusted. It is possible for damped oscillations to be set up after the load change; this will be discussed in chapter 7.

An important factor regarding turbines is the possibility of overspeeding, when the load on the shaft is lost completely, with possible drastic mechanical breakdown. To avoid this special valves are incorporated to automatically cut off the energy supply to the turbine. In a turbo-generator normally running at 3000 rev/min this overspeed protection operates at about 3300 rev/min.

*Example 3.1* A 75 MVA synchronous generator operates initially at no-load at 3000 rev/min, 50 Hz. A 20 MW load is suddenly applied to the machine and the steam valves to the turbine commence to open after 0·5 seconds due to the time-lag in the governor system. Calculate the frequency to which the generated voltage drops before the steam flow commences to increase to meet the new load. The value of $H$ (stored energy) for the machine is 4 kW seconds per kVA of generator capacity.

*Solution* For this machine the stored energy at 3000 rev/min

$$= 4 \times 75,000 = 300,000 \text{ kW sec.}$$

Before the steam valves start to open the machine loses $20,000 \times 0·5 = 10,000$ kW sec of the stored energy in order to supply the load.

The stored energy $\propto$ (speed)$^2$

$\therefore$ the new frequency

$$= \sqrt{\left(\frac{300,000 - 10,000}{300,000}\right)} \times 50 \text{ Hz}$$

$$= 49·2 \text{ Hz.}$$

## 3.3 Control Loops

The machine and its associated governor and voltage-regulator control systems may be represented by the block diagram shown in figure 3.5. The nature of the voltage control loop has been discussed already in chapter 2, accurate machine representation is required involving two-axis expressions. Here again the scope of the problem will be outlined and the reader referred to references 5 and 12 for further information.

Two factors have a large influence on the dynamic response of the prime mover, (a) entrained steam between the inlet valves and the first stage of the turbine (in large machines this can be sufficient to cause loss of synchronism after the valves have closed), (b) the storage action in the reheater which causes the output of the low-pressure turbine to lag behind that of the high-pressure side. The transfer function $\dfrac{\text{prime mover torque}}{\text{valve opening}}$

accounting for both these effects is shown in reference 12 to be,

$$\frac{G_1 G_2}{(1 + \tau_t p)(1 + \tau_r p)} \qquad (3.2)$$

where

$G_1$ = entrained steam constant,
$G_2$ = reheater gain constant,
$\tau_t$ = entrained steam time constant,
$\tau_r$ = reheater time constant.

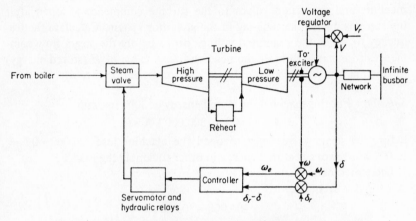

Figure 3.5   Block diagram of complete turbo-alternator control systems. The governor system is more sophisticated than that shown in figure 3.1 owing to the inclusion of the load-angle δ in the control loop. Suffixes *r* refer to reference quantities and suffixes *e* refer to error quantities. The controller modifies the error signal from the governor by taking into account the load angle.

The transfer function relating steam-valve opening to changes in speed due to the governor feedback loop is,

$$\frac{\Delta d}{\Delta \omega}(p) = \frac{G_3 G_4 G_5}{(1 + \tau_g p)(1 + \tau_1 p)(1 + \tau_2 p)} \qquad (3.3)$$

where

$\tau_g$ = governor-relay time constant (0·047 sec)
$\tau_1$ = primary-relay time constant (0·06 sec)
$\tau_2$ = secondary-valve-relay time constant (0·024 sec)
$(G_3 G_4 G_5)$ = constant relating steam-valve lift to speed change.

By a consideration of the transfer function of the synchronous generator with the above expressions the dynamic response of the overall system may be obtained.

### 3.4 Division of Load between Generators

The use of the speed changer enables the steam input and electrical power output at a given frequency to be changed as required. The effect of this on two machines can be seen in figure 3.6. The output of each machine

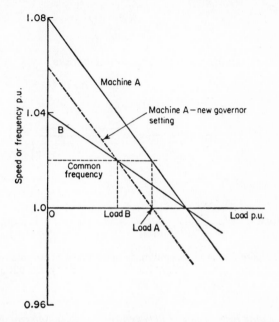

Figure 3.6 Two machines connected to an infinite busbar. The speeder gear of machine A is adjusted so that the machines load equally.

is not therefore determined by the governor characteristics but can be varied to meet economical and other considerations by the operating personnel. The governor characteristics only completely decide the outputs of the machines when a sudden change in load occurs or when machines are allowed to vary their outputs according to speed within a prescribed range in order to keep the frequency constant. This latter mode of operation is known as *free-governor action*.

It has been shown in chapter 2 that the voltage difference between the two ends of an interconnector of total impedance $R+jX$ is given by

$$\Delta V = E - V = \frac{RP+XQ}{V} \tag{2.6}$$

Also the angle between the voltage phasors (i.e. the transmission angle) $\delta$ is given by

$$\sin^{-1}\frac{\delta V}{E}$$

where

$$\delta V = \frac{XP - RQ}{V} \qquad\qquad (2.7)$$

When $X \gg R$, i.e. for most transmission networks,

$$\delta V \propto P \quad \text{and} \quad \Delta V \propto Q.$$

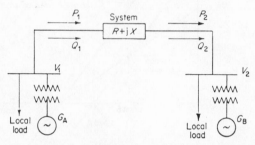

Figure 3.7   Two generating stations linked by an interconnector of impedance $(R+jX)\Omega$. The rotor of A is in phase advance of B and $V_1 > V_2$.

Hence, (a) *the flow of power between two nodes is determined largely by the transmission angle*, (b) *the flow of reactive power is determined by the scalar voltage difference between the two nodes.*

These two facts are of fundamental importance to the understanding of the operation of power systems.

The angular advance of $G_A$ (figure 3.7) is due to a greater relative energy input to turbine A than to B. The provision of this extra steam (or water) to A is possible because of the action of the speeder gear without which the power outputs of A and B would be determined solely by the nominal governor characteristics. The following simple example illustrates these principles.

*Example 3.2*   Two synchronous generators operate in parallel and supply a total load of 200 MW. The capacities of the machines are 100 MW and 200 MW and both have governor droop characteristics of 4 per cent from no load to full load. Calculate the load taken by each machine assuming free governor action.

*Solution* Let $x$ MW be the power supplied from 100 MW generator. Referring to figure 3.8,

$$\frac{4}{100} = \frac{\alpha}{x}$$

For the 200 MW machine,

$$\frac{4}{200} = \frac{\alpha}{200 - x}$$

$$\therefore \quad \frac{4x}{100} = \frac{800 - 4x}{200}$$

and $x = 66.6$ MW = load on the 100 MW machine. The load on the 200 MW machine = 133.3 MW.

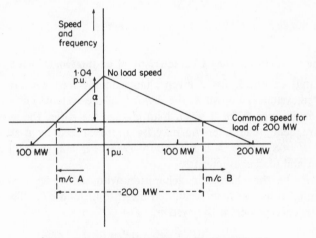

Figure 3.8   Speed-load diagram for Example 3.3.

It will be noticed that when the governor droops are the same the machines share the total load in proportion to their capacities or ratings. Hence it is advantageous for the droops of all turbines to be equal.

*Example 3.3* Two units of generation maintain 66 kV and 60 kV (line) at the ends of an interconnector of inductive reactance per phase of 40 Ω and with negligible resistance and shunt capacitance. A load of 10 MW is to be transferred from the 66 kV unit to the other end. Calculate the necessary conditions between the two ends, including the power factor of the current transmitted.

*Solution*

$$\delta V = \frac{XP - RQ}{V_R} = \frac{40 \times 3 \cdot 33 \times 10^6}{60,000/\sqrt{3}} = 3840 \text{ V}$$

Also,

$$\frac{\delta V}{66,000/\sqrt{3}} = \sin \delta = 0 \cdot 101$$

$$\therefore \qquad\qquad \delta = 5^\circ \ 44'$$

Hence the 66 kV busbars are 5° 44′ in advance of the 60 kV busbars.

$$\Delta V = \frac{6000}{\sqrt{3}} = \frac{RP + XQ}{V_R} = \frac{40 \ Q}{60,000/\sqrt{3}}$$

Hence

$$Q = 3 \text{ MVAr per phase.}$$

The p.f. angle $\phi = \tan^{-1} Q/P = 42^\circ$ and hence the p.f. $= 0 \cdot 74$.

### 3.5  The Power/Frequency Characteristic of an Interconnected System

The change in power for a given change in the frequency in an inter-connected system is known as the *stiffness* of the system. The smaller the change in frequency for a given load change the stiffer the system. The power-frequency characteristic may be approximated by a straight line and $dP/df = K$, where $K$ is a constant (MW per Hz) depending on the governor and load characteristics.

Let $dP_G$ be the change in generation with the governors operating 'free acting' resulting from a sudden increase in load $dP_L$. The resultant out of balance power in the system,

$$dP = dP_L - dP_G \qquad\qquad\qquad (3.4)$$

and therefore

$$K = \frac{dP_L}{df} - \frac{dP_G}{df} \qquad\qquad\qquad (3.5)$$

$dP_L/df$ measures the effect of the frequency characteristics of the load and $dP_G \propto (P_T - P_G)$, where $P_T$ is the turbine capacity connected to the network and $P_G$ the output of the associated generators. When steady conditions are again reached the load $P_L$ is equal to the generated power $P_G$ (neglecting losses), hence, $K = K_1 P_T - K_2 P_L$ where $K_1$ and $K_2$ are the coefficients relevant to the turbines and load respectively.

$K$ can be determined experimentally by connecting two large separate systems by a single link, breaking the connexion and measuring the frequency change. For the British system, tests show that $K = 0.8P_T - 0.6P_L$ and lies between 2000 and 5500 MW per Hz, i.e. a change in frequency of 0.1 Hz requires a change in the range 200 to 550 MW depending on the amount of plant connected. In smaller systems the change in frequency for a reasonable load change is relatively large and electrical governors have been introduced to improve the power/frequency characteristic.

### 3.6 Systems Connected by Lines of Relatively Small Capacity

Let $K_A$ and $K_B$ be the respective power-frequency constants of two separate power systems A and B. Let the change in the power transferred from A to B when a change resulting in an out of balance power $dP$ occurs in system B, be $dP_t$ where $dP_t$ is positive when power is transferred from A to B. The change in frequency in system B due to an extra load $dP$ and an extra input of $dP_t$ from A, is $-(dP - dP_t)/K_B$ (the negative sign indicates a fall in frequency). The drop in frequency in A due to the extra load $dP_t$ is $-dP_t/K_A$, but the changes in frequency in each system must be equal as they are electrically connected.

Hence,

$$\frac{-(dP - dP_t)}{K_B} = \frac{-dP_t}{K_A}$$

$$\therefore dP_t = +\left(\frac{K_A}{K_A + K_B}\right)dP \qquad (3.6)$$

Next consider the two systems operating at a common frequency $f$ with A exporting $dP_t$ to B. The connecting link is now opened and A is relieved of $dP_t$ and assumes a frequency $f_A$ and B has $dP_t$ more load and assumes $f_B$.

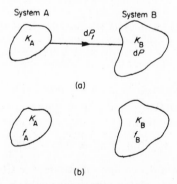

(a)

(b)

Figure 3.9a Two interconnected power systems connected by a tie line, b The two systems with the tie line open.

Hence,

$$f_A = f + \frac{dP_t}{K_A} \quad \text{and} \quad f_B = f - \frac{dP_t}{K_B}$$

from which

$$\frac{dP_t}{f_A - f_B} = \frac{K_A K_B}{K_A + K_B} \tag{3.7}$$

Hence, by opening the link and measuring the resultant change in $f_A$ and $f_B$ the values of $K_A$ and $K_B$ can be obtained.

When large interconnected systems are linked electrically to others by means of tie-lines the power transfers between them are usually decided by mutual agreement and the power controlled by regulators in accordance. As the capacity of the tie-lines is small compared with the systems care must be taken to avoid excessive transfers of power.

*Effect of Governor Characteristics*
A fuller treatment of the performance of two interconnected systems in the steady state requires a further consideration of the control aspects of the generation process.

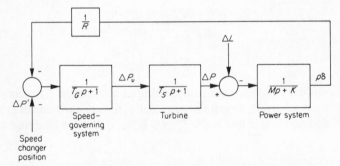

Figure 3.10   Block diagram for turbine-generator connected to a power system. (From *Economic Control of Interconnected Systems* by L. K. Kirchmayer. Copyright © 1959 John Wiley & Sons Ltd. By permission of John Wiley & Sons, Inc.)

A more complete block diagram for the steam turbine-generator connected to a power system is shown in figure 3.10. For this system the following equation holds:

$$M p^2 \delta + K p \delta = \Delta P - \Delta P_L$$

where $M$ is a constant depending on inertia (see Chapter 7) in radians.

$K$ is the stiffness or damping coefficient (i.e. change of power with speed), in MW/Hz or MW per rad/s.

$$= \frac{\partial \,(\text{Load power})}{\partial \,(p\delta)} - \frac{\partial \,(\text{Turbine power})}{\partial \,(p\delta)}$$

$p \equiv \mathrm{d}/\mathrm{d}t.$

$\Delta P$ and $\Delta P_L$ = change in prime mover and load powers.

$\delta$ = change from initial angular position,

$R$ = preset speed regulation (or governor droop), i.e. drop in speed or frequency when combined machines of an area change from no-load to full-load, expressed as p.u. or Hz or rad/s per MW.

$\Delta P^1$ = change in speed-changer setting.

$\therefore$ change from normal speed or frequency,

$$p\delta = \frac{1}{Mp + K}(\Delta P - \Delta P_L).$$

This analysis holds for steam turbine generation; for hydro turbines, the large inertia of the water must be accounted for and the analysis is more complicated (see reference 2).

The representation of two systems connected by a tie line is shown in figure 3.11. The general analysis is as before except for the additional power terms due to the tie line. The machines in the individual power systems are considered to be closely coupled and to possess one equivalent rotor.

For system (1), $M_1 p^2\delta_1 + K_1 p\delta_1 + T_{12} (\delta_1 - \delta_2)$
$$= \Delta P_1 - \Delta P_{L1} \tag{3.8}$$

where $T_{12}$ is the synchronizing torque coefficient of the tie line.

For system (2), $M_2 p^2\delta_2 + K_2 p\delta_2 + T_{12} (\delta_2 - \delta_1)$
$$= \Delta P_2 - \Delta P_{L2} \tag{3.9}$$

The steady-state analysis of two interconnected systems may be obtained from the transfer functions given in the block diagram.

The speed governor general response is given by

$$\Delta P_V = \frac{-1}{(T_G p + 1)} \left(\frac{1}{R}p\delta + \Delta P'\right) \tag{3.10}$$

In the steady state from (3.10), $\Delta P_1 = -\dfrac{1}{R_1} p\delta_1$ (3.11)

and $\Delta P_2 = -\dfrac{1}{R_2} p\delta_2$ (3.12)

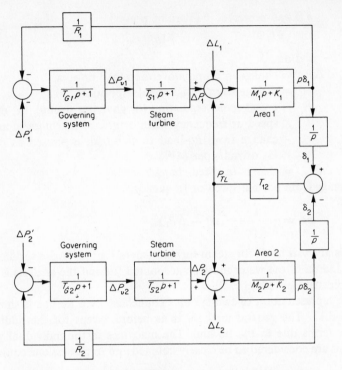

Figure 3.11   Block control diagram of two power systems connected by a tie line.   (From *Economic Control of Interconnected Systems* by L. K. Kirchmayer. Copyright © 1959 John Wiley & Sons Ltd. By permission of John Wiley & Sons, Inc.)

Similarly, from equations (3.8) and (3.9) in the steady state,

$$\left(K_1 + \frac{1}{R_1}\right) p\delta_1 + T_{12}\left(\delta_1 - \delta_2\right) = -\Delta P_{L1} \qquad (3.13)$$

and

$$\left(K_2 + \frac{1}{R_2}\right) p\delta_2 + T_{12}\left(\delta_2 - \delta_1\right) = -\Delta P_{L2} \qquad (3.14)$$

Adding (3.13) and (3.14),

$$\left(K_1 + \frac{1}{R_1}\right) p\delta_1 + \left(K_2 + \frac{1}{R_2}\right) p\delta_2 = -\Delta P_{L1} - \Delta P_{L2} \qquad (3.15)$$

In a synchronous system

$$p\delta_1 = p\delta_2 = p\delta = \text{angular frequency and (3.15) becomes}$$

$$\left[ K_1 + K_2 + \left( \frac{1}{R_1} + \frac{1}{R_2} \right) \right] p\delta = -\Delta P_{L1} - \Delta P_{L2}$$

and $$p\delta = \frac{-\Delta P_{L1} - \Delta P_{L2}}{(K_1 + K_2) + (1/R_1 + 1/R_2)} \qquad (3.16)$$

From (3.13) and (3.14)

$$T_{12} (\delta_1 - \delta_2) = \frac{-\Delta P_{L1} [K_2 + (1/R_2)] + \Delta P_{L2} [K_1 + (1/R_1)]}{[K_2 + (1/R_2) + K_1 + (1/R_1)]} \qquad (3.17)$$

It is normally required to keep the system frequency constant and to maintain the interchange through the tie line at its scheduled value. To achieve this, additional controls are necessary to operate the speed-changer settings as follows.

For area (1), $p \, \Delta P_1^1 \, \alpha \, \gamma_{1t} \, T_{12} (\delta_1 - \delta_2) + \gamma_{1f} \, p\delta_1$

$$\alpha \, \gamma_{1t} \left[ T_{12} (\delta_1 - \delta_2) + \frac{\gamma_{1f}}{\gamma_{1t}} p\delta_1 \right] \qquad (3.18)$$

Similarly for area (2),

$$p \, \Delta P_2^1 \, \alpha \, \gamma_{2t} \left[ T_{12} (\delta_2 - \delta_1) + \frac{\gamma_{2f}}{\gamma_{2t}} p\delta_2 \right] \qquad (3.19)$$

where $\gamma_t$ and $\gamma_f$ refer to the control constants for power transfer and frequency, respectively. When a load change occurs in a given area the changes in tie-line power and frequency have opposite signs, i.e. the frequency falls for a load increase and the power transfer increases and vice versa. In the interconnected area, however, the changes have the same sign. Typical system parameters have the following orders of magnitude:

$K$, 0·75 p.u. on system capacity base.
$T_{12}$, 0·1 p.u. (10 per cent of system capacity results in 1 radian displacement between areas (1) and (2)).
$R$, 0·04 p.u. on a base of system capacity.
$\gamma_f$, 0·005.
$\gamma_t$, 0·0009.

*Example 3.4* Two power systems, A and B, each have a regulation $(R)$ of 0·1 p.u. (on respective capacity bases) and a stiffness $K$ of 1 p.u. The capacity of system A is 1500 MW and of B 1000 MW. The two systems are interconnected through a tie line and are initially at 60 Hz. If there is a 100 MW load change in system A, calculate the change in the steady-state values of frequency and power transfer.

*Solution*

$$K_A = 1 \times 1500 \text{ MW per Hz}$$
$$K_B = 1 \times 1000 \text{ MW per Hz}$$
$$R_A = \frac{\Delta f \text{ (No load to full load)}}{\text{Full load capacity}} = \frac{0 \cdot 1 \times 60}{1500}$$
$$= 6/1500 \frac{\text{Hz}}{\text{MW}}$$

$$R_B = (6/1000) \frac{\text{Hz}}{\text{MW}}$$

From (3.16),

$$\Delta f = \frac{-\Delta P_1}{(K_1 + 1/R_1) + (K_2 + 1/R_2)}$$

$$= \frac{-100}{1500 + \dfrac{1500}{6} + 1000 + \dfrac{1000}{6}}$$

$$= \frac{-600}{17500} = -0 \cdot 034 \text{ Hz}$$

$$P_{12} = T_{12} (\delta_1 - \delta_2) = \frac{-\Delta P_1 (K_2 + 1/R_2)}{(K_1 + 1/R_1) + (K_2 + 1/R_2)}$$

$$= \frac{-100 \left( \dfrac{7000}{6} \right)}{10500/6} = \frac{-7000}{105}$$

$$= -6 \text{ MW}$$

Note without the participation of governor control,

$$P_{12} = \left( \frac{-K_B}{K_A + K_B} \right) dP = -\frac{1000}{2500} \times 100$$

$$= -40 \text{ MW}$$

### Frequency-bias–tie-line control

Consider three interconnected power systems A, B and C, as shown in figure 3.12, the systems being of similar size. Assume that initially A and B export to C their previously agreed power transfers. If C has an increase in load the overall frequency tends to decrease and hence the generation in A, B and C increases. This results in increased power transfers from A and B to C. These transfers, however, are limited by the tie-line power controller to the previously agreed values and therefore instructions are given for A and B to reduce generation and hence C is not

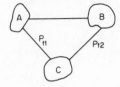

Figure 3.12    Three power systems connected by tie lines.

helped. This is a severe drawback of what is known as straight tie-line control, which can be overcome if the systems are controlled by using a consideration of both load transfer and frequency, such that the following equation holds,

$$\sum \Delta P + K \Delta f = 0. \qquad (3.20)$$

$\sum \Delta P$ is the nett transfer error and depends on the size of the system and the governor characteristic and $\Delta f$ is the frequency error and is positive for high frequency. In the case above, after the load change in C the frequency error is negative (i.e. low frequency) for A and B and the sum of $\Delta P$ for the line AC and BC is positive. For correct control,

$$\sum \Delta P_A + K_A \Delta f = \sum \Delta P_B + K_B \Delta f = 0.$$

Systems A and B take no regulating action despite their fall in frequency. In C, $\sum \Delta P_C$ is negative as it is importing from A and B and therefore the governor speeder motors in C operate to increase output and restore frequency. This system is known as frequency-bias, tie-line control.

### 3.7    Phase-Shift Transformers

Although the generators constitute the only source of power in the system the distribution of this power around the network can be influenced by the provision of phase changing equipment such as booster transformers. Often in practice lines or cables operating in parallel have currents which are unrelated to their thermal ratings; an overhead line and underground cable in parallel is a typical example.

The distribution of power and reactive power in the network is decided by the impedances of the various paths. As has been already seen, voltage magnitudes decide the reactive power flow but not, to any large extent, the flow of power. To influence the power in a line a phase shift is needed and this can be obtained by the connexions shown in figure 3.13a. The

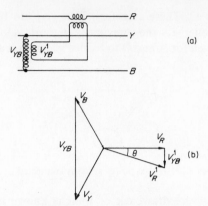

Figure 3.13a   Connexions for one phase of a phase-shifting transformer. Similar connexions for the other two phases. b Corresponding phasor diagram.

booster arrangement shows only the injection of voltage into one phase; it is repeated for the other two phases. In figure 3.13b the corresponding phasor diagram is shown and the nature of the angular shift of the voltage boost $V'_{YB}$ indicated. By the use of tappings on the energizing transformer several values of phase shift may be obtained.

## 3.8   Optimization of Power System Operation

*Introduction*

The extensive interconnexion of power sources has made the operation of a system in the most economical manner a complex subject. Economy must be balanced against considerations such as security of supply. The use of the *merit order* ensures that as far as possible the most economical generating sets are used. However, a lack of precise knowledge of the flows of real and reactive power and other parameters in the network and effective means of dealing with the analysis of large systems have prevented the operation from attaining a real economic optimum although experienced operators certainly approach this aim.

The rapid expansion in the use of digital computers for load flows and fault calculations and the development of optimization techniques in control theory has resulted in much attention being given to this topic.

Apart from financial considerations, it is becoming difficult for operators to cope with the information produced by large complex systems in times of emergency such as with major faults. Computers with on-line facilities can more readily digest this information and take correct action by instructing control gear or by the suitable display of relevant information to enable human operators to take correct action. In the attempt to obtain an economic optimization the limitations of the system such as plant ratings and stability limits must be obeyed.

Optimization may be considered in a number of ways according to the time scale involved, namely: daily, yearly (especially with hydro stations) and over much longer periods when planning for future developments, although this latter is not strictly operational optimization. In an existing system the various factors involved are the fixed and variable costs. The former includes labour, administration, interest and depreciation, etc., and the latter mainly fuel. A major problem is the effective prediction of the future load whether it occurs in ten minutes, a few hours or in several years time. It is not intended here to describe the mathematical theory of the solutions so far achieved as this would be beyond the scope and level intended in this text, but the problem will be outlined and several references given and further discussed in Appendix 4.

The following information is of interest.

*For each generator:*

  (a) maximum and economic output capacities,
  (b) fixed and incremental heat rates,
  (c) minimum shut-down time,
  (d) minimum stable outputs, maximum run-up and run-down rates.

*For each station:*

  (a) cost and calorific value of fuel (thermal stations),
  (b) factor reflecting recent operational performance of the station,
  (c) minimum time between loading and unloading successive generators,
  (d) constraints in station output.

*For the system:*

  (a) load demand at given intervals for the specified period,
  (b) specified constraints imposed by transmission capability,
  (c) running-spare requirements,
  (d) effect of transmission losses.

The input-output characteristic of a turbine is of great importance when economical operation is considered. A typical characteristic is shown in figure 3.14. The *incremental heat rate* is defined as the slope of the input-output curve at any given output. The graph of the incremental heat-rate against output is known as the Willans line. For large turbines with a single valve the incremental heat-rate is approximately constant over the operating range (most steam turbines in Britain are of this type); with multi-valve turbines (as used in the U.S.A.) the Willans line is not horizontal but curves upwards and is often represented by the closest linear law.

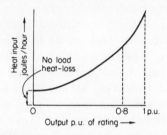

Figure 3.14 Input–output characteristic of a turbo-generator set. Often the curve above an output of 0·8 p.u. is steeper than the remainder and the machine has a maximum rating at 1 p.u. and a maximum economic rating at, say, 0·8 p.u.

The value taken for the incremental heat-rate of a generating set is sometimes complicated because if only one or two shifts are being operated (there are normally three shifts per day) heat has to be expended in banking boilers when they are not required.

Instead of plotting incremental heat-rate or fuel-consumption against power output for the turbine-generator the incremental fuel-cost may be used. This is advantageous when allocating load to generators for optimum economy as it incorporates differences in the fuel costs of the various generating stations. Usually the graph of incremental fuel-cost against power output can be approximated by a straight line (figure 3.15). Consider two turbine-generator sets having the following different incremental fuel costs, $dC_1/dP_1$ and $dC_2/dP_2$; where $C_1$ is the cost of the fuel input to unit number 1 for a power output of $P_1$ and similarly $C_2$ and $P_2$ relate to unit number 2. It is required to load the generators to meet a given requirement in the most economical manner. Obviously the load on the machine with the higher $dC/dP$ will be reduced by increasing the load taken by the machine with the lower $dC/dP$. This transfer will be beneficial until the values of $dC/dP$ for both sets are equal, after which the machine with the previously higher $dC/dP$ now becomes the one with the lower value and vice versa. There is no economic advantage in further transfer of load and the condition when $dC_1/dP_1 = dC_2/dP_2$ therefore gives optimum economy; this can be seen by considering figure 3.15. The above argument can be extended to several machines supplying a load. Generally for optimum economy the *incremental fuel-cost should be identical for all contributing turbine-generator sets.*

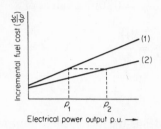

Figure 3.15 Idealized graphs of incremental fuel cost against output for two machines sharing a load equal to $P_1$ and $P_2$.

The above reasoning must be modified when the distances of generating stations from the common loads are different; here the cost of different transmission losses will affect the argument.

As important as the transmission losses, is the optimum method of transporting fuel from the production centres to the generating stations. The transport of both electrical energy and fuel in the optimum manner forms *Transportation Problems* which may be dealt with by special techniques (e.g. the northwest rule) or by the general method of linear programming. A good introduction to these methods is given in reference 6.

*Basic formulation of the short-term optimization problem for thermal stations*

Kirchmeyer[4] uses Lagrange multipliers in formulating equations including transmission losses.

Let;  $P_i$  = power output of unit i (MW),

$P_R$  = total load on system (MW),

$P_L$  = transmission losses (MW),

$F_T$  = total cost of units (money/hour),

$\lambda$  = Lagrange multiplier (money per MW-hour),

$n$  = number of generating units.

The total input to the system from all generators $= P_T = \sum_{i=1}^{n} P_i$, and

$$\left( \sum_{1}^{n} P_i \right) - P_L - P_R = 0$$

Using Lagrangian multipliers the expression,

$$\gamma = F_T - \lambda \left( \sum_1^n P_1 - P_L - P_R \right) \text{ is obtained.}$$

For minimum cost $(F_T)$, $\dfrac{\partial \gamma}{\partial P_1} = 0$ for all values of $i$.

$$\therefore \qquad \frac{\partial \gamma}{\partial P_1} = \frac{dF_1}{dP_1} - \lambda + \lambda \frac{\partial P_L}{\partial P_1} = 0$$

i.e. $\qquad\qquad \dfrac{dF_1}{dP_1} + \lambda \dfrac{\partial P_L}{\partial P_1} = \lambda \qquad\qquad\qquad (3.21)$

In equation (3.21), $dP_L/dP_1$ is the incremental transmission loss. One way of solving the equations described by (3.21) is known as the penalty factor method in which (3.21) is rewritten as

$$\frac{dF_1}{dP_1} L_1 = \lambda \qquad\qquad\qquad (3.22)$$

where $L_1 = \left( \dfrac{1}{1 - \dfrac{\partial P_L}{\partial P_1}} \right) = $ Penalty Factor of plant $i$.

An approximate penalty factor, $L_1 = 1 + \dfrac{\partial P_L}{\partial P_1}$

Using this and (3.22)

$$\frac{dF_1}{dP_1} + \left( \frac{dF_1}{dP_1} \cdot \frac{\partial P_L}{\partial P_1} \right) = \lambda$$

where $i = 1 \ldots n$ (number of plants).
In practice the determination of $dP_L/dP_1$ is difficult and the use of the so-called loss or 'B' coefficients is made, i.e.

$$\frac{\partial P_L}{\partial P_1} = \Sigma_1 \, 2B_{m1} \, P_m + B_{1o},$$

where the $B$ coefficients are determined from the network (references 1 and 4).

There are many drawbacks to the above treatment, e.g. limitations on power flows by equipment ratings, transformer settings and maximum phase angles allowable across transmission lines on stability grounds. Also it is only concerned with active power, reactive power being neglected. A brief introduction to a more comprehensive approach using linear programming is given in Appendix 4.

### 3.9  Computer Control of Load and Frequency

*Control of tie lines*

Automatic control of area power-systems connected by tie lines has been already discussed. The methods used will now be extended to include optimum economy as well as power transfer and frequency control. The basic systems described are typical of United States practice and have been comprehensively discussed by Kirchmeyer[4]. In the previous section, methods for economic analysis and optimization as developed by Kirchmeyer[2] have been summarized. The choice of generating units to be operated is largely decided by spinning reserve, voltage control, stability and protection requirements. The methods discussed decide the allocation of load to particular machines.

If transmission losses be neglected, it has been shown that optimum economy results when $dF_n/dP_n = \lambda$. Control equipment to adjust the governor speed-change settings such that all units comply with the appropriate value of $dF_n/dP_n$ is incorporated in the control loops for frequency and power-transfer adjustment, as shown in figure 3.16. The frequency and load-transfer control acts quickly, and once these quantities have been decided the slower acting economic controls act. For example, if an increase in load occurs in the controlled area, a signal requiring increased

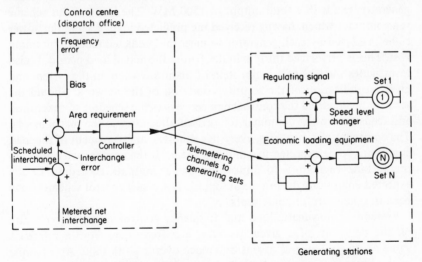

Figure 3.16  Schematic diagram of automatic control arrangements covering frequency, tieline power flows and economic loading of generating sets. (From *Economic Control of Interconnected Systems* by L. K. Kirchmayer. Copyright © 1959 John Wiley and Sons Ltd. By permission of John Wiley & Sons, Inc.)

generation transmits through the control system. These changes alter the value of $\lambda$ and cause the economic control apparatus to call for generation to be operated at the same incremental cost. Eventually the system is again in the steady state, the load change having been absorbed and all units operate at an identical value of incremental loss.

If transmission losses are included, the basic economic requirement calls for, $dF_n/dP_n = \lambda/L_n$, where $L_n$ is the penalty factor. $(1/L_n)$ signals are generated by a computer from a knowledge of system parameters in the form of the so-called $B$ coefficients. A complete and detailed account of United States practice in the economic control of interconnected systems is given in references 2 and 4.

In Britain, using the present non-automatic methods, there is a frequency error of less than 0·1 Hz for 90 per cent of the day. For the frequency to be automatically controlled a range of controlled generation of $\pm(300-500)$ MW would be required to correct the frequency error and this must be spread over several generating stations to avoid uneconomic power flows. A possible scheme would be for the selected machines to be fitted with turbine regulators which control the speeder motor gear and hence the power output. These regulators would be controlled by means of telemetering pulses transmitted from a central system controller.

A British scheme which has been in operation controls 31 turbo-generator sets with a total output of 1500 MW. The heart of this scheme is a computer which, having received the predicted load, decides the power to be produced by each generator to meet the predicted load at minimum cost. This is performed thirty minutes before the actual load period. It also investigates the loss of a major item of plant anywhere in the system and gives an indication of the security condition of the network. Should the security be impaired the control engineer revises the schedule of generation and feeds this to the computer for revised allocation to the machine sets. The computer also provides, every five minutes, an assessment of the load conditions to be expected five minutes ahead. It then sends instructions to the turbine regulators to adjust the power outputs according to the expected conditions. In France automatic frequency control schemes have been in operation for some years.

Systems of automatic load and frequency control may either operate on the deviation of a given quantity, e.g. frequency, from a reference value or on the integral of this difference over a given time. An example of the latter is known as *load-phase-energy regulation*. A proposed comprehensive approach to automatic control has been reported in reference 14 and is summarized by the block diagram in figure 3.17. There are three sub-systems each of which can function alone or be replaced by a human operator. The plant section produces three hours in advance a security checked order, in which plant is to be connected to the system. Access to a large fast digital computer is required and the computation gives the merit order, load flow and security check.

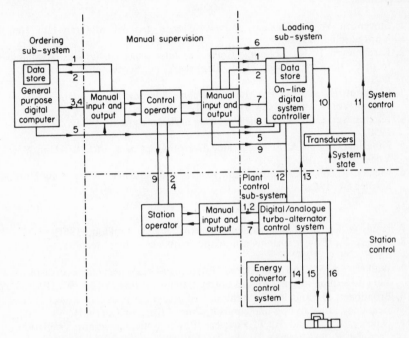

Figure 3.17 Functional scheme of automatic power-system control. Information and command flow. 1. Modification of parameters, 2. Modification of constraints, 3. Load data; configuration, 4. Available plant, 5. Security-checked optimum schedule of active and reactive power, 6. Processed operational data, 7. Alarms and decision demands, 8. Modifications to schedules, 9. Emergency manual instructions, 10. Load; frequency; voltage; configuration, 11. Supplementary information, 12. Required output; future output; required regulation, 13. Power output, 14. Power output; future power output, 15. Energy control; excitation control, 16. MW; frequency; voltage. (*Permission of Institution of Electrical Engineers.*)

The loading sub-system receives the plant order, stores it and when required sends loading instructions (digitally coded) to the turbo-alternator control system. Input information is obtained by telemetry from the generating stations and by manual in-feed. A limiting factor to such schemes is the length of time for the load flows (see chapter 5) to be computed; computation times which appear fast in the normal context may be long in terms of a completely automatic control system. As the evaluation of powers, voltages and currents represents the core of the control system this rapidity of load flow computation is of vital importance.

## References

BOOKS

1. Stevenson, W. D., *Elements of Power System Analysis*, McGraw-Hill, New York, 2nd ed., 1962.
2. Kirchmeyer, L. K., *Economic Control of Interconnected Systems*. Available from C. K. Kirchmeyer, 27 St. Stephens Lane, Scotia, New York, 12302, U.S.A.
3. Chestnut, H., and R. W. Mayer, *Servomechanisms and Regulating System Design*, Vol. 1, Wiley, New York, 1951.
4. Kirchmeyer, L. K., *Economic Operation of Power Systems*, Wiley, New York, 1958.
5. Kingsley, C., and A. E. Fitzgerald, *Electric Machinery*, McGraw-Hill, New York, 2nd ed., 1961.
6. Metzer, K. W., *Elementary Mathematical Programming*, Wiley, New York, Science ed., 1963.

ARTICLES AND PAPERS

7. Kidd, W. L., 'Load and frequency control', *Elec.J.*, London, (1961).
8. Moran, F., 'Power systems automatic frequency control techniques', *Proc. I.E.E.*, **106A**, (1959).
9. Walker, P. A. W., and A. S. Aldred, 'Frequency—response analysis of displacement governing in synchronous power systems', *Proc.I.E.E.*, **108C**, (1961).
10. Broadbent, D., and K. N. Stanton, 'An analytical review of power-system frequency, time and tie-line control', *Proc. I.E.E.*, **108C**, (1961).
11. Dineley, J. L., and E. T. Powner, 'Power-system governor simulation', *Proc. I.E.E.*, **111**, (1964).
12. Coles, H. E., 'Effects of prime-mover governing and voltage regulation on turbo-alternator performance', *Proc. I.E.E.*, **112**, (1965).
13. Wardrop, K. W., 'An analogue computer to evaluate the cost of changes in power on a power system', *Proc. I.E.E.*, **106**, 285, (1959).
14. 'Automation in the electricity supply industry', *I.E.E. Conf. Rept.*, (1962).
15. Carpentier, J., 'Contribution à l'étude du dispatching economique', *Bulletin de la Société Francaise des Electriciens*, Ser. 8, **3**, (1962).
16. Carpentier, J., and J. Sirioux, 'L'optimization de la production à l'électricité de France', *Bulletin de la Société Francaise des Electriciens*, (1963).
17. Dopazo, J. F., O. A. Klitin, G. W. Stagg, and M. Watson, 'An optimization technique for real and reactive power allocation', *Trans. I.E.E.E., P.A. & S.*, **55**, No. 11, (1967).
18. Kuhn, H. W., and A. W. Tucker, 'Non-linear programming', *Proc. of Second Berkeley Symposium on Math. Stat. & Prob.* University of California Press, Berkeley, U.S.A., (1951).
19. Moskalev, A. G., 'Principles of the most economical distribution of the active and reactive loads in automatically controlled power systems', *Elektrichestvo*, No. 12, 24–33, (1963).

20. Peschon, J., D. S. Piercy, W. F. Tinney, O. J. Tviet, and M. Cuenard, 'Optimal control of reactive power flow', *I.E.E.E. Transactions on Power Apparatus and Systems*, **PAS-8**, No. 1, (1968).
21. Brewer, C., G. F. Charles, C. C. M. Parish, J. N. Prewett, D. C. Salthouse, and B. J. Terry, 'The performance of a predictive automatic load dispatching system', *Proc. I.E.E.*, **115**, No. 10, 1577, (1968).
22. Benthall, T. P., 'Automatic load scheduling in a multi-area power system', *Proc. I.E.E.*, **115**, No. 4, 592, (1968).
23. Pettit, P., 'Digital computer programs for an experimental automatic load dispatching system', *Proc. I.E.E.*, **115**, No. 4, 597, (1968).
24. Wells, D. W., 'Method for economic secure loading of a power system', *Proc. I.E.E.*, **115**, No. 8, 1190, (1968).
25. Long, R. W., and J. R. Barrios, 'A parametric equations approach to the economic dispatch problem', *I.E.E.E. Trans., P.A. & S.*, **86**, 141, (1967).
26. Caprez, A., and D. Caha, 'Optimized long-range power scheduling for a hydro reservoir system', *I.E.E.E. Trans., P.A. & S.*, **86**, 150, (1967).
27. Kennedy, T., and A. G. Hoffman, 'On-line digital computer application techniques for complex electric systems dispatch', *I.E.E.E. Trans., P.A. & S.*, **87**, 67, (1968).
28. Sasson, A. M., 'Non-linear programming solutions for load-flow, minimum-loss, and economic dispatching problems', *I.E.E.E. Trans., P.A. & S.*, **88**, 399, (1969).
29. El-Abiad, A. H., and F. J. Jaimes, 'A method for optimum scheduling of power and voltage magnitude', *I.E.E.E. Trans., P.A. & S.*, **88**, 413, (1969).
30. Rindt, L. J., 'Economic scheduling of generation for planning studies', *I.E.E.E. Trans., P.A. & S.*, **88**, 801, (1969).
31. Cohn, N., S. B. Biddle, Jr., R. G. Lex, Jr., E. H. Preston, C. W. Ross, and D. R. Whitten, 'On-line computer applications in the electric power industry', *Proc. I.E.E.E.*, **58**, No. 1, 78, (1970).
32. Stagg, G. W., J. F. Dopazo, O. A. Klitin, and L. S. VanSlyck, 'Techniques for the real-time monitoring of power system operations', *I.E.E.E. Trans.*, Paper No. 69TP657-PWR, presented at the I.E.E.E. Summer Power Meeting (1969).

## Problems

3.1. Two identical 60 MW synchronous generators operate in parallel. The governor settings on the machines are such that they have 4 per cent and 3 per cent droops (no load to full load percentage speed drop). Determine (a) the load taken by each machine for a total load of 100 MW, (b) the percentage adjustment in the no-load speed to be made by the speeder motor if the machines are to share the load equally. (42·8 and 57·2 MW, 0·83 per cent increase in no load speed on the 4 per cent droop machine.)

3.2. Two generating stations A and B are linked by a line and two transformers of total reactance 20 Ω referred to 132 kV and negligible resistance. A load of 100 MW, 0·9 p.f. lagging is taken from the busbars of A and 200 MW, 0·85 p.f. lagging from B. Determine the phase angle between the busbars of A and B and the voltage adjustment required, to equalize the load on each station.

Initially both stations have busbar voltages of 11 kV which are in phase. (3° 9′; 5·3 kV increase on 132 kV side of A.)

3.3. The incremental fuel costs of two units in a generating station are as follows:

$$\frac{dF_1}{dP_1} = 0.003P_1 + 0.7$$

$$\frac{dF_2}{dP_2} = 0.004P_2 + 0.5$$

where $F$ is in £/hr and $P$ is in MW.

Assuming continuous running with a total load of 150 MW calculate the saving per hour obtained by using the most economical division of load between the units as compared with loading each equally. The maximum and minimum operational loadings are the same for each unit and are 125 MW and 20 MW. ($P_1 = 57$ MW, $P_2 = 93$ MW saving £1·14 per hour.)

3.4. For the two generating units in problem 3.3 plot incremental fuel cost for maximum economy against total load, over the range 40 to 250 MW. (Note the maximum and minimum limits to machine outputs.)

3.5. Two power systems A and B are inter-connected by a tie line and have power-frequency constants $K_A$ and $K_B$ MW per Hz. An increase in load of 500 MW on system A causes a power transfer of 300 MW from B to A. When the tie line is open the frequency of system A is 49 Hz and of system B 50 Hz. Determine the values of $K_A$ and $K_B$ deriving any formulae used. ($K_A$ 500 MW/Hz, $K_B$ 750 MW/Hz.)

3.6. Two power systems, A and B, having capacities of 3000 and 2000 MW, respectively, are interconnected through a tie-line and both operate with frequency-bias–tie-line control. The frequency bias for each area is 1 per cent of the system capacity per 0·1 Hz frequency deviation. If the tie-line interchange for A is set at 100 MW and for B set (incorrectly) at 200 MW, calculate the steady-state change in frequency.

Answer:            0·6 Hz.  Use $\Delta P_A + \sigma_A \Delta f = \Delta P_B + \sigma_B \Delta f$

# Control of Voltage and Reactive Power

## 4.1 Introduction

The approximate relationship between the scalar voltage difference between two nodes in a network and the flow of reactive power was shown in chapter 2 to be

$$\Delta V = \frac{RP_2 + XQ_2}{V_2} \tag{2.6}$$

Also it was shown that the transmission angle is proportional to

$$\frac{XP_2 - RQ_2}{V_2} \tag{2.7}$$

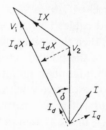

Figure 4.1 Phasor diagram for system shown in figure 3.7; $V_1 > V_2$. Resistance of line zero, inductive reactance $X\,\Omega$. $I_d$ and $I_q$ in phase and quadrature components of the current $I$.

For networks where $X \gg R$, i.e. most power circuits, $\Delta V$, the voltage difference, determines $Q$. Consider the simple interconnector linking two generating stations A and B as shown in figure 3.7. The machine at A is in phase advance of that at B and $V_1$ is greater than $V_2$, hence there is a flow of power and reactive power from A to B. This can be seen from the phasor diagram shown in figure 4.1. It is seen that $I_d$ and hence $P$ is determined by $\angle\delta$ and $I_q$ and hence $Q$ mainly by $V_1 - V_2$. In this case $V_1 > V_2$ and reactive power is transferred from A to B; if by varying the generator excitations such that $V_2 > V_1$, the direction of the reactive power is reversed as shown in figure 4.2. Hence power can be sent from A to B or B to A by suitably adjusting the amount of steam (or water) admitted to the turbine and reactive power càn be sent in either direction by adjusting the voltage magnitudes. These two operations are approximately independent of each other if $X \gg R$ and the flow of reactive power can be studied almost independently of the power flow. The phasor diagrams show that if a scalar voltage difference exists across a largely reactive link the reactive power flows towards the node of lower voltage. From another point of

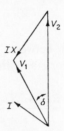

Figure 4.2   Phasor diagram for system in figure 3.7. $V_2 > V_1$.

view, if in a network there is a deficiency of reactive power at a point, this has to be supplied from the connecting lines and hence the voltage at that point falls. Conversely if there is a surplus of reactive power generated (e.g. lightly loaded cables absorb leading or negative vars and hence generate positive vars) then the voltage will rise. This is a convenient way of expressing the effect of the power factor of the transferred current and although it may seem unfamiliar initially, the ability to think in terms of var flows instead of exclusively with power factors and phasor diagrams, will make the study of power networks much easier.

If it can be arranged that $Q_2$ in the system in figure 3.7 is zero, then there will be no voltage drop between A and B, a very satisfactory state of affairs. Assuming that $V_1$ is constant, consider the effect of keeping $V_2$ and hence the voltage drop $\Delta V$ constant. From equation (2.6)

$$Q_2 = \frac{V_2 \cdot \Delta V - R \cdot P_2}{X} = K - \frac{R}{X}P_2 \qquad (4.1)$$

where $K$ is a constant.

If this value of $Q_2$ does not exist naturally in the circuit then it will have to be obtained by artificial means such as the connexion at B of capacitors or inductors. If the value of the power changes from $P_2$ to $P_2^1$ and if $V_2$ remains constant then the reactive power at B must change to $Q_2^1$ such that,

$$Q_2^1 - Q_2 = \frac{R}{X}(P_2^1 - P_2)$$

i.e. an increase in power causes an increase in reactive power. The change however is proportional to $(R/X)$ which is normally small. It is seen that voltage can be controlled by the injection into the network of reactive power of the correct sign. Other methods of a more obvious kind for controlling voltage are the use of tap-changing transformers and voltage boosters.

## 4.2 The Generation and Absorption of Reactive Power

In view of the findings in the previous section a review of the characteristics of a power system from the viewpoint of reactive power is now appropriate.

### Synchronous generators

These can be used to generate or absorb reactive power. The limits on the capability for this can be seen in figure 2.21 of chapter 2. The ability to supply reactive power is determined by the short-circuit ratio (1/synchronous reactance) as the distance between the power axis and the theoretical stability-limit line in figure 2.21 is proportional to the short circuit ratio. In modern machines the value of this ratio is made low for economic reasons and hence the inherent ability to operate at leading-power-factors is not large. For example a 200 MW, 0·85 p.f. machine with a 10 per cent stability allowance has a capability of 45 MVAr at full power outputs. The var capacity can however be increased by the use of continuously-acting voltage regulators as explained in chapter 2. An over-excited machine, i.e. one with greater than normal excitation generates reactive power whilst an under-excited machine absorbs it (or generates negative or leading vars). The generator is the main source of supply to the system of both positive and negative vars.

### Overhead lines and transformers

When fully loaded, lines absorb reactive power. With a current $I$ A for a line of reactance per phase $X$ $\Omega$ the vars absorbed are $I^2 X$ per phase. On light loads the shunt capacitances of longer lines may become predominant and the lines become var generators.

Transformers always absorb reactive power. A useful expression for the quantity may be obtained for a transformer of reactance $X_T$ p.u. and a full-load rating of $3\,V\,.\,I_{\text{rated}}$.

The ohmic reactance

$$= \frac{V\,.\,X_T}{I_{\text{rated}}}$$

Therefore the vars absorbed

$$= 3\,.\,I^2\,.\,\frac{V\,.\,X_T}{I_{\text{rated}}}$$

$$= 3\,.\,\frac{I^2 V^2}{(IV)_{\text{rated}}}\,.\,X_T = \frac{(VA \text{ of load})^2}{\text{Rated } VA}\,.\,X_T$$

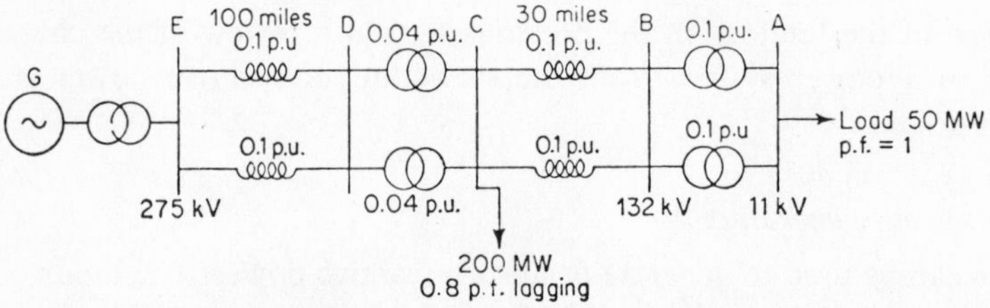

Figure 4.3   Radial transmission system with intermediate loads. Calculation of reactive-power requirement.

## Cables

Cables are generators of reactive power owing to their high capacitance. A 275 kV, 240 MVA cable produces 10 to 12 MVAr per mile; a 132 kV cable roughly 3 MVAr per mile and a 33 kV cable, 0·2 MVAr per mile.

## Loads

A load at 0·95 power factor implies a reactive power demand of 0·33 kVAr per kw of power, which is more appreciable than the mere quoting of the power factor would suggest. In planning a network it is desirable to assess the reactive power requirements to ascertain whether the generators are able to operate at the required power factors for the extremes of load to be expected. An example of this is shown in figure 4.3 where the reactive losses are added for each item until the generator power factor is obtained.

*Example 4.1*   In the radial transmission system shown in figure 4.3 all per unit values are referred to the voltage bases shown and 100 MVA. Determine the power factor at which the generator must operate.

*Solution* Voltage drops in the circuits will be neglected and the nominal voltages assumed.

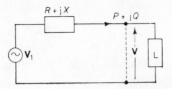

Figure 4.4 Single-phase equivalent circuit of a line supplying a load of $P+jQ$ from an infinite busbar of voltage $V_1$.

Busbar A,

$$P = 0.5 \text{ p.u.}, \quad Q = 0$$

$I^2X$ loss in 132 kV lines and transformers

$$= \frac{P^2+Q^2}{V^2}X_{CA} = \frac{0.5^2}{1^2} \cdot 0.1$$

$$= 0.025 \text{ p.u.}$$

Busbar C,

$$P = 2+0.5 = 2.5 \text{ p.u.}$$
$$Q = 1.5+0.025 \text{ p.u.}$$
$$= 1.525 \text{ p.u.}$$

$I^2X$ loss in 275 kV lines and transformers

$$= \frac{2.5^2+1.525^2}{1^2} \, 0.07$$

$$= 0.6 \text{ p.u.}$$

The $I^2X$ loss in the large generator-transformer will be negligible so that the generator must deliver $P = 2.5$ and $Q = 2.125$ p.u. and operate at a power factor of 0.76 lagging. It is seen in this example that starting with the consumer load the vars for each circuit in turn are added to obtain the total.

### 4.3 Relation Between Voltage, Power, and Reactive Power at a Node

The phase voltage $V$ at node is a function of $P$ and $Q$ at that node, i.e.

$$V = \phi(P,Q).$$

The total differential of $V$,

$$dV = \frac{\partial V}{\partial P} \cdot dP + \frac{\partial V}{\partial Q} \cdot dQ$$

and using

$$\frac{\partial P}{\partial V} \cdot \frac{\partial V}{\partial P} = 1 \quad \text{and} \quad \frac{\partial Q}{\partial V} \cdot \frac{\partial V}{\partial Q} = 1,$$

$$dV = \frac{dP}{(\partial P/\partial V)} + \frac{dQ}{(\partial Q/\partial V)} \tag{4.2}$$

It can be seen from equation (4.2) that the change in voltage at a node is defined by the two quantities,

$$\left(\frac{\partial P}{\partial V}\right) \quad \text{and} \quad \left(\frac{\partial Q}{\partial V}\right).$$

As an example consider a line with series impedance $(R+jX)\Omega$ and zero shunt admittance. From equation (2.6),

$$(V_1 - V)V - PR - XQ = 0 \tag{4.3}$$

where $V_1$, the sending-end voltage, is constant and $V$ the receiving-end voltage depends on $P$ and $Q$ (figure 4.4).

From equation (4.3)

$$\frac{\partial P}{\partial V} = \frac{V_1 - 2V}{R} \tag{4.4}$$

Also,

$$\frac{\partial Q}{\partial V} = \frac{V_1 - 2V}{X} \tag{4.5}$$

Hence,

$$dV = \frac{dP}{\partial P/\partial V} + \frac{dQ}{\partial Q/\partial V}$$

$$= \frac{dP \cdot R + dQ \cdot X}{V_1 - 2V} \tag{4.6}$$

For constant $V$ and $\Delta V$, $R dP + X dQ = 0$ and $dQ = -(R/X)dP$ which is obtainable directly from (4.1).

Normally $\partial Q/\partial V$ is the quantity of greater interest. It can be found experimentally using a Network Analyser (see chapter 5) by the injection of a known quantity of vars at the node in question and measuring the difference in voltage produced. From the results obtained,

$$\frac{\Delta Q}{\Delta V} = \frac{Q_{after} - Q_{before}}{V_{after} - V_{before}}$$

$\Delta V$ should be small for this test, a few percent of the normal voltage.

From the expression,

$$\frac{\partial Q}{\partial V} = \frac{V_1 - 2V}{X}$$

proved for the line, it is evident that the smaller the reactance associated with a node the larger the value of $\partial Q/\partial V$ for a given voltage drop, i.e. the voltage drop is inherently small. The greater the number of lines meeting at a node the smaller the resultant reactance and the larger the value of $\partial Q/\partial V$. $\partial Q/\partial V$ obviously depends on the network configuration but a high value would lie in the range 10–15 MVAr/kV. If the natural voltage drop at a point without the artificial injection of vars is, say 5 kV and the value of $\partial Q/\partial V$ at this point is 10 MVAr/kV, then to maintain the voltage at its no-load level would require $5 \times 10$ or 50 MVAr. Obviously the greater the value of $\partial Q/\partial V$ the more expensive it becomes to maintain voltage levels by injection of reactive power.

$\partial Q/\partial V$ *and the short-circuit current at a node*   It has been shown that for a connector of reactance $X$ $\Omega$ with a sending-end voltage $V_1$ and a received voltage $V$ per phase,

$$\frac{\partial Q}{\partial V} = \frac{V_1 - 2V}{X} \tag{4.5}$$

If the three phases of the connector are now short-circuited at the receiving end (i.e. three-phase symmetrical short-circuit applied) the current flowing in the lines

$$= \frac{V_1}{X}\text{A, assuming } R \ll X.$$

With the system on no-load

$$V = V_1 \quad \text{and} \quad \frac{\partial Q}{\partial V} = -\frac{V_1}{X}$$

Hence the magnitude of $(\partial Q/\partial V)$ is equal to the short-circuit current; the sign decides the nature of the reactive power. With normal operation $V$ is within a few per cent of $V_1$ and hence the value of $\partial V/\partial Q$ at $V = V_1$ gives useful information regarding reactive power/voltage characteristics for small excursions from the nominal voltage. This relationship is especially useful as the short-circuit current will normally be known at all substations.

*Example 4.2*   Three supply points A, B and C are connected to a common busbar M. Supply point A is maintained at a nominal 275 kV and is connected to M through a 275/132 kV transformer (0·1 p.u. reactance) and a 132 kV line of reactance 50 $\Omega$. Supply point B is nominally at 132 kV and is connected to M through a 132 kV line of 50 $\Omega$ reactance. Supply point C is nominally at 275 kV and is connected to M by a 275/ 132 kV transformer* (0·1 p.u. reactance) and a 132 kV line of 50 $\Omega$ reactance.

If at a particular system load the line-voltage at M falls below its nominal value by 5 kV, calculate the magnitude of the reactive volt ampere injection required at M to re-establish the original voltage.

The p.u. values are expressed on a 500 MVA base and resistance may be neglected throughout.

*Solution*  The line diagram and equivalent single-phase circuit are shown in figures 4.5 and 4.6.

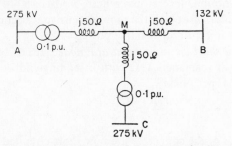

Figure 4.5  Schematic diagram of the system for Example 4.2.

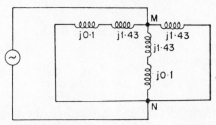

Figure 4.6  Equivalent single-phase network with the node *M* short-circuited to neutral (refer to chapter 6 for full explanation on the derivation of this circuit).

It is necessary to determine the value of $\partial Q/\partial V$ at the node or busbar M; hence the current flowing into a three-phase short-circuit at M is required. The base value of reactance in the 132 kV circuit

$$= \frac{132^2 \times 1000}{500,000} = 35\,\Omega$$

∴  the line reactances

$$= \frac{j50}{35} = j1{\cdot}43 \text{ p.u.}$$

The equivalent reactance from M to N = j0·5 p.u.

Hence the fault MVA at M

$$= \frac{500}{0 \cdot 5} = 1000 \text{ MVA}$$

and the fault current

$$= \frac{1000 \times 10^6}{\sqrt{(3)} \times 132{,}000}$$

$$= 4380 \text{ A at zero power factor lagging.}$$

It has been shown that $\partial Q_M / \sqrt{(3)} \partial V_M$ = three-phase short-circuit current, when $Q_M$ and $V_M$ are three-phase and line values

$$\therefore \qquad\qquad \frac{\partial Q_M}{\partial V_M} = 7 \cdot 6 \text{ MVAr/kV.}$$

The natural voltage drop at M = 5 kV

∴ the value of the injected vars required, $\Delta Q_M$, to offset this drop

$$= 7 \cdot 6 \times 5 = 38 \text{ MVAr.}$$

### 4.4 Methods of Voltage Control (i)—Injection of Reactive Power

The background to this method has been given in the previous sections. This is the most fundamental method but in transmission systems it lacks the flexibility and economy of transformer tap changing. Hence it is only used in schemes when transformers alone will not suffice. The provision of static capacitors to improve the power factors of factory loads has been long established. The capacitance required for the power-factor improvement of loads for optimum economy is determined as follows.

Let the tariff of a consumer be

$$£A \times kVA + B \times kWh$$

A load of $P_1$ kW at power factor $\phi_1$ lagging has a kVA of $P_1/\cos \phi_1$. If this power factor is improved to $\cos \phi_2$, the new kVA is $P_1/\cos \phi_2$. The saving is therefore

$$£P_1 A \left( \frac{1}{\cos \phi_1} - \frac{1}{\cos \phi_2} \right).$$

The reactive power required from the correcting capacitors

$$= P_1 \tan \phi_1 - P_1 \tan \phi_2 \text{ kVAr}$$

Let the cost per annum in interest and depreciation on the capacitor installation be $£C$ per kVAr or

$$£C(P_1 \tan \phi_1 - P_1 \tan \phi_2).$$

The net saving

$$= \pounds\left[AP_1\left(\frac{1}{\cos\phi_1} - \frac{1}{\cos\phi_2}\right) - CP_1(\tan\phi_1 - \tan\phi_2)\right]$$

This saving is a maximum when

$$\frac{d\,(\text{saving})}{d\phi_2} = 0,$$

i.e. when $\sin\phi_2 = C/A$.

It is interesting to note that the optimum power factor is independent of the original one. The improvement of load power factors in such a manner will obviously alleviate the whole problem of var flow in the transmission system.

The main effect of transmitting power at non-unity power factors is as follows. It is evident from equation (2.6) that the voltage drop is largely determined by the reactive power ($Q$). The line currents are larger giving increased $I^2R$ losses and hence reduced thermal capability. One of the obvious places for the artificial injection of reactive power is at the loads themselves. In general three methods of injection are available, involving the use of,

(a) static shunt capacitors

(b) static series capacitors

(c) synchronous compensators

*Shunt capacitors and reactors*

Shunt capacitors are used for lagging power-factor circuits whereas reactors are used on those with leading power-factors such as created by lightly loaded cables. In both cases the effect is to supply the requisite reactive power to maintain the values of the voltage. Capacitors are connected either directly to a busbar or to the tertiary winding of a main transformer and are disposed along the route to minimize the losses and voltage drops. Unfortunately as the voltage falls the vars produced by a shunt capacitor or reactor fall, thus when needed most their effectiveness falls. Also on light loads when the voltage is high the capacitor output is large and the voltage tends to rise to excessive levels.

*Series capacitors*

These are connected in series with the line conductors and are used to reduce the inductive reactance between the supply point and the load. One major drawback is the high overvoltage produced when a short-circuit current flows through the capacitor and special protective devices are incorporated (e.g. spark gaps). The phasor diagram for a line with a series capacitor is shown in figure 4.7b.

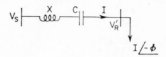

Figure 4.7a Line with series capacitor.

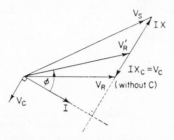

Figure 4.7b Phasor diagram.

The relative merits between shunt and series capacitors may be summarized as follows:

(a) If the load var requirement is small, series capacitors are of little use.

(b) With series capacitors the reduction in line current is small, hence if thermal considerations limit the current little advantage is obtained and shunt compensation should be used.

(c) If voltage drop is the limiting factor, series capacitors are effective; also voltage fluctuations due to arc furnaces, etc., are evened out.

(d) If the total line reactance is high, series capacitors are very effective and stability is improved.

*Synchronous compensators*

A synchronous compensator is a synchronous motor running without a mechanical load and depending on the value of excitation it can absorb or generate reactive power. As the losses are considerable compared with static capacitors the power-factor is not zero. When used with a voltage regulator the compensator can automatically run overexcited at times of high load and underexcited at light load. A typical connexion of a synchronous compensator is shown in figure 4.8 and the associated voltage-var output characteristic in figure 4.9. The compensator is run up as an induction motor in 2·5 minutes and then synchronized.

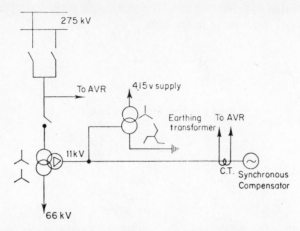

Figure 4.8  Typical installation with synchronous compensator connected to tertiary (delta) winding of main transformer. A neutral point is provided by the earthing transformer shown. The automatic voltage regulator on the compensator is controlled by a combination of the voltage on the 275 kV system and the current output; this gives a droop to the voltage-var output curve which may be varied as required.

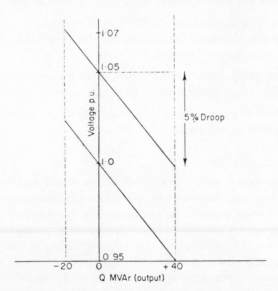

Figure 4.9  Voltage—reactive power output of a typical synchronous compensator.

A great advantage is the flexibility of operation for all load conditions. Although the cost of such installations is high in some circumstances it is justified, for example, at the receiving-end busbar of a long high-voltage line where transmission at power factors less than unity cannot be tolerated. Static shunt capacitors would then be used on the lower voltage network as required.

*Reactive-power requirements for the voltage control of long lines*
It is advantageous to use the generalized line equation,

$$\mathbf{V}_S = \mathbf{A}\mathbf{V}_r + \mathbf{B}\mathbf{I}_r$$

which takes into account the presence of transformers.

Let the received load-current $\mathbf{I}_r$, lag $\mathbf{V}_r$ by $\phi_r$ where $\mathbf{V}_r$ is the reference phasor

$$\mathbf{A} = A\underline{/\alpha} \quad \text{and} \quad \mathbf{B} = B\underline{/\beta}$$

$$V_s\underline{/\delta_s} = AV_r\underline{/\alpha} + BI_r\underline{/\beta-\phi_r} \tag{4·7}$$

$$\therefore \quad V_s\underline{/\delta_s} = AV_r \cos \alpha + jAV_r \sin \alpha + BI_r \cos (\beta-\phi_r) + jBI_r \sin (\beta-\phi_r)$$

As the modulus of the left hand side is equal to that of the right

$$V_s^2 = A^2 V_r^2 + B^2 I_r^2 + 2ABV_r I_r \cos (\alpha - \beta + \phi_r)$$
$$= A^2 V_r^2 + B^2 I_r^2 + 2ABV_r I_r [\cos (\alpha - \beta) \cos \phi_r$$
$$- \sin (\alpha - \beta) \sin \phi_r] \tag{4.8}$$

$$P_r = V_r I_r \cos \phi_r \quad \text{and} \quad Q_r = V_r I_r \sin \phi_r.$$

Hence equation 4.8 becomes

$$V_s^2 = A^2 V_r^2 + B^2 I_r^2 + 2ABP_r \cos (\alpha - \beta) - 2ABQ_r \sin (\alpha - \beta)$$

Also,

$$\mathbf{I}_r = I_p - jI_q \quad \text{and} \quad I_r^2 = I_p^2 + I_q^2 \quad \text{and} \quad I_p = \frac{P_r}{V_r} \quad \text{and} \quad I_q = \frac{Q_r}{V_r}$$

$$\therefore \quad V_s^2 = A^2 V_r^2 + B^2 \left( \frac{P_r^2}{V_r^2} + \frac{Q_r^2}{V_r^2} \right) + 2ABP_r \cos (\alpha - \beta)$$

$$- 2ABQ_r \sin (\alpha - \beta) \tag{4.9}$$

For a given network $P_r$, $\mathbf{A}$ and $\mathbf{B}$ will be known. The magnitude of $Q_r$ such that $V_r$ is equal to, or a specified ratio of, $V_S$ can be determined by the use of equation (4.9).

## 4.5 Methods of Voltage Control (ii)—Tap Changing Transformers

The basic operation of the tap-changing transformer has been discussed in chapter 2; by changing the transformation ratio the voltage in the secondary circuit is varied and voltage control obtained. This constitutes the most popular and widespread form of voltage control at all voltage levels.

Consider the operation of a radial transmission system with two tap-changing transformers, as shown in the equivalent single-phase circuit of figure 4.10. $t_s$ and $t_r$ are fractions of the nominal transformation ratios, i.e. the tap ratio/nominal ratio. For example, a transformer of nominal ratio 6·6 kV to 33 kV when tapped to give 6·6 to 36 kV has a $t_s = 36/33 = 1·09$. $V_1$ and $V_2$ are the nominal voltages at the ends of the line and the actual voltages being $t_s V_1$ and $t_r V_2$. It is required to determine the tap changing ratios required to completely compensate for the voltage drop in the line. The product $t_s t_r$ will be made unity; this ensures that the overall voltage-level remains in the same order and that the minimum range of taps on both transformers is used.

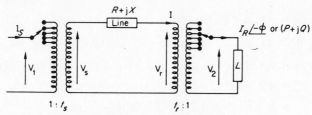

Figure 4.10   Co-ordination of two tap-changing transformers in a radial transmission link.

From figure 4.10,
$$t_s V_1 = t_r V_2 + IZ$$

Using equation (2.6),
$$IZ \doteq \Delta V = \frac{RP + XQ}{t_r V_2}$$

$\therefore$
$$t_s V_1 = t_r V_2 + \frac{RP + XQ}{t_r V_2}$$

As $t_s t_r = 1$,
$$t_s = \frac{1}{V_1}\left(\frac{1}{t_s}V_2 + \frac{RP + XQ}{V_2/t_s}\right)$$

$\therefore$
$$t_s^2 = \frac{V_2}{V_1} + \frac{RP + XQ}{V_1 V_2} \cdot t_s^2$$

$\therefore$
$$t_s^2\left(1 - \frac{RP + XQ}{V_1 V_2}\right) = \frac{V_2}{V_1} \qquad (4.10)$$

For complete line voltage drop compensation $V_1 = V_2$.

*Example 4.3* A 132 kV line is fed through an 11/132 kV transformer from a constant 11 kV supply. At the load end of the line the voltage is reduced by another transformer of nominal ratio 132/11 kV. The total impedance of the line and transformers at 132 kV is $(25+j66)$ $\Omega$. Both transformers are equipped with tap-changing facilities which are so arranged that the product of the two off-nominal settings is unity. If the load on the system is 100 MW at 0·9 p.f. lagging calculate the settings of the tap-changers required to maintain the voltage of the load busbar at 11 kV.

*Solution* The line diagram is shown in figure 4.11. As the line voltage-drop is to be completely compensated $V_1 = V_2 = 132$ kV.

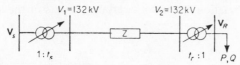

Figure 4.11 Schematic diagram of system for Example 4.3.

Also $t_s \times t_r = 1$.
The load is 100 MW, 48·3 MVAr.
It has been shown that,

$$t_s^2\left(1 - \frac{RP+XQ}{V_1V_2}\right) = \frac{V_2}{V_1}$$

∴ $$t_s^2\left(1 - \frac{25 \times (100 \times 10^6/3) + 66 \times (48·3 \times 10^6/3)}{(132/\sqrt{3})^2 \times 10^6}\right) = 1$$

Hence

$$t_s = 1·22 \text{ and therefore } t_r = 0·82.$$

These settings are large for the normal range of tap-changing transformers (usually not more than $\pm 20$ per cent tap range). It would be necessary in this system to inject vars at the load end of the line to maintain the voltage at the required value.

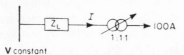

V constant

Figure 4.12 Effect of tap changing on line current and voltage drop.

It is important to note that the transformer does not improve the var-flow position and also that the current in the supplying line is increased if the ratio is increased. From figure 4.12 it can be seen that initially the line current $(I) = 100$ A (ratio $1:1$). After the tap change the current $(I) = 100 \times 1\cdot1$, and the volt drop in $Z_L = 1\cdot1 \times$ the original value. Hence the secondary voltage increase is somewhat offset by the increased drop in the line. Obviously if the line impedance is high it is possible for the voltage drop to be too large for the transformer to maintain voltages with the tap range available.

### 4.6   Combined Use of Tap-Changing Transformers and Reactive Power Injection

The usual practical arrangement is shown in figure 4.13 where the tertiary winding of a three-winding transformer is connected to a synchronous compensator. For given load conditions it is proposed to determine the necessary transformation ratios with certain outputs of the compensator.

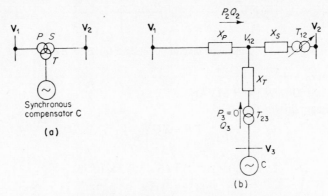

Figure 4.13a   Schematic diagram of system with combined tap changing and synchronous compensation.
Figure 4.13b   Equivalent network.

The transformer is represented by the equivalent star connexion and any line impedances from $V_1$ or $V_2$ to the transformer can be lumped together with the transformer branch impedances. $V_n$ is the phase voltage at the star-point of the equivalent circuit in which the secondary impedance $(Z_s)$ is usually approaching zero and hence neglected. Resistance and losses are neglected. The allowable ranges of voltage for $V_1$ and $V_2$ are specified and the values of $P_2$, $Q_2$, $P_3$ and $Q_3$ given, $P_3$ is usually taken as zero.

The volt drop $V_1$ to $V_n$

$$= \Delta V \doteqdot X_p \frac{Q_2/3}{V_n}$$

or

$$X_p \frac{Q_2}{V_L\sqrt{3}},$$

where $V_L$ is the line voltage $= \sqrt{(3)}V_n$ and $Q_2$ is the total vars. Also,

$$\delta V = X_p \frac{P_2}{V_L\sqrt{3}}$$

$\therefore$ $(V_n + \Delta V)^2 + (\delta V)^2 = V_1^2$

(see phasor diagram of 2.7; phase values used) and

$$\left(V_n + X_p \frac{Q_2}{V_L\sqrt{3}}\right)^2 + X_p^2\left(\frac{P_2^2}{3V_L^2}\right) = V_1^2$$

$\therefore$ $(V_L^2 + X_p Q_2)^2 + X_p^2 P_2^2 = V_{1L}^2 V_L^2$

where $V_{1L}$ is the line voltage $= \sqrt{(3)}V_1$

$\therefore$ $$V_L^2 = \frac{V_{1L}^2 - 2X_p Q_2}{2} \pm \tfrac{1}{2}\sqrt{[V_{1L}^2(V_{1L}^2 - 4X_p Q_2) - 4 \cdot X_p^2 P_2^2]}$$

Once $V_L$ is obtained, the transformation ratio is easily found. The procedure is best illustrated by an example.

*Example 4.4* A three-winding grid transformer has windings rated as follows: 132 kV (line), 75 MVA, star connected; 33 kV (line), 60 MVA, star connected; 11 kV (line), 45 MVA, delta connected. A synchronous compensator is available for connexion to the 11-kV winding.

The equivalent circuit of the transformer may be expressed in the form of three windings, star connected, with an equivalent 132 kV primary reactance of 0·12 p.u., negligible secondary reactance and an 11 kV tertiary reactance of 0·08 p.u., both values expressed on a 75 MVA base.

In operation, the transformer must deal with the following extremes of loading:

(a) Load of 60 MW, 30 MVAr with primary and secondary voltages governed by the limits 120 kV and 34 kV; synchronous compensator disconnected.

(b) No load. Primary and secondary voltage limits 143 kV and 30 kV; synchronous compensator in operation and absorbing 20 MVAr.

Calculate the range of tap changing required. Ignore all losses.

*Solution* The value of $X_p$ the primary reactance in ohms $=$

$$0\cdot12 \times 132^2 \times 1000/75 \times 1000 = 27\cdot8 \ \Omega.$$

Similarly the effective reactance of the tertiary winding is 18·5 Ω. The equivalent star circuit is shown in figure 4.14.

The first operating conditions are as follows:

$$P_1 = 60 \text{ MW}, \quad Q_1 = 30 \text{ MVAr}, \quad V_{1L} = 120 \text{ kV}.$$

Hence,

$$V_L^2 = \tfrac{1}{2}(120,000^2 - 2 \times 27\cdot8 \times 30 \times 10^6) \pm$$

$$\tfrac{1}{2}\sqrt{[120,000^2(120,000^2 - 4 \times 27\cdot8 \times 30 \times 10^6) - 4 \times 27\cdot8^2 \times 60^2 \times 10^{12}]}$$

$$= \left(63\cdot61 \pm \frac{122}{2}\right)10^8 = 124\cdot4 \times 10^8$$

∴     $V_L = 111 \text{ kV}.$

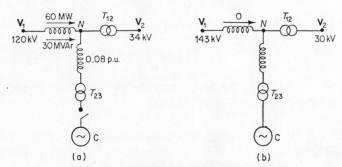

Figure 4.14a   Example 4.4. System with loading condition (a).
Figure 4.14b   System with loading condition (b).

The second set of conditions are:

$$V_{1L} = 143 \text{ kV}, \quad P_2 = 0 \text{ and } Q_2 = 20 \text{ MVAr}.$$

Again using the formula for $V_L$,

$$V_L = 138\cdot5 \text{ kV}.$$

The transformation ratio under the first condition

$$= 111/34 = 3\cdot27$$

and for the second condition,

$$\frac{138\cdot5}{30} = 4\cdot61.$$

The actual ratio will be taken as the mean of these extremes, i.e. 3·94, varying by ±0·67 or 3·94±17 per cent. Hence the range of tap changing required is ±17 per cent.

### 4.7 Booster Transformers

It may be desirable on technical or economic grounds to increase the voltage at an intermediate point in a line rather than at the ends as with tap-changing transformers, or the system may not warrant the expense of tap changing. Here booster transformers are used as shown in figure 4.15. The booster can be brought into the circuit by the closure of relay B and the opening of A and vice versa. The mechanism by which the relays are operated can be controlled either from a change in the voltage or in the

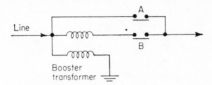

Figure 4.15   Connexion of in-phase booster transformer. One phase only shown.

current. The latter method is the more sensitive, as from no-load to full-load represents a 100 per cent change in current, but only in the order of a 10 per cent change in voltage. This booster gives an in-phase boost as does a tap-changing transformer. An economic advantage is that the rating of the booster is the product of the current and the injected voltage and is hence only about 10 per cent of that of a main transformer. Boosters are often used in distribution feeders where the cost of tap-changing transformers is not warranted.

*Example 4.5*   In the system shown by the line diagram in figure 4.16 each transformer $T_A$ and $T_B$ have tap ranges of $\pm 10$ per cent in 14 steps of 1·43 per cent. Initially $V_A^1 = V_B^1$ and hence no power or var transfer takes place through the line. It is required to calculate the magnitude of the circulating current resulting from an in-phase boost of 8·75 per cent on $T_A$, with the busbar voltages maintained constant by automatic voltage regulators on generators $G_A$ and $G_B$. The plant data are as follows expressed on a 20 MVA base, resistance ignored. Generators $G_A$ and $G_B$; 20 MVA, $X = 0·2$ p.u. Transformers $T_A$ and $T_B$; 20 MVA, 132/33, $X = 0·1$ p.u.; 132 kV line, 5 miles, 0·175 in², $X = 3·85\ \Omega = 0·061$ p.u.

*Solution*   The equivalent circuits of the network are shown in figure 4.16. The only voltage source is the boost $E_x = 0·0875$ p.u. which produces a current of 0·336 p.u. As the busbar voltages are maintained constant at 33 kV by voltage regulators no circulating current flows in the loads.

The reactive power in generator $G_A$ is now $12+(0.336 \times 1 \times 20)$, i.e. 18·7 MVAr and in $G_B$ is $12+(-0.336 \times 1 \times 20)$, i.e. 5·3 MVAr.

It is seen that $G_A$ is heavily loaded with vars whilst $G_B$ is loaded lightly; this could lead to a loss of stability. In phase boosts redistribute currents in a network and this has been used to de-ice lines in the winter, the extra current produces sufficient $I^2 R$ loss heating.

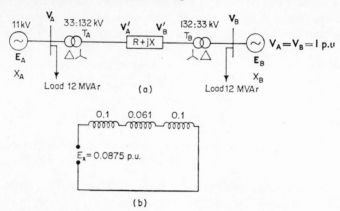

Figure 4.16a   Line diagram of system for Example 4.5.   b   Equivalent network with voltage boost $E_x$ acting.

Circulating current is also used to automatically make tap-changing transformers in parallel, return to the same tap point if they tend to operate several taps apart.

*Example 4.6*   In the system shown in figure 4.17 it is required to keep the nominally 11 kV busbar at constant voltage. The range of taps is not

*Three winding transformer*

| Winding | MVA | Voltage kV | p.u. $Z$ referred to nameplate MVA | p.u. $Z$ on 15 MVA base | $(Z(\Omega))$ referred to 33 kV | Equivalent $(Z(\Omega))$ referred to 33 kV |
|---|---|---|---|---|---|---|
| P–S | 15 | 33/11 | 0·008+j0·1 | 0·008+j·1 | 0·57+j7·3 | $Z_1 =$ 0·214+j8·2 |
| P–T | 5 | 33/1·5 | 0·0035+j·0595 | 0·0105+j·179 | 0·76+j4·32 | $Z_2 =$ 0·363−j·78 |
| S–T | 5 | 11/1·5 | 0·0042+j0·0175 | 0·0126+j·0525 | 0·915+j1·27 | $Z_3 =$ 0·545+j4·77 |

sufficient and it is proposed to use shunt capacitors connected to the tertiary winding. The data are as follows, p.u. quantities being referred to a 15 MVA base; line 10 miles, 0·2 in², OHL, 33 kV, $Z = (0·0304 + j0·0702)$; $Z$ referred to 33 kV side $= (2·2 + j5·22)$ Ω.

For the three winding transformer the measured impedances between the windings and the resulting equivalent star impedances $Z_1$, $Z_2$ and $Z_3$ are given. The equivalent circuit referred to 33 kV is shown in figure 4.17b.

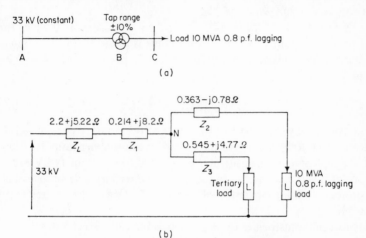

(a)

(b)

Figure 4.17a   Line diagram for Example 4.6.
Figure 4.17b   Equivalent network—referred to 33 kV.

The voltage across the receiving-end load

$$= \frac{33,000}{\sqrt{3}} - \Delta V,$$

where

$$\Delta V \doteqdot \frac{RP + XQ}{V_C}$$

∴

$$\Delta V \doteqdot \frac{2·77 \times 8/3 \times 10^6 + 12·64 \times 6/3 \times 10^6}{33,000/\sqrt{3}}$$

As $V_C$ referred to the 33 kV base is not known because of the system volt drop, 33 kV is assumed initially. The revised value is then used and the process repeated.

$$\Delta V = \frac{7·4 + 25·28}{19} \text{ kV} = 1·715 \text{ kV} \quad \text{and} \quad V_C = 17·285 \text{ kV}.$$

Repeating the calculation for $\Delta V$ with the new $V_C$,

$$\Delta'V = \frac{2 \cdot 77 \times 8/3 \times 10^6 + 12 \cdot 64 \times 6/3 \times 10^6}{17 \cdot 285} = 1 \cdot 89 \text{ kV}.$$

Hence

$$V_C' = 19 - 1 \cdot 89 = 17 \cdot 11 \text{ kV}.$$

$$\Delta''V = 1 \cdot 9 \quad \text{and} \quad V_C'' = 17 \cdot 1 \text{ kV}.$$

This will be the final value of $V_C$.

$V_C$ referred to $11 \text{ kV} = 17 \cdot 1/3 = 5 \cdot 7 \text{ kV}$ (phase) or $9 \cdot 9 \text{ kV}$ (line). In order to maintain $11 \text{ kV}$ at C the voltage is raised by tapping down on the transformer. Using the full range of 10 per cent, i.e. $t_r = 0 \cdot 9$ the voltage at C is

$$\frac{29 \cdot 7}{(33 \times 0 \cdot 9)/11} = 11 \text{ kV}$$

The true voltage will be less than this as the primary current will have increased by $(1/0 \cdot 9)$ because of the change in transformer ratio. It is evident that the tap-changing transformer is not able to maintain $11 \text{ kV}$ at C and the use of a static capacitor connected to the tertiary will be investigated.

Consider a shunt capacitor of capacity $5 \text{ MVAr}$ (the capacity of the tertiary).

Assume the transformer to be at its nominal ratio $33/11 \text{ kV}$. The voltage drop

$$= \frac{2 \cdot 414 \times 8/3 \times 10^6 + 13 \cdot 42 \times 1/3 \times 10^6}{V_N (\doteqdot 19 \text{ kV})}$$

$$= 0 \cdot 587 \text{ kV}$$

$$V_N' = 19 - 0 \cdot 587 = 18 \cdot 413 \text{ kV (phase)}$$

$$\therefore \qquad \Delta'V_N = 0 \cdot 606 \quad \text{and} \quad V_N' = 18 \cdot 394 \text{ kV}.$$

$\therefore$   The volt drop N to C

$$\Delta V_C = \frac{0 \cdot 363 \times 8/3 - 0 \cdot 78 \times 6/3}{18 \cdot 394} \text{ kV}$$

$$= -0 \cdot 032 \text{ kV}$$

$$\therefore \qquad V_C = 18 \cdot 394 + 0 \cdot 032$$

$$= 18 \cdot 426 \text{ kV (phase)}.$$

As $\Delta V_C$ is so small there is no need to iterate further.

Referred to $11 \text{ kV}$, $V_C = 10 \cdot 55 \text{ kV}$ (line). Hence to have $11 \text{ kV}$ the transformer will tap such that $t_r = (1 - 0 \cdot 35/11) = 0 \cdot 97$, i.e. a 3 per cent tap change, which is well within the range and leaves room for load increases. On *no load*

$$\Delta V = \frac{2 \cdot 959 \times 0 + 18 \cdot 19 \times (-5/3)}{19} \text{ kV}$$

$$= -\frac{30 \cdot 3}{19} = -1 \cdot 595 \text{ kV (phase)}.$$

The shunt capacitor is a constant impedance load and hence as $V_N$ rises the current taken increases causing further volt increase.

Ignoring this effect initially,

$$V_C = 10 + 1 \cdot 6 = 20 \cdot 6 \text{ kV (phase); at 11 kV,}$$

$$V_C = 11 \cdot 85 \text{ kV (line)}$$

therefore the tap change will have to be at least 7·15 per cent which is well within the range. Further refinement in the value of $V_C$ will be unnecessary. If an accurate value of $V_C$ is required then the reactance of the capacitor must be found and the current evaluated.

### 4.8 Voltage Stability

Voltage stability is essentially an aspect of load stability to be discussed in chapter 7. As the voltages to be maintained in a system are influenced by voltage stability it is appropriate to discuss this subject here.

Consider the circuit shown in figure 4.18a. If $V_S$ is fixed (i.e. an infinite busbar supply) the graph of $V_R$ against $P$ for given power factors is as shown in figure 4·18b. In figure 4.18b, $Z$ represents the series impedance of a 100 mile long, double circuit, 400 kV, 0·4 in² conductor overhead line. The fact that two values of voltage exist for each value of power is easily demonstrated by considering the analytical solution of this circuit. At the lower voltage a very high current is taken to produce the power. The seasonal thermal ratings of the line are also shown and it is apparent that for loads of power factor less than unity the possibility exists that before the thermal rating is reached the operating power may be on that part of the characteristic where small changes in load cause large voltage changes and *voltage instability* will have occurred. In this condition the action of tap changing transformers is interesting. If the receiving end transformers 'tap up' to maintain the load voltage the line current increases thereby causing further increase in the voltage drop. It would in fact be more profitable to 'tap down', thereby reducing the current and voltage drop. It is feasible therefore for a 'tapping-down' operation to result in increased secondary voltage, and vice-versa.

The possibility of an actual voltage collapse depends upon the nature of the load. If this is stiff (constant power), e.g. induction motors, the collapse is aggravated. If the load is soft, e.g. heating, the power falls off rapidly with voltage and the situation is alleviated. Referring to figure 4.18 it is evident that a critical quantity is the power factor; at full load a change in lagging power factor from 0·99 to 0·90 will precipitate voltage collapse. On long lines therefore, for reasonable power transfers it is necessary to keep the power factor of transmission approaching unity, certainly above 0·97 lagging and it is economically justifiable to employ var injection by static capacitors or synchronous compensators at the load.

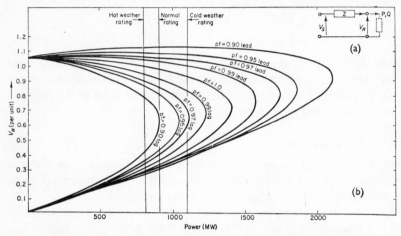

Figure 4.18a   Equivalent circuit of a line supplying a load $P+jQ$.
b Relation between load voltage and received power at constant power factor for a 400 kV, $2\times0\cdot4$ in$^2$ conductor line, 100 miles in length. Thermal ratings of the line are indicated.

A problem arises with the operation of two or more lines in parallel, for example the system shown in figure 4.19 in which the shunt capacitance has been represented as in a $\pi$ section. If one of the three lines is removed from the circuit due to a fault, the total system reactance will increase from $X/3$ to $X/2$, and the capacitance, which normally improves the power factor decreases to $2C$ from $3C$. Thus the overall voltage drop is greatly increased and owing to the increased $I^2X$ loss of the lines and the decreased generation of vars by the shunt capacitances the power factor decreases; hence the possibility of voltage instability. The same argument will of course apply to two lines in parallel.

Usually there will be local generation feeding the receiving-end busbars at the end of long lines. If this generation is electrically close to the load busbars, i.e. low connecting impedance $Z_g$, a fall in voltage will automatically increase the local var generation and this may be sufficient to keep the reactive power transmitted low enough to avoid large voltage drops in the long lines. Often, however, the local generators supply lower voltage networks and are electrically remote from the high-voltage busbar of figure 4·19 and $Z_g$ is high. The fall in voltage now causes little change in the local generator var output and the use of capacitors, static or rotary, at the load may be required. As $Z_g$ is inversely proportional to the three-phase short-circuit level at the load busbar due to the local generation; the reactive-power contribution of the local machines is proportional to this fault level.

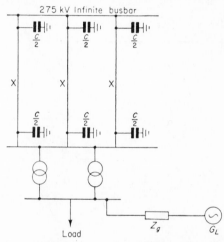

Figure 4.19   Line diagram of three long lines in parallel—effect of the loss of one line. $G_L$ = local generators.

At the time of writing about 40 per cent of the power generated in the British system is fed to load centres less than 20 miles in distance. This is changing, however, owing to the construction of the 400 kV network and the greater distance between generation and loads. It has been estimated that the reactive demand in Britain at peak loads is 0·4 kVAr/kW and the nett var loss in the lines, etc., 0·1 kVAr/kW, giving a total of 0·5 kVAr/kW. This is supplied almost entirely by generators operating with an average power-factor of 0·9 lagging. At minimum load the demand is 0·5 kVAr/kW with a nett generation of vars in the network. For modern machines with short-circuit ratios of 0·55, operation at unity power-factor is close to the theoretical stability limit with a 20 per cent margin. With an extensive very-high-voltage network however, the vars absorbed by the network on load will amount to 0·3 kVAr/kW and on light load the network will generate 0·25 kVAr/kW. Hence the voltage regulation problem will be greatly aggravated.

### 4.9 Voltage Control in Distribution Networks

Single phase supplies to houses and other small consumers are tapped off from three-phase feeders. Although efforts are made to allocate equal loads to each phase the loads are not applied at the same time and some unbalance occurs. An empirical method for modifying the voltage drop obtained by assuming balanced operation to allow for the average degree of unbalance is reported in reference 6.

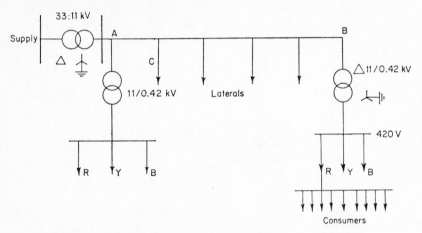

Figure 4.20   Line diagram of typical radial distribution scheme.

In the distribution network (British practice) shown in figure 4.20 an 11 kV distributor supplies a number of lateral feeders in which the voltage is 420 V and then each phase loaded separately (i.e. a, b, c).

The object of design is to keep the consumers' nominal 415 volt supply within the statutory ±6% of the declared voltage. The main 33/11 kV transformer gives a 5% rise in voltage between no load and full load. The distribution transformers have fixed taps and a secondary phase voltage of 250 volts which is 4% high on the nominal value of 240 volts. A typical distribution of voltage drops would be as follows: main distributor 6%, 11/0·42 kV transformer 3%, 420 volt circuit 7%, consumer circuit 1·5% giving a total of 17·5%. On very light load (10% of full load) the corresponding drop may be 1·5%. To offset these drops various voltage boosts are employed as follows: main transformer, +5% (zero on light load); distribution transformer, inherent boost of $+4\%$ (i.e. 250 volt secondary) plus a fixed $2.5\%$ boost. These add to give a total boost on full load of $11.5\%$ and on light load of $6.5\%$. Hence the consumers' voltage varies between $(-17.5+11.5)$, i.e. $-6\%$ and $6.5-1.5$, i.e. $+5\%$, which is just permissible. There will also be a difference in consumer voltage depending upon the position of the lateral feeder on the main distributor, obviously a consumer supplied from C will have a higher voltage than one from B.

Uniformly loaded distribution feeder. In areas such as towns with high load densities a large number of tappings are made from feeders and a uniform load along the length of a feeder may be considered to exist. Consider the load taken from a length d$x$ of the feeder distant $x$ metres from the supply end. Let i$A$ be the current tapped per unit length and $r$ and $x$ the resistance and reactance per phase per unit length.

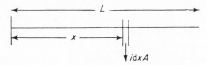

Figure 4.21   Uniformly loaded distributor.

The voltage drop at $x = rix\,\mathrm{d}x\cos\theta + xix\mathrm{d}x\sin\theta$, where $\cos\theta$ is the power factor of the uniformly distributed load.

The total voltage drop

$$= r\int_0^L ix\,\mathrm{d}x\cos\theta + x\int_0^L ix\,\mathrm{d}x\sin\theta$$

$$= ri\frac{L^2}{2}\cos\theta + xi\frac{L^2}{2}\sin\theta$$

$$= \frac{Lr}{2}I\cos\theta + \frac{Lx}{2}I\sin\theta$$

where $I = Li$ the total current load. Hence the uniformly distributed load may be represented by the total load tapped at the centre of the feeder length.

## References

**BOOKS**

1. *Electrical Transmission and Distribution Reference Book*, Westinghouse Electric. Corp., East Pittsburgh, Pennsylvania, 1964.
2. Waddicor, H., *Principles of Electric Power Transmission*, Chapman and Hall, London, 5th ed., 1964.

3. Mortlock, J. R., and M. W. Humphrey Davies, *Power System Analysis*, Chapman and Hall, London, 1952.
4. Dahl, O. G., *Electric Power Circuits*, Vol. II, McGraw-Hill, New York, 1938.

PAPERS

5. Smith, H. B., 'Principle of AC power system control for operating personnel', *Trans. A.I.E.E. Power Apparatus and Systems*, April (1949).
6. Egginton, J. L., 'Power systems planning and economics, Part VI, Low Voltage Networks', *Elec. J., London*, June (1956).
7. Salzmann, A., 'Voltage Levels in Multi-Voltage Power Systems', *Elec. Distbn.*, September (1961).
8. Casson, W., and H. J. Sheppard, 'Technical and Economic Aspects of the Supply of Reactive Power in England and Wales', *Proc. I.E.E.*, 108A (1961).
9. Booth, E. S., et al., 'The 400 kV Grid System for England and Wales', *Proc. I.E.E. Part A* (1962).
10. Csicy, J., 'The influence of the regulating transformer and the excitation of alternators on system voltages and stability', *C.I.G.R.E.*, No. 323 (1962).
11. Miller, G. M., and C. W. Robbins, 'Fundamental relation of system voltage drop and system loads', *Trans. A.I.E.E. Power Apparatus and Systems*, April (1955).
12. Venikov, V. A., 'The influence of power factor correction on load stability', (Russian) *Elektrichestro*, No. 6 (1960).
13. I.E.E.E. Committee Report, 'Bibliography on power capacitors 1959–1962', *I.E.E.E., P.A. & S.*, 83, 1110, (1964).
14. Breuer, G. D., H. M. Rustebakke, R. A. Gibley, and H. O. Simmons, Jr., 'The use of series capacitors to obtain maximum EHV transmission capability', *I.E.E.E., P.A. & S.*, 83, 1090, (1964).
15. Schmill, J. V., 'Optimum size and location of shunt capacitors on distribution feeders', *I.E.E.E., P.A. & S.*, 84, 825, (1965).
16. Baum, W. V., and W. A. Frederick, 'A method of applying switched and fixed capacitors for voltage control', *I.E.E.E., P.A. & S.*, 84, 42, (1965).
17. Kimbark, E. W., 'Improvement of system stability by switched series capacitors', *I.E.E.E., P.A. & S.*, 85, 180, (1966).
18. Duran, H., 'Optimum number, location, and size of shunt capacitors in radial distribution feeders—A dynamic programming approach', *I.E.E.E., P.A. & S.*, 87, 1769, (1968).
19. Kumai, K., and K. Ode, 'Power system voltage control using a process control computer', *I.E.E.E., P.A. & S.*, 87, 1985, (1968).
20. Chang, N. E., 'Locating shunt capacitors on primary feeder for voltage control and loss reduction', *I.E.E.E., P.A. & S.*, 88, 1574, (1969).
21. Boehne, E. W., and S. S. Low, 'Shunt capacitor energization with vacuum interrupters—A possible source of overvoltage', *I.E.E.E., P.A. & S.*, 88, 1424, (1969).
22. Kumar, B. S. A., K. Parthasarathy, F. S. Prabhakara, and H. P. Khincha, 'Effectiveness of series capacitors in long distance transmission lines', *I.E.E.E., P.A. & S.*, 89, 941, (1970).

**Problems**

4.1. An 11 kV supply busbar is connected to an 11/132 kV, 100 MVA, 10 per cent reactance, transformer. The transformer feeds a 132 kV transmission link consisting of an overhead line of impedance $(0.014+j0.04)$ p.u. and a cable of impedance $(0.03+j0.01)$ p.u. in parallel. If the receiving end is to be maintained at 132 kV when delivering 80 MW, 0.9 p.f. lagging calculate the power and re-active power carried by the cable and the line. All p.u. values relate to 100 MVA and 132 kV bases. (Line $(23+j38)$ MVA, cable $(57+j1)$ MVA.)

4.2. A three-phase induction motor delivers 500 h.p. at an efficiency of 0.91, the operating power factor being 0.76 lagging. A loaded synchronous motor with a power consumption of 100 kW is connected in parallel with the induction motor. Calculate the necessary kVA and the operating power factor of the synchronous motor if the overall power factor is to be unity. (365 kVA, 0.274.)

4.3. The load at the receiving end of a three-phase, overhead line is 25 MW, power factor 0.8 lagging, at a line voltage of 33 kV. A synchronous compensator is situated at the receiving end and the voltage at both ends of the line is main-tained at 33 kV. Calculate the MVAr of the compensator. The line has resistance 5 Ω per phase and inductive reactance (line to neutral) 20 Ω per phase. (33 MVAr.)

4.4. In the previous question calculate the maximum value of power that can be transmitted if the thermal rating of the line is not to be exceeded. Assume that without compensation the line was fully loaded, hence the current under the new conditions is unchanged. (36 MW.)

4.5. A three-phase transmission line has resistance and inductive reactance (line to neutral) of 25 Ω and 90 Ω respectively. With no load at the receiving end but with a synchronous compensator there taking a current lagging by 90°, the voltage at the sending end is 145 kV and 132 kV at the receiving end. Calculate the value of the current taken by the compensator.

When the load at the receiving end is 50 MW, it is found that the line can operate with unchanged voltages at sending and receiving ends, provided that the compensator takes the same current as before but now leading by 90°. Calculate the power factor of the load. (83.5 A, $Q_L$ 24.2 MVAr.)

4.6. Repeat question 4.3 making use of $\partial Q/\partial V$ at the receiving end.

4.7. In Example No. 4.3 determine the tap ratios if the receiving end voltage is to be maintained at 0.9 p.u. of the sending end voltage. ($1.19\ t_s$, $0.84\ t_r$.)

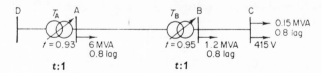

Figure 4.22    Line diagram for system in Problem 4.9.

4.8. In the transmission system in Example 4.4 a synchronous compensator is connected to the tertiary winding and produces 10 MVAr when a secondary load of 50 MVA 0.8 p.f. lagging is taken. On zero secondary load the compensator is adjusted to absorb 2 MVAr. Calculate the range of voltage at the load busbar if the supply is constant at 132 kV. (Lower limit 127 kV.)

4.9. In the system shown in Fig. 4.22 determine the supply voltage necessary at D to maintain a phase voltage of 240 V at the consumer's terminals at C. The following data apply:

(33 kV.)

| Line Trans- formers | Rated voltage kV | Line: length of cross- sectional area | | | Impedance ohms |
|---|---|---|---|---|---|
| | | Trans. | Rating | Nominal tap ratio | |
| BC | 0·415 | 0·25 in² | U/G | 300 ft | 0·0127+j0·00909 |
| AB | 11 | 0·15 in² | O.H.L. | 5 miles | 1·475+j2·75 |
| DA | 33 | 0·15 in² | O.H.L. | 5 miles | 1·475+j2·75 |
| $T_A$ | 33/11 | 10 MVA | 33/11 | 30·69/11 | 1·09+j9·8 referred to 33 kV |
| $T_B$ | 11/0·415 | 2·5 MVA | 11/0·415 | 10·450/0·415 | 0·24+j1·95 referred to 33 kV |

4.10. A load is supplied through a 275 kV link of total reactance 50 $\Omega$ from an infinite busbar at 275 kV. Plot the receiving-end voltage against power graph for a constant load power-factor of 0·95 lagging. The system resistance may be neglected.

4.11. In the system shown in figure 4.19 each line is 100 miles long and is rated at 275 kV with $2 \times 0·175$ in² conductors. The lines are carrying their rated (normal weather) MVA at 0·95 power factor lagging. Calculate the voltage at the load busbar, (a) initially and (b) when one line is switched out, taking into account the shunt capacitances of the lines. Examine the effect on the load voltage of generation in the load area which may be represented by a voltage source in series with a reactance of (a) 2 p.u. and (b) 0·5 p.u. connected to the load busbar. The local generator may be assumed to export zero current when all three lines are in service. (When $V_S = 1$ p.u., $V_R = 0·92$ p.u.)

(*Note.* Use Table 2·2a for line ratings. Initially $P_g$ and $Q_g$ from local generator $= 0$ and generator e.m.f. $(E_g) =$ received busbar voltage $(V_R)$. After outage, if $V_R^1$ is new received voltage and $V_s$ sending voltage,

$$V_s - V_R^1 = \frac{RP + XQ_r}{V_R^1}$$

Also $E_g - V_R = X_g Q_g / V_R^1$ as $P_g$ is still zero).

4.12. Explain the limitations of tap-changing transformers. A transmission link (figure 4.23a) connects an infinite busbar supply of 400 kV to a load busbar supplying 1000 MW, 400 MVAr. The link consists of lines of effective impedance $(7 + j70)$ $\Omega$ feeding the load busbar via a transformer with a maximum tap ratio of 0·9 : 1. Connected to the load busbar is a synchronous compensator. If the maximum overall voltage drop is to be 10 per cent with the transformer taps fully utilized, calculate the reactive power requirement from the compensator. (148 MVAr).

*Note.* Refer voltage and line $Z$ to load side of transformer as in figure 4.23b from which

$$V_R = \frac{V_s}{t} - \left( \frac{\frac{RP}{t^2} + \frac{X}{t^2} Q}{V_R} \right)$$

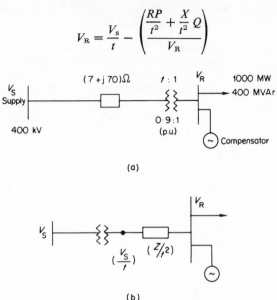

(a)

(b)

Figure 4.23   Circuits for Problem 4.12

4.13. A generating station consists of four 500-MW, 20-kV, 0·95-p.f. (lag) generators, each feeding through a 525-MVA, 0·1-p.u. reactance transformer on to a common busbar. It is required to transmit 2000 MW at 0·95 p.f., lagging, to a substation maintained at 500 kV in a power system at a distance of 500 km from the generating station. Design a suitable transmission link of nominal voltage 500 kV to achieve this, allowing for a reasonable margin of stability and a maximum voltage drop of 10 per cent. Each generator has synchronous and transient reactances of 2 p.u. and 0·3 p.u., respectively, and incorporates a fast-acting automatic voltage regulator. The 500-kV transmission lines have an inductive reactance per phase of 0·4 $\Omega$/km and a shunt capacitive reactance per phase of 0·3 $\times$ 10$^6$ $\Omega$/km. Both series and shunt capacitors may be used if desired and the number of 3-phase lines used should be not more than three—less if feasible. Use approximate methods of calculation, ignore resistance and state clearly any assumption made. Assume shunt capacitance to be lumped at the receiving end only. (Use two 500 kV lines with series capacitors compensating to 70 per cent of series inductance.)

4.14. It is required to transmit power from a hydro-electric station to a load centre 300 miles away using two lines in parallel for security reasons.

Assume sufficient bundle conductors are used such that there are no thermal limitations and that the effective reactance per phase per mile is 0·7 $\Omega$ and the resistance negligible. The shunt capacitive reactance of each line is 0·2 M$\Omega$ per phase per mile, and each line may be represented by the nominal $\pi$-circuit with half the capacitance at each end. The load is 2000 MW at 0·95 lagging and is independent of voltage over the permissible range.

Investigate the feasibility and performance of the link if the sending end voltage is (a) 345 kV, and (b) 500 kV, from the point of view of stability and voltage drop.

The lines may be compensated up to 70 per cent by series capacitors and at the load end synchronous compensators of 120 MVAr capacity are available. The maximum permissible voltage drop is 10 per cent. As two lines are provided for security reasons, your studies should include the worst operating case of only one line in use.

Calculate the voltage at the receiving end on no-load with the synchronous compensators not in operation. (Use 500 kV; 525 kV.)

# 5

# Load Flows

## 5.1 Introduction

A load flow is power system parlance for the steady-state solution of a net-work. This does not essentially differ from the solution of any other type of network except that certain constraints are peculiar to power supply. In previous chapters the manner in which the various components of a power system may be represented by equivalent circuits has been demonstrated. It should be stressed that the simplest representation should always be used consistent with the accuracy of the information available. There is no merit in using very complicated machine and line models when the load and other data are known only to a limited accuracy, for example the long-line representation should only be used where absolutely necessary. Similarly synchronous-machine models of more sophistication than given in this text are only needed for very specialized purposes, for example in some stability studies. Usually the size and complexity of the network itself provides more than sufficient intellectual stimulus without undue refinement of the components. Often resistance may be neglected with little loss of accuracy and an immense saving in computation.

The following combinations of quantities are usually specified at the system busbars for load-flow studies.

(a) *Slack or floating busbar*   One node is always specified by a voltage, constant in magnitude and phase. The effective generator at this node supplies the losses to the network; this is necessary because the magnitude of the losses will not be known until the calculation of currents is complete and this cannot be achieved unless one busbar has no power constraint and can feed the required losses into the system. The location of the slack node can influence the complexity of the calculations; the node approach-ing most closely an infinite busbar should be used.

(b) *Load nodes*   The complex power $S = P \pm jQ$ is specified.

(c) *Generation nodes*  The voltage magnitude and power are usually specified. Often limits to the value of the reactive power are given depending upon the characteristics of individual machines.

Load-flow studies are performed to investigate the following.

1. Flow of MW and MVAr in the branches of the network.
2. Busbar voltages.
3. Effect of rearranging circuits and incorporating new circuits on system loading.
4. Effect of temporary loss of generation and transmission circuits on system loading.
5. Effect of injecting in-phase and quadrature boost voltages on system loading.
6. Optimum system running conditions and load distribution.
7. Optimum system losses.
8. Optimum rating and tap range of transformers.
9. Improvement from change of conductor size and system voltage.

Studies will normally be performed for minimum load conditions (possibility of instability due to high voltage-levels and self-excitation of induction machines) and maximum load conditions (possibility of synchronous instability). Having ascertained that a network behaves reasonably under these conditions further load flows will be performed to attempt to optimize various quantities. The design and operation of a power network to obtain optimum economy is of paramount importance and the furtherance of this ideal will be greatly advanced by the growing use of centralized automatic-control of generating stations.

Although the same approach can be used to solve all problems, e.g. the nodal voltage method, the object should be to use the quickest and most efficient method for the particular type of problem. Radial networks will require less sophisticated methods than closed loops. In very large networks the problem of organizing the data is almost as important as the method of solution and the calculation must be carried out on a systematic basis and the nodal-voltage method is the most advantageous. Such methods as network reduction combined with the Thevenin or Superposition theorems are at their best with smaller networks. In the nodal method, greater accuracy is required in the computation as the currents in the branches are derived from the voltage differences between the ends. These differences are small in well designed networks and numerical accuracy of a high order is necessary. Hence this method is ideally suited for computation using digital computers.

## 5.2 Radial and Simple Loop Networks

In radial networks the phase shifts due to transformer connexions along the circuit are not usually important. The following examples illustrate the solution of this type of network.

*Example 5.1*  Distribution feeders with several tapped loads.

A distribution feeder with several tapped inductive loads (or laterals) and fed at one end only, is shown in figure 5.1a. Determine the total voltage drop.

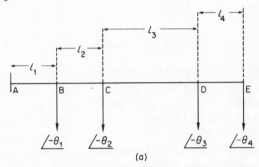

(a)

Figure 5.1a  Feeder with several load tappings along its length.

*Solution*   The current in $AB = (I_1 \cos \theta_1 + I_2 \cos \theta_2 + I_3 \cos \theta_3 + I_4 \cos \theta_4) - j(I_1 \sin \theta_1 + I_2 \sin \theta_2 + I_3 \sin \theta_3 + I_4 \sin \theta_4)$. Similarly the currents in the other section of the feeder are obtained. The approximate voltage drop is obtained from $\Delta V = RI \cos \theta + XI \sin \theta$ for each section. That is

$$R_1(I_1 \cos \theta_1 + I_2 \cos \theta_2 + I_3 \cos \theta_3 + I_4 \cos \theta_4) +$$
$$R_2(I_2 \cos \theta_2 + I_3 \cos \theta_3 + I_4 \cos \theta_4) +$$
$$R_3(I_3 \cos \theta_3 + I_4 \cos \theta_4) + R_4(I_4 \cos \theta_4) +$$
$$X_1(I_1 \sin \theta_1 + I_2 \sin \theta_2 + I_3 \sin \theta_3 + I_4 \sin \theta_4)) \text{ and so on.}$$

Rearranging and letting the resistance per loop metre be $r\ \Omega$ and the reactance per loop metre be $x\ \Omega$ (the term loop metre refers to single-phase circuits and includes the go and return conductors),

$$\Delta V = r[I_1 . l_1 \cos \theta_1 + I_2 \cos \theta_2 . (l_1 + l_2) +$$
$$I_3 \cos \theta_3(l_1 + l_2 + l_3) +$$
$$I_4 \cos \theta_4(l_1 + l_2 + l_3 + l_4)] +$$
$$x[I_1 l_1 \sin \theta_1 + I_2 \sin \theta_2(l_1 + l_2) +$$
$$I_3 \sin \theta_3(l_1 + l_2 + l_3) + I_4 \sin \theta_4(l_1 + l_2 + l_3 + l_4)].$$

In the system shown in figure 5.1b calculate the size of cable required if the voltage drop at the end load on the feeder must not exceed 12 V (line value). Let the resistance and reactance per metre per phase be $r\,\Omega$ and $x\,\Omega$ respectively. Then, referring to figure 5.1c,

$$r[100 \times 40 + 250 \times 20 + 330 \times 25] + x[100 \times 30 + 250 \times 0 + 330 \times 25] = 12/\sqrt{3}$$

i.e.

$$r(4000 + 5000 + 8250) + x(3000 + 0 + 8250) = 12/\sqrt{3}$$

i.e.

$$17{,}250r + 11{,}250x = 12/\sqrt{3}$$

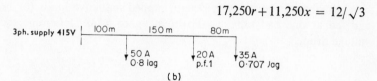

Figure 5.1b    Line diagram of feeder in Example 5.1.

The procedure now is to consult the appropriate overhead line or cable tables to select a cross-section which gives values of $r$ and $x$ which best fit the above equation. Usually if underground cables are used the size of cross-section is selected on thermal considerations and then the voltage drop calculated. With overhead lines the volt drop is the prime consideration and the conductor size will be determined accordingly. Thus in this example an overhead line with a 7/0·16 conductor is adequate.

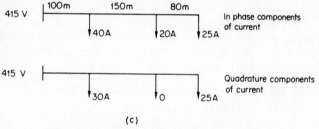

Figure 5.1c    Inphase and quadrature quantities.

*Example 5.2*    The system shown in figure 5.2a feeds two distinct loads, a group of domestic consumers and a group of induction-motors which on starting takes $5 \times$ full-load current at zero power-factor lagging. The induction motor voltage is nominally 6 kV. Calculate the dip in voltage on the domestic-load busbar when the induction motor is started.

*Solution*    The problem of the drop in the voltage to other consumers when an abnormally large current is taken for a brief period from an interconnected busbar is often serious. The present problem poses a typical situation. The 132 kV system is not an infinite busbar and is represented by a voltage source in series with a 4 per cent reactance on a

50 MVA base. Using a 50 MVA base the equivalent single-phase circuit
is as shown in figure 5.2b.

Starting current of the induction motors

$$= -j\frac{15,000 \times 10^3}{\sqrt{(3)} \times 6000} \times 5 \text{ A}$$

$$= -j7210 \text{ A}$$

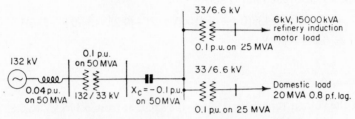

Figure 5.2a   Line diagram of system in Example 5.2.

Domestic load voltage before the induction-motor load is switched on

$$= 1 - \mathbf{IZ}$$

where

$$\mathbf{Z} = j0 \cdot 24 \text{ p.u.}$$

and

$$\mathbf{I} = \frac{20 \times 10^6}{\sqrt{(3)} \times 6000} = 1920 \text{ A}, 0 \cdot 8 \text{ p.f. lagging.}$$

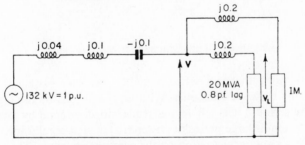

Figure 5.2b   Equivalent circuit of system in Example 5.2.

Base current in the 6·6 kV circuit

$$= \frac{50 \times 10^6}{\sqrt{(3)} \times 6600} = 4380 \text{ A}$$

∴   $\mathbf{I}$ p.u. $= 0 \cdot 438$ p.u. at $0 \cdot 8$ p.f. lagging (i.e. 1920/4380)

$\mathbf{V}_L = 1 - 0 \cdot 438(0 \cdot 8 - j0 \cdot 6)(j0 \cdot 24)$ p.u.

$= 0 \cdot 9366 - j0 \cdot 0843$ p.u.

and

$$V_L = 0.935 \text{ p.u.}$$

Starting current of the induction motors

$$= \frac{-j7210}{4380} = -j1.645 \text{ p.u.}$$

Voltage $V'_L$ when the motors start

$$= 1 - j0.04[0.438(0.8 - j0.6) - j1.645] - j0.2[0.438(0.8 - j0.6)]$$
$$= 0.871 - j0.084 \text{ p.u.}$$

and

$$V'_L = 0.874 \text{ p.u.}$$

Hence the voltage dips from 0.935 to 0.87 p.u. or from 6.175 kV to 5.82 kV.

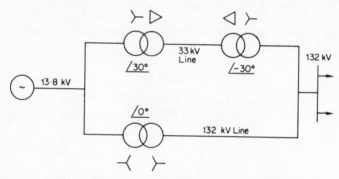

Figure 5.3   Loop with transformer phase shifts.

It will be noticed that a series capacitor has been installed to partly neutralize the network reactance. Without this capacitor the dip will be much more serious; it is left to the reader to determine by how much. Often in steel mills large pulses of power are taken at regular intervals and the result is a regular fluctuation on the consumer busbars. This is known as *voltage flicker* owing to the effect on electric lights.

*Load flows in closed loops*   In radial networks any phase shifts due to transformer connexions are not usually of importance as the currents and voltages are shifted by the same amount. In a closed loop to avoid circulating currents the product of the transformer transformation ratios (magnitudes) round the loop should be unity and the sum of the phase shifts in a common direction round the loop should be zero. This will be illustrated by the system shown in figure 5.3.

Neglecting phase shifts due to the impedances of components, for the closed loop formed by the two lines in parallel the total shift and transformation ratio is,

$$\left(\frac{33}{13\cdot8}\underline{/30°}\right)\left(\frac{132}{33}\underline{/-30°}\right)\left(\frac{1}{132/13\cdot8\underline{/0°}}\right) = 1\underline{/0°}.$$

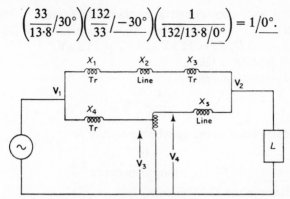

Figure 5.4 Equivalent circuit of network in figure 5.3 showing autotransformer.

In practice the transformation ratios of transformers are frequently changed by the provision of tap-changing equipment. This results in the product of the ratios round a loop being no longer unity although the phase shifts are still equal to zero. To represent this condition a fictitious autotransformer is connected as shown in figure 5.4. The increase in voltage on the 132 kV line by tap changing has the effect, in a largely

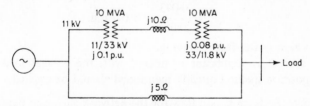

Figure 5.5 Line diagram of system for Example 5.3.

reactive path, of changing the flow of reactive power and thus the power factor of the current. An undesirable effect is the circulating current set up around the loop.

Often the out of balance or remnant voltage represented by the autotransformer can be neglected. If this is not the case the best method of calculation is to determine the circulating current and consequent voltages due to the remnant voltage acting alone, and then superpose these values on those obtained for operation with completely nominal voltage ratios.

*Example 5.3*   In figure 5.5 is shown a system in which the receiving end transformer is tapped from its nominal 33/11 kV to 33/11·8 kV. Calculate the circulating current.

*Solution*   The equivalent circuit is as shown in figure 5.4, where $V_1 = 11$ kV (base), $V_2 = 11\cdot8$ kV (base), $X_1 = j0\cdot1$ p.u., $X_2 = j10\ \Omega$, $X_3 = j0\cdot08$ p.u., $X_4 = j5\ \Omega$, $V_3 = 11$ kV and $V_4 = 11\cdot8$ kV.

A base of 10 MVA is taken.

Ratio of the autotransformer

$$= \frac{11\cdot8}{11} = 1\cdot073.$$

A first approximation to the circulating current is

$$\frac{1\cdot073 - 1}{\text{loop impedance}}\ \text{p.u.}$$

Base impedance in the 33 kV circuit

$$= \frac{33^2 \times 10^6}{10 \times 10^6} = 109\ \Omega$$

$$\therefore \qquad\qquad X_2 = j\frac{10}{109} = j0\cdot09\ \text{p.u.}$$

Similarly in the 11 kV circuit

$$X_4 = j0\cdot413\ \text{p.u.}$$

The circulating current

$$= \frac{0\cdot073}{j0\cdot683} = 0\cdot107\ \text{p.u.}$$

or 52 A at zero power-factor lagging.

If this current is significant compared with the normal load-current the superposition method already mentioned should be applied.

*Example 5.4*   Perform a load flow for the network shown in figure 5.6.

*Solution*   From Table 2.2 for a 132 kV, 0·175 in² conductor single-circuit line, $R = 0\cdot25\ \Omega$/mile, $X = 0\cdot66\ \Omega$/mile. Also for a 132 kV cable of 0·55 in² conductor, $R = 0\cdot104\ \Omega$/mile and $X = 0\cdot206\ \Omega$/mile. The actual impedance values are shown in brackets in figure 5.6.

The loads consume constant power and reactive power regardless of the line voltages and an iterative procedure must be used. It is proposed to use Thevenin's Theorem although care must be taken because this applies only to linear circuits and because of the loads, the present problem is non-linear. However if the effect of the constant-power constraint is small the errors will be negligible. The network between B and C will be replaced by a voltage source and equivalent impedance. Link BC is open circuited and the voltage across it found. Voltage drop A to B (refer to figure 2.7).

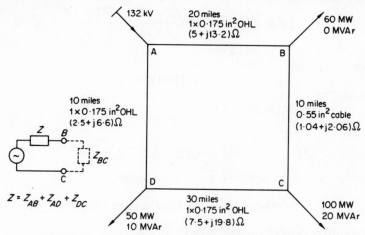

Figure 5.6 Line diagram of loop network for Example 5.4.

$$\Delta V = \frac{PR + XQ}{V_B} = \frac{5 \times 20 \times 10^6}{76 \cdot 2 \times 10^3} = 1 \cdot 315 \text{ kV (using values per phase)}$$

and

$$\delta V = \frac{13 \cdot 2 \times 20 \times 10^6}{76 \cdot 2 \times 10^3} = 3 \cdot 46 \text{ kV (assuming initially } V_B = V_A)$$

$\therefore$

$$(V_B + 1 \cdot 315)^2 + 3 \cdot 46^2 = 76 \cdot 2^2$$

and

$$V_B = 74 \cdot 885 \text{ kV}$$

If

$$V_B \doteqdot V_A - \Delta V \text{ is used,}$$

$$V_B \doteqdot 76 \cdot 2 - 1 \cdot 315 = 74 \cdot 885 \text{ kV,}$$

i.e. no error is involved.

As $V_B$ falls the current in AB increases to maintain the load, further decreasing $V_B$. The new value of

$$V_B = V'_B, \quad \text{and} \quad \Delta' V = \frac{5 \times 20 \times 10^6}{74 \cdot 885 \times 10^3} = 1 \cdot 335 \text{ kV}$$

$\therefore$

$$V'_B = 76 \cdot 2 - 1 \cdot 335 \text{ kV}$$

$$= 74 \cdot 865 \text{ kV.}$$

Similarly

$$V_B'' = 74 \cdot 8 \text{ kV.}$$

Obviously the stiff load makes little difference to the voltage at B.
Voltage drop A to D,

$$\Delta V = \frac{2 \cdot 5 \times (150/3) \times 10^6 + 6 \cdot 6 \times (30/3) \times 10^6}{76 \cdot 2 \times 10^3}$$

$$= 2 \cdot 51 \text{ kV.}$$

Hence

$$V_D \doteqdot 76 \cdot 2 - 2 \cdot 51 = 73 \cdot 69 \text{ kV.}$$

Similarly

$$V_D' = 73 \cdot 61 \text{ kV.}$$

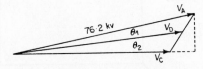

Figure 5.7   Phasor diagram of voltages at busbars in Example 5.4.

Voltage drop D to C, initially take $V_C = V_D'$,

$$\therefore \qquad \Delta V = \frac{7 \cdot 5 \times (100/3) \times 10^6 + 19 \cdot 8 \times (20/3) \times 10^6}{73 \cdot 61 \times 10^3}$$

$$= 5 \cdot 18 \text{ kV}$$

$$V_C = 73 \cdot 61 - 5 \cdot 18 = 68 \cdot 43 \text{ kV}$$

$$\Delta' V = \frac{382}{68 \cdot 43} = 5 \cdot 58 \text{ kV}$$

$$\therefore \qquad V_C' = = 73 \cdot 61 - 5 \cdot 58 = 68 \cdot 03 \text{ kV}$$

and

$$V_C'' = 67 \cdot 99 \text{ kV } (\delta'' V = 8 \cdot 95 \text{ kV).}$$

Overall voltage drop ADC,

From figure 5.7

$$\sin \theta_1 = \frac{4 \cdot 14}{76 \cdot 2}, \quad \theta_1 = 3° \, 7'$$

$$\sin \theta_2 = \frac{8 \cdot 95}{73 \cdot 61}, \quad \theta_2 = 6° \, 58'$$

$$\theta_1 + \theta_2 = 10° \, 5'$$

$$\delta V_{AC} = 76 \cdot 2 \sin 10° \, 5' = 13 \cdot 32 \text{ kV}$$

$$\Delta V_{AC} = 76 \cdot 2 \cos 10° \, 5' - 67 \cdot 99 = 7 \cdot 01 \text{ kV.}$$

Also,

$$\sin \theta_B = \frac{3 \cdot 62}{76 \cdot 2}, \quad \theta_B = 2° \, 43'$$

$$V_B - V_C = 74 \cdot 87 \, (\cos 2° \, 43' - j \sin 2° \, 43')$$
$$-67 \cdot 99 \, (\cos 10° \, 5' - j \sin 10° \, 5')$$
$$= 7 \cdot 87 + j8 \cdot 35 \text{ kV}$$

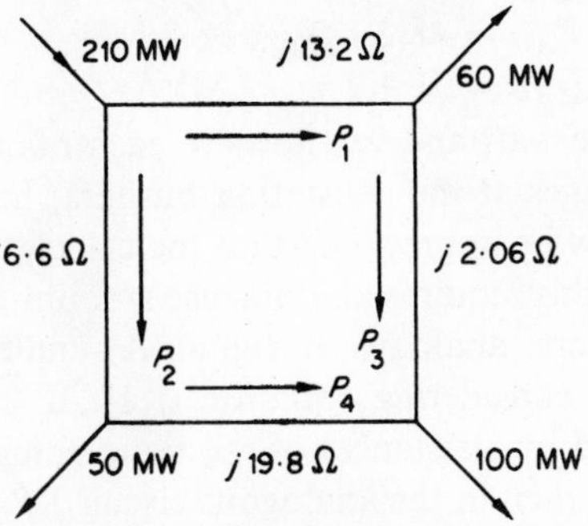

Figure 5.8   Line diagram showing powers in system of Example 5.4
—real-power analysis.

$$\therefore \qquad I_{BC}(\text{kA}) = \frac{7 \cdot 87 + j8 \cdot 35}{16 \cdot 04 + j41 \cdot 66} = (237 \cdot 5 - 97 \cdot 3j) \text{ A}$$

$$(P + jQ)_{BC} = (67 - j11 \cdot 9)(237 \cdot 5 + j97 \cdot 3)10^{-3} \text{ MVA}$$

$$\therefore \qquad P = 17 \cdot 10 \text{ MW per phase}$$

and

$$Q = 3 \cdot 68 \text{ MVAr per phase.}$$

An approximate estimate for $P_{BC}$ may be obtained by assuming zero resistance in the network and using the equation

$$\sum P_{ij} X_{ij} = 0 \qquad (5.1)$$

where $i$ and $j$ refer to the branch between nodes $i$ and $j$ of the network. This follows from the fact that $\delta V \propto \underline{/\delta} \doteqdot (XP/V)$ and the sum of the angular shifts round the loop is zero. Labelling the powers as in figure 5.8 the following equations hold:

$$13 \cdot 2 P_1 + 2 \cdot 06 P_3 - 19 \cdot 8 P_4 - 6 \cdot 6 P_2 = 0$$
$$P_4 + P_3 = 100 \text{ or } P_4 = 100 - P_3$$
$$P_2 - P_4 = 50; \quad P_1 = 60 + P_3$$
$$P_2 + P_1 = 210; \quad P_2 = 210 - 60 - P_3 = 150 - P_3$$
$$\therefore \quad 13 \cdot 2(60 + P_3) + 2 \cdot 06(P_3) - 19 \cdot 8(100 - P_3) - 6 \cdot 6(150 - P_3) = 0$$

and        $792+13\cdot2P_3+2\cdot06P_3-1980+19\cdot8P_3-990+6\cdot6P_3 = 0$.

From which $P_3 = 52\cdot2$ MW, which is a reasonably accurate estimate.

Once $P$ and $Q$ in BC are known the flows in the other branches follow readily, i.e.

$$P_{DC} = 100-51\cdot8 = 48\cdot2 \text{ MW}$$
$$Q_{DC} = 20-11 = 9 \text{ MVAr}$$
$$P_{AB} = 60+51\cdot8 = 111\cdot8 \text{ MW}$$
$$Q_{AB} = 0+11 = 11 \text{ MVAr}$$
$$P_{AD} = 48\cdot2+50 = 98\cdot2 \text{ MW}$$
$$Q_{AD} = 10+9 = 19 \text{ MVAr}$$

In example 5·4 the watt and var flows in each branch have been found along with the voltages at the substation busbars; hence the load-flow is complete. It will now be appreciated that the calculation of load flows on larger systems than this requires an immense amount of labour and hence the need for computers, analogue or digital. A simple analogue computer can be designed by considering equation (5.1). If the reactances of the network are replaced by resistances of the same ohmic values and powers replaced by currents then in the analogous circuit $\Sigma R_{ij}I_{ij} = 0$. An analogous network of resistors is constructed and suitably energized. The currents which represent power flows are then measured. Suitable scaling factors between $P$ and $I$, and $X$ and $R$ can be used to advantage.

## 5.3  Large Systems

The calculations of load flows carried out so far have been on small systems. Similar methods on large networks would be extremely tedious if not virtually impossible. One method of approaching the analyses of large systems is to represent them by model lumped-constant networks at low voltages ($< 100$ V) and currents (few milliamperes). Lines and cables are represented by the $\pi$ model, generators by the internal voltage (E) in series with the appropriate impedance and transformers by their series impedance. The effects of tap changing are simulated by the use of miniature autotransformers. These equivalent-circuit models are usually known as Alternating Current Analysers and for several years formed the main approach to large-system studies. They are steady-state models and often operate at higher than mains frequency to reduce the physical size of the components to give the actual system reactances. The many variable resistors, inductors and capacitors which comprise analysers are scaled in p.u. values so that the actual power-system components (also in p.u.) can be represented directly. One step further towards the physical system is the micro-reseaux system (reference 3) developed in France. In this the generators are represented by actual machines scaled down to laboratory proportions. This arrangement is not economical for large systems and is confined entirely to the study of a few interconnected machines. Special-purpose electronic analogue computers have also been used for the assessment of the effect of prime mover and generator characteristics on synchronous stability.

With the rapid development of digital computers in the 1950's the attention of power engineers has been directed towards the use of numerical methods for analysis. This has advantages over the use of alternating-current analysers in that analysis can be placed on a systematic basis once suitable programmes have been developed, thus requiring less trained-engineer time. The various components of the network are uniquely labelled and these designations together with the per-unit values of admittance, power, reactive power and voltage are fed into the computer which then performs the analysis to prescribed values of accuracy.

With digital methods, in both load-flow and fault analysis, the nodal-voltage method of network solution is preferred; this method is described in Appendix 2 with illustrative examples. The admittance matrix is formed, the constraints of the network expressed, and then a suitable method of solution used. Usually the voltage magnitudes at the nodes are evaluated and from these the current, watt and var flows obtained. In a well-designed network the voltage magnitudes at the various nodes will be close and high accuracy required in computation. Before discussing in detail the methods of solution, a simple problem will be solved to illustrate the nodal voltage method.

*Example 5.5*  Perform a load flow on the interconnected busbars shown in figure 5.9.

*Solution*  Remembering the manner in which the self and mutual admittances are formed from Appendix 2, the equations may be written directly as follows:

$$2V_1 - 0{\cdot}5V_2 - 1{\cdot}5V_3 = I_1$$
$$-0{\cdot}5V_1 + 1{\cdot}25V_2 - 0{\cdot}75V_3 = I_2$$
$$-1{\cdot}5V_1 - 0{\cdot}75V_2 + 2{\cdot}25V_3 = I_3$$

In matrix form,

$$\begin{bmatrix} 2 & -0{\cdot}5 & -1{\cdot}5 \\ -0{\cdot}5 & 1{\cdot}25 & -0{\cdot}75 \\ -1{\cdot}5 & -0{\cdot}75 & +2{\cdot}25 \end{bmatrix} \quad \begin{bmatrix} V_1 \\ V_2 \\ V_3 \end{bmatrix} = \begin{bmatrix} I_1 \\ I_2 \\ I_3 \end{bmatrix}$$

It is usual to write the admittance matrix direct. If the values in the current column vector are specified it is required to determine the values of $V_1$, $V_2$ and $V_3$. [Y] may be inverted or the equations solved by elimination. As $\qquad$ [Y][V] = [I] then [Y]$^{-1}$[I] = [V] hence it is required to invert the admittance matrix. A simple method will be used in this case,

$$V_1 = 0\cdot5I_1 + 0\cdot25V_2 + 0\cdot75V_3$$

Substitute for $V_1$ in equations for $I_2$ and $I_3$,

$$-0\cdot25I_1 + 1\cdot125V_2 - 1\cdot125V_3 = I_2$$
$$-0\cdot75I_1 - 1\cdot125V_2 + 1\cdot125V_3 = I_3$$

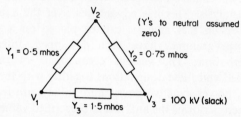

Figure 5.9    Three-busbar system—load flow using matrix inversion (resistance zero).

Also from the equation for $I_2$,

$$V_2 = 0\cdot222I_1 + 0\cdot888I_2 + V_3$$

Substitute for $V_2$ in the other equations,

$$0\cdot555I_1 + 0\cdot222I_2 + V_3 = V_1$$
$$0\cdot222I_1 + 0\cdot888I_2 + V_3 = V_2$$
$$-I_1 \quad -I_2 - 0V_3 = I_3$$

i.e.
$$\begin{bmatrix} 0\cdot555+0\cdot222+1\cdot0 \\ 0\cdot222+0\cdot888+1\cdot0 \\ -1 \quad -1 \quad 0\cdot0 \end{bmatrix} \begin{bmatrix} I_1 \\ I_2 \\ V_3 \end{bmatrix} = \begin{bmatrix} V_1 \\ V_2 \\ I_3 \end{bmatrix}$$

The node at which the voltage is specified, the slack busbar, is node (3), hence the manipulation of the matrix into the form shown. Next some loadings at the nodes will be specified.

$$I_1 = 1000 \text{ A (Generator node)}$$
$$I_2 = -1500 \text{ A (Load node)}$$
$$V_3 = 100 \text{ kV (Phase value), slack busbar.}$$

Dealing in kiloamperes and kilovolts,

$$V_1 = 0\cdot555(1) + 0\cdot222(-1\cdot5) + (+1)(100)$$
$$= 100\cdot222 \text{ kV}$$
$$V_2 = 0\cdot222(1) + 0\cdot888(-1\cdot5) + (1)(100)$$
$$= 98\cdot89 \text{ kV}$$
$$I_3 = -(1)(1) + (1)(1\cdot5) + 0(100)$$
$$= +0\cdot5 \text{ kA, i.e. a generator node.}$$

Note that $I_1 + I_2 + I_3 = 0$

In this example the loads have been specified as currents and the solution is obtained directly after the matrix manipulation. In practice however, power, reactive power and voltage are specified. The normal combination of quantities at a busbar have already been discussed and it is evident that the following equations require solution:

$$[Y][V] = [I] \tag{5.2}$$

and

$$S = VI^* \tag{5.3}$$

The problem is non-linear and iterative procedures must be used either completely or combined with matrix inversion.

### 5.4 Methods of Solution for Large Systems

A comprehensive summary of the many numerical methods available is given in reference 4.

(a) DIRECT METHODS  These basically invert the admittance matrix, a process consuming both in time and computer storage. Many methods are available and a text on numerical methods should be consulted. These methods include Gaussian elimination.

*Triangulation and Partitioning*  Direct methods only solve linear systems, i.e. $[Y][V] = [I]$ where $[I]$ is specified. The fact that powers are specified in practice makes the problem non-linear. A new value for $I$ must be obtained from $S = VI^*$ after each direct solution and this value used to obtain a new one.

Partitioning is useful for handling large problems on small computers, for network reduction, and the elimination of unwanted nodes.

A matrix
$$[Y] = \begin{bmatrix} Y_{11} & Y_{12} & Y_{13} \\ Y_{21} & Y_{22} & Y_{23} \\ \hline Y_{31} & Y_{32} & Y_{33} \end{bmatrix} \tag{5.4}$$

can be partitioned into four submatrices by the dividing lines shown in equation (5.4). Hence in this typical admittance matrix for a network,

$$[Y] = \begin{bmatrix} B & C \\ C^t & D \end{bmatrix}$$

where $C^t$ is the transpose of $C$,

$$B = \begin{bmatrix} Y_{11} & Y_{12} \\ Y_{21} & Y_{22} \end{bmatrix} \quad \text{and so on.}$$

A saving in computation time and storage is often obtained by manipulating the submatrices instead of the main one and in the equation $[I] = [Y][V]$ it is possible to eliminate nodes at which $I$ is specified.

The matrix [Y] is partitioned so that the nodes to be eliminated are grouped together in a submatrix, e.g.

$$\begin{bmatrix} \mathbf{I}_w \\ \mathbf{I}_u \end{bmatrix} = \begin{bmatrix} \mathbf{B} & \mathbf{C} \\ \mathbf{C}^t & \mathbf{D} \end{bmatrix} \begin{bmatrix} \mathbf{V}_w \\ \mathbf{V}_u \end{bmatrix} \tag{5.5}$$

where the suffix $u$ indicates unwanted and $w$ wanted nodes. For example if the currents at the unwanted nodes are zero,

$$\mathbf{I}_u = 0, \quad -\mathbf{C}^t\mathbf{V}_w = \mathbf{D}\mathbf{V}_u \quad \text{and} \quad -\mathbf{D}^{-1}\mathbf{C}^t\mathbf{V}_w = \mathbf{V}_u$$

Substituting this value of $\mathbf{V}_u$ in the expression for $\mathbf{I}_w$,

$$\mathbf{I}_w = \mathbf{B}\mathbf{V}_w - \mathbf{C}\mathbf{D}^{-1}\mathbf{C}^t\mathbf{V}_w$$

which gives a new admittance matrix containing only wanted nodes, i.e.

$$[\mathbf{Y}] = \mathbf{B} - \mathbf{C}\mathbf{D}^{-1}\mathbf{C}^t \tag{5.6}$$

Methods are available for modifying the admittance matrix to allow for changes in the configuration of the network, say, due to line outages. These modifications can be performed such that a completely fresh inversion of the new admittance matrix is not necessary, see reference 4.

Consider the solution of the large (say 50 by 50) system,

$$[\mathbf{Y}] \, [\mathbf{V}] = [\mathbf{I}]$$

[Y] can be partitioned into 9 submatrices:

$$\begin{bmatrix} \mathbf{Y}_{11} & \mathbf{Y}_{12} & \mathbf{Y}_{13} \\ \mathbf{Y}_{21} & \mathbf{Y}_{22} & \mathbf{Y}_{23} \\ \mathbf{Y}_{31} & \mathbf{Y}_{32} & \mathbf{Y}_{33} \end{bmatrix} \cdot \begin{bmatrix} \mathbf{V}_1 \\ \mathbf{V}_2 \\ \mathbf{V}_3 \end{bmatrix} = \begin{bmatrix} \mathbf{I}_1 \\ \mathbf{I}_2 \\ \mathbf{I}_3 \end{bmatrix}$$

where, if $\mathbf{Y}_{11}$ is a 20 by 20 matrix and $\mathbf{Y}_{22}$ a 20 by 20 matrix, $\mathbf{Y}_{33}$ will be 10 by 10, $\mathbf{Y}_{23}$ 20 by 10 and both $\mathbf{I}_3$ and $\mathbf{V}_3$ are 10 by 1.

The solution is carried out by elimination in stages as follows.

In the [Y] matrix containing the submatrices first elimate $\mathbf{V}_1$ by the following process. Multiply row 1 by $(\mathbf{Y}_{21}/\mathbf{Y}_{11})$ and subtract it from row 2. Also, row 1 of [I] is multiplied by $(\mathbf{Y}_{21}/\mathbf{Y}_{11})$ and subtracted from row 2. Next rows 1 of the [Y] and [I] matrices are multiplied by $(\mathbf{Y}_{31}/\mathbf{Y}_{11})$ and subtracted from row 3.

Hence the original [Y] and [I] matrices become

$$\begin{bmatrix} \mathbf{Y}_{22}^1 & \mathbf{Y}_{23}^1 \\ \mathbf{Y}_{32}^1 & \mathbf{Y}_{33}^1 \end{bmatrix} \quad \text{and} \quad \begin{bmatrix} \mathbf{I}_2^1 \\ \mathbf{I}_3^1 \end{bmatrix}$$

where, $(\mathbf{Y}_{22}^1, \mathbf{Y}_{23}^1, \mathbf{I}_2^1) = (\mathbf{Y}_{22}, \mathbf{Y}_{23}, \mathbf{I}_2) - \mathbf{Y}_{21}\,\mathbf{Y}_{11}^{-1}\,(\mathbf{Y}_{12}, \mathbf{Y}_{13}, \mathbf{I}_1),$

$(\mathbf{Y}_{32}^1, \mathbf{Y}_{33}^1, \mathbf{I}_3^1) = (\mathbf{Y}_{32}, \mathbf{Y}_{33}, \mathbf{I}_3) - \mathbf{Y}_{31}\,\mathbf{Y}_{11}^{-1}\,(\mathbf{Y}_{12}, \mathbf{Y}_{13}, \mathbf{I}_1),$

Finally, rows 1 of the new reduced matrices are multiplied by $(\mathbf{Y}^1_{32}/\mathbf{Y}^1_{22})$ and $\mathbf{Y}^{11}_{33}$ . $\mathbf{V}_3 = \mathbf{I}^{11}_3$ from which $\mathbf{V}_3$ is found remembering that $\mathbf{Y}^{11}_{33}$, $\mathbf{V}_3$ and $\mathbf{I}_3$ are matrices,

and $\quad (\mathbf{Y}^{11}_{33}, \mathbf{I}^{11}_3) = (\mathbf{Y}^1_{33}, \mathbf{I}^1_3) - \mathbf{Y}^1_{32} (\mathbf{Y}^1_{22})^- (\mathbf{Y}^1_{23}, \mathbf{I}^1_2)$

$\mathbf{V}_1$ and $\mathbf{V}_2$ are obtained by back substitution.

$\mathbf{V}_3 = (\mathbf{Y}^{11}_{33})^{-1} \mathbf{I}^{11}_3 \, ; \, \mathbf{V}_2 = (\mathbf{Y}^1_{22})^{-1} (\mathbf{I}^1_2 - \mathbf{Y}^1_{23} \, \mathbf{V}_3)$, etc.

Most power system matrices are symmetric, i.e. $\mathbf{Y}_{ij} = \mathbf{Y}_{ji}$ and it is only required to store elements on and above the diagonal. Frequently all the non-zero elements lie within a narrow band about the diagonal and again considerable savings in computer storage may be attained.

*Accuracy* When solving the set of equations [A] [x] = [b] the residual vector r = [b] − [A] [x] is not zero because of rounding errors. Troubles arise when ill-conditioned equations are obtained in which although the residual is small the solution may be inaccurate. Such matrices are often large and sparse and for most rows the diagonal element is equal to the sum of the non-diagonal elements and of opposite sign.

(b) ITERATIVE METHODS The two sets of equations (5.2) and (5.3) are solved simultaneously as the iteration proceeds. A number of methods are available and as two of these have gained wide adoption they will be described in detail. The first is the Gauss-Seidel method which has been widely used for many years and is simple in approach, the second is the Newton-Raphson method which although more complex has certain advantages. The speed of convergence of these methods is of extreme importance as apart from the cost of computer time the use of these methods in schemes for the automatic control of power systems requires very fast load-flow solutions.

*The Gauss-Seidel method*

In this method the unknown quantities are initially assumed and the value obtained from the first equation for say $V_1$ is then used when obtaining $V_2$ from the second equation and so on. Each equation is considered in turn and then the complete set solved again until the values obtained for the unknowns converge to within required limits.

Application of the method to the simple network of Example 5.5. The nodal equations are,

$$2V_1 - 0.5V_2 - 1.5V_3 = I_1 = 1$$
$$-0.5V_1 + 1.25V_2 + 0.75V_3 = I_2 = -1.5$$
$$-1.5V_1 - 0.75V_2 + 2.25V_3 = I_3$$

where the voltages are in kV and the currents in kA.

As $V_3$ is known, i.e. 100 kV (slack busbår voltage) it is only necessary to solve the first two equations. Initially make $V_2 = 100$ kV.

$$\therefore \qquad {}^1V_1 = 0{\cdot}5(1+150+50) = 100{\cdot}5000.$$

Using this value to evaluate $V_2$,

$$ {}^1V_2 = \frac{1}{1{\cdot}25}(-1{\cdot}5+75+50{\cdot}25) = 99{\cdot}0000. $$

Using this value of $V_2$ to evaluate $V_1$,

$$ {}^2V_1 = 0{\cdot}5(1+150+49{\cdot}5) = 100{\cdot}2500 $$

$$ {}^2V_2 = \frac{1}{1{\cdot}25}(-1{\cdot}5+75+50{\cdot}125) = 98{\cdot}9000 $$

$$ {}^3V_1 = 100{\cdot}2250; \quad {}^3V_2 = 98{\cdot}8900; $$

$$ {}^4V_1 = 100{\cdot}2225; \quad {}^4V_2 = 98{\cdot}8890; $$

$$ {}^5V_1 = 100{\cdot}22225; \quad {}^5V_2 = 98{\cdot}8888. $$

Iterations are now producing changes only in the fourth place of decimals and the process may be stopped.

$$I_3 = -1{\cdot}5 \times 100{\cdot}22225 - 0{\cdot}75 \times 98{\cdot}8888 + 2{\cdot}25 \times 100 = 0{\cdot}5 \text{ kA}.$$

The solution has been obtained with much less computation than the more direct method previously used.

In the three-node system of Example 5.5 the iterative form of the three nodal equations with the nodal constraints that $\mathbf{S} = \mathbf{VI}^*$ is obtained as follows ($p$ indicates the iteration number and is *not* a power):
for node 1,

$$ \mathbf{I}_1 = \mathbf{V}_1\mathbf{Y}_{11} + \mathbf{V}_2\mathbf{Y}_{12} + \mathbf{V}_3\mathbf{Y}_{13} $$

$$ \therefore \qquad \mathbf{V}_1 = -\mathbf{V}_2\frac{\mathbf{Y}_{12}}{\mathbf{Y}_{11}} - \mathbf{V}_3\frac{\mathbf{Y}_{13}}{\mathbf{Y}_{11}} + \frac{\mathbf{I}_1}{\mathbf{Y}_{11}} $$

$$ \therefore \qquad \mathbf{V}_1^* = -\frac{\mathbf{Y}_{12}^*}{\mathbf{Y}_{11}^*}\mathbf{V}_2^* - \frac{\mathbf{Y}_{13}^*}{\mathbf{Y}_{11}^*}\mathbf{V}_3^* + \frac{\mathbf{I}_1^*}{\mathbf{Y}_{11}^*} $$

Substituting $\mathbf{I}_1^* = \mathbf{S}_1/\mathbf{V}_1$ and writing in the iterative form,

$$ \mathbf{V}_1^{*p+1} = -\frac{\mathbf{Y}_{12}^*}{\mathbf{Y}_{11}^*}\mathbf{V}_2^{*p} - \frac{\mathbf{Y}_{13}^*}{\mathbf{Y}_{11}^*}\mathbf{V}_3^{*p} + \frac{\mathbf{S}_1}{\mathbf{Y}_{11}^{*p}}\frac{1}{\mathbf{V}_1^p} $$

Similarly,

$$ \mathbf{V}_2^{*p+1} = -\frac{\mathbf{Y}_{21}^*}{\mathbf{Y}_{22}^*}\mathbf{V}_1^{*p+1} - \frac{\mathbf{Y}_{23}^*}{\mathbf{Y}_{22}^*}\mathbf{V}_3^{*p} + \frac{\mathbf{S}_2}{\mathbf{Y}_{22}^*}\frac{1}{\mathbf{V}_2^p} $$

and

$$V_3^{*p+1} = -\frac{V_{31}^*}{Y_{33}^*}V_1^{*p+1} - \frac{V_{32}^*}{Y_{33}^*}V_2^{*p+1} + \frac{S_3}{Y_{33}^*}\frac{1}{V_3^p}$$

Generally,

$$(V_i^{p+1})^* = +\frac{S_i}{Y_{ii}^*}\cdot\frac{1}{V_i^p} - \sum\frac{Y_{ij}^*}{Y_{ii}^*}V_j^{*p} \tag{5.7}$$

In this particular case node three is the slack-bus and as $V_i$ is known the equation for it is not required. It is seen in the above equations that the new value of $V_l$ in the preceding equation is immediately used in the next equation, i.e. $V_1^{*p+1}$ is used in the $V_2$ equation. Each node is scanned in turn over a complete iteration.

Equation (5.7) refers to a busbar with $P$ and $Q$ specified. At a generator node $(i)$, usually $V_i$ and $P_i$ are specified with perhaps upper and lower limits to $Q_i$.

The magnitude of $V_i$ is fixed but its phase depends on $Q_i$. The values of $V_i$ and $Q_i$ from the previous iteration are not related by (5.7) because $V_i$ has been modified to give a constant value of magnitude. It is necessary at the next iteration to calculate the value of $Q_i$ corresponding to $V_i$ from equation (5.7),

i.e. $Q_i =$ imaginary part of $S_i$, i.e. of

$$Y_{ii}^* V_i^p \left[ V_i^{*\,p+1} + \sum_{\substack{j=1 \\ j\neq i}}^{N} \frac{Y_{ij}^*}{Y_{ii}^*}V_j^{*p} \right]$$

This value of $Q_i$ holds for the existing value of $V_i$ and is then substituted into

$$-\sum_{\substack{j=1 \\ j\neq i}}^{N} \frac{Y_{ij}^*}{Y_{ii}^*}V_j^{*p} + \frac{P_i+jQ_i}{Y_{ii}^*V_i^p}$$

to obtain $V_i^{p+1}$.

The real and imaginary components of $V_i^{p+1}$ are then multiplied by the ratio $V_i/V_i^{p+1}$ thus complying with the constant $V_i$ constraint. This final step is a slight approximation but the error involved is small and the saving in computation large. The phase of $V_i$ is thus found and the iteration can proceed to the next node. The process continues until the value of $V^{*p+1}$ at any node differs from $V^{*p}$ by a specified amount, a common figure is 0·0001 per unit. The study is commenced by assuming 1 per unit voltage at all nodes except one; this exception is necessary in order that current flows may be obtained in the first iteration.

*Acceleration factors* The number of iterations required to reach the specified convergence can be greatly reduced by the use of acceleration factors (reference 5). The correction in voltage from $V^p$ to $V^{p+1}$ is multiplied by such a factor so that the new voltage is brought closer to its final value, i.e.

$$^1V^{p+1} = V^p + \omega(V^{p+1} - V^p)$$
$$= V^p + \omega\Delta V^p$$

where $^1V^{p+1}$ is the accelerated new voltage. It can be shown that a complex value of $\omega$ reduces the number of iterations more than a real value. However up to the present real values only seem to be used; the actual value depends on the nature of the system under study but a value of 1·6 is widely used. The Gauss-Seidel method with the use of acceleration factors is known as the method of Successive Over Relaxation.

*Transformer tap-changing*

Further changes which must be accommodated in the admittance matrix are those due to transformer tap-changing. When the ratio is at the nominal value the transformer is represented by a single series impedance but when off-nominal, adjustments have to be made as follows.

Consider a transformer of ratio $t : 1$ connected between two nodes $i$ and $j$; the series admittance of the transformer is $Y_t$. Referring to figure 5.10 the following nodal-voltage equation holds.

$$I_j = Y_j(Y_{jr} + Y_{j0} + Y_{js} + Y_t) - (V_r Y_{jr} + V_x Y_t + V_s Y_{js} + V_0 Y_{j0})$$

As $V_0 = 0$ and $V_x = V_i/t$,

$$I_j + V_s Y_{js} = V_j(Y_{jr} + Y_{j0} + Y_{js} + Y_t) - \left(Y_{jr}V_r + \frac{Y_t V_i}{t}\right)$$

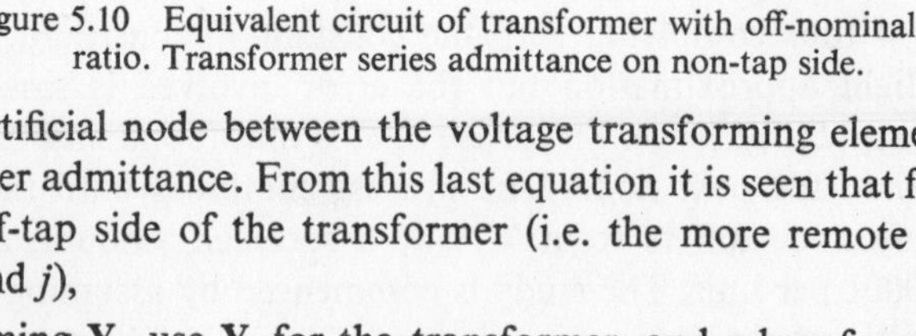

Figure 5.10 Equivalent circuit of transformer with off-nominal tap ratio. Transformer series admittance on non-tap side.

$x$ is an artificial node between the voltage transforming element and the transformer admittance. From this last equation it is seen that for the node on the off-tap side of the transformer (i.e. the more remote of the two nodes $i$ and $j$),

when forming $Y_{jj}$ use $Y_t$ for the transformer, and when forming $Y_{ij}$ use $Y_t/t$ for the transformer.

It can be similarly shown for the tap-side node that,

when forming $Y_{ii}$ use $Y_t/t^2$, and when forming $Y_{ij}$ use $Y_t/t$.

These conditions may be represented by the $\pi$ section shown in figure 5.11, although it is probably easier to modify the mutual and self admittances directly.

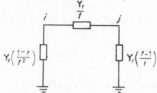

Figure 5.11   $\pi$-Section to represent transformer with off-nominal tap ratio.

*The Newton-Raphson method*

Although the Gauss-Seidel method is well established, more recently the Newton-Raphson method has received much attention. With some systems it gives a greater assurance of convergence and is at the same time economical in computer time.

The basic iterative procedure is given by

$$\text{value at new iteration, } x^{p+1} = x^p - \frac{f(x^p)}{f'(x^p)}$$

Extending this to a multi-equation system,

$$\mathbf{x}^{p+1} = \mathbf{x}^{(p)} - J^{-1}(\mathbf{x}^p) f(\mathbf{x}^p)$$

where $\mathbf{x}$ and $\mathbf{f}$ are column vectors and $J(\mathbf{x}^p)$ is a matrix known as the Jacobian matrix, of the form

$$
\begin{bmatrix}
\dfrac{\partial f_1}{\partial x_1} & \cdot & \cdot & \cdot & \dfrac{\partial f_1}{\partial x_n} \\
\cdot & \cdot & \cdot & \cdot & \cdot \\
\cdot & \cdot & \dfrac{f_k}{x_k} & \cdot & \cdot \\
\cdot & \cdot & \cdot & \cdot & \cdot \\
\dfrac{\partial f_n}{\partial x_1} & \cdot & \cdot & \cdot & \dfrac{\partial f_n}{\partial x_n}
\end{bmatrix} \quad p^{\text{th}} \text{ iteration.}
$$

Consider now the application to an $n$-node power system, for a link connecting nodes $k$ and $j$ of admittance $Y_{kj}$,

$$P_k + jQ_k = V_k I_k^* = V_k \sum_{j=1}^{n-1} (Y_{kj} V_j)^*$$

Let $\qquad \mathbf{V}_k = a_k + jb_k$ and $\mathbf{Y}_{kj} = G_{kj} - jB_{kj}$

Then, $\quad P_k + jQ_k = (a_k + jb_k) \sum_{1}^{n-1} [(G_{kj} - jB_{kj})(a_j + jb_j)]^*$ $\qquad$ (5.8)

from which,

$$P_k = \sum_{j=1}^{n-1} [a_k (a_j G_{kj} + b_j B_{kj}) + b_k (b_j G_{kj} - a_j B_{kj})] \qquad (5.9)$$

$$Q_k = \sum_{j=1}^{n-1} [b_k (a_j G_{kj} + b_j B_{kj}) - a_k(b_j G_{kj} - a_j B_{kj})] \qquad (5.10)$$

Hence, there are two non-linear simultaneous equations for each node. Note that $(n - 1)$ nodes are considered because the slack node $n$ is completely specified.

Changes in $P$ and $Q$ are related to changes in $a$ and $b$ by equations (5.9) and (5.10), e.g.

$$\Delta P_1 = \frac{\partial P_1}{\partial a_1} \Delta a_1 + \frac{\partial P_1}{\partial a_2} \Delta a_2 + \ldots \frac{\partial P_1}{\partial a_{n-1}} \Delta a_{n-1}$$

Similar equations hold in terms of $\Delta P$ and $\Delta b$, and $\Delta Q$ in terms of $\Delta a$ and $\Delta b$.

The equations may be expressed generally in the following manner.

$$
\begin{bmatrix}
\Delta P_1 \\
\cdot \\
\cdot \\
\cdot \\
\Delta P_{n-1} \\
\Delta Q_1 \\
\cdot \\
\cdot \\
\cdot \\
\Delta Q_{n-1}
\end{bmatrix}
=
\begin{bmatrix}
\frac{\partial P_1}{\partial a_1} & \cdots & \frac{\partial P_1}{\partial a_{n-1}} & \frac{\partial P_1}{\partial b_1} & \cdots & \frac{\partial P_1}{\partial b_{n-1}} \\
 & \cdots & & & \cdots & \\
 & \cdots & & & \cdots & \\
 & \cdots & & & \cdots & \\
\frac{\partial P_{n-1}}{\partial a_1} & \cdots & \frac{\partial P_{n-1}}{\partial a_{n-1}} & \frac{\partial P_{n-1}}{\partial b_1} & \cdots & \frac{\partial P_{n-1}}{\partial b_{n-1}} \\
\frac{\partial Q_1}{\partial a_1} & \cdots & \frac{\partial Q_1}{\partial a_{n-1}} & \frac{\partial Q_1}{\partial b_1} & \cdots & \frac{\partial Q_1}{\partial b_{n-1}} \\
 & \cdots & & & \cdots & \\
 & \cdots & & & \cdots & \\
 & \cdots & & & \cdots & \\
\frac{\partial Q_{n-1}}{\partial a_1} & \cdots & \frac{\partial Q_{n-1}}{\partial a_{n-1}} & \frac{\partial Q_{n-1}}{\partial b_1} & \cdots & \frac{\partial Q_{n-1}}{\partial b_{n-1}}
\end{bmatrix}
\cdot
\begin{bmatrix}
\Delta a_1 \\
\cdot \\
\cdot \\
\cdot \\
\Delta a_{n-1} \\
\Delta b_1 \\
\cdot \\
\cdot \\
\cdot \\
\Delta b_{n-1}
\end{bmatrix}
$$

$$\ldots (5.11)$$

For convenience denote the Jacobian matrix by

| $J_A$ | $J_B$ |
|-------|-------|
| $J_C$ | $J_D$ |

The elements of the matrix are evaluated for the values of $P$, $Q$ and $V$ at each iteration as follows.

For the submatrix $J_A$ and from equation (5.9),

$$\frac{\partial P_k}{\partial a_j} = a_k\, G_{kj} - b_k\, B_{kj} \tag{5.12}$$

where $k \neq j$, i.e. off-diagonal elements.

Diagonal elements,

$$\frac{\partial P_k}{\partial a_k} = 2a_k\, G_{kk} + b_k\, B_{kk} - b_k\, B_{kk} + \sum_{\substack{j=1 \\ j \neq k}}^{n-1} (a_j\, G_{kj} + b_j\, B_{kj}) \tag{5.13}$$

This element may be more readily obtained by expressing some of the quantities in terms of the current at node $k$, $\mathbf{I}_k$ which can be determined separately at each iteration.

Let $\quad \mathbf{I}_k = c_k + jd_k = (G_{kk} - jB_{kk})\,(a_k + jb_k)$

$$+ \sum_{\substack{j=1 \\ j \neq k}}^{n-1} (G_{kj} - jB_{jk})\,(a_j + jb_j)$$

from which,

$$c_k = a_k\, G_{kk} + b_k\, B_{kk} + \sum_{\substack{j=1 \\ j \neq k}}^{n-1} (a_j\, G_{kj} + b_j\, B_{kj})$$

and

$$d_k = b_k\, G_{kk} - a_k\, B_{kk} + \sum_{\substack{j=1 \\ j \neq k}}^{n-1} (b_j\, G_{kj} - a_j\, B_{jk})$$

So that, $\quad \dfrac{\partial P_k}{\partial a_k} = a_k\, G_{kk} - b_k\, B_{kk} + c_k$

For $J_B$, $\quad \dfrac{\partial P_k}{\partial b_k} = a_k\, B_{kk} + b_k\, G_{kk} + d_k$

and $\quad \dfrac{\partial P_k}{\partial b_j} = a_k\, B_{kj} + b_k\, G_{kj} \quad (k \neq j)$

For $J_C$, $\quad \dfrac{\partial Q_k}{\partial a_k} = a_k\, B_{kk} + b_k\, G_{kk} - d_k$

and $\quad \dfrac{\partial Q_k}{\partial a_j} = a_k B_{kj} + b_k G_{kj} \quad (k\neq j)$

For $J_D$, $\quad \dfrac{\partial Q_k}{\partial b_j} = -a_k G_{kj} + b_k B_{kj} \quad (k\neq j)$

and $\quad \dfrac{\partial Q_k}{\partial b_k} = -a_k G_{kk} + b_k B_{kk} + c_k$

The process commences with the iteration counter '$p$' set to zero and all the nodes except the slack bus being assigned voltages, usually 1 p.u.

From these voltages, $P$ and $Q$ are calculated from equations (5.9) and (5.10). The changes are then calculated, $\Delta P_k^p = P_k$ (specified) $- P_k^p$ and $\Delta Q_k^p = Q_k$ (specified) $- Q_k^p$, where $p$ is the iteration number.

Next the node currents are computed,

$$I_k^p = \left(\frac{P_k^p + jQ_k^p}{V^p}\right)^* = c_k^p + jd_k^p$$

The elements of the Jacobian matrix are then formed and from (5.11),

$$\begin{array}{|c|} \hline \Delta a \\ \hline \Delta b \\ \hline \end{array} = \begin{array}{|c|c|} \hline J_A & J_B \\ \hline J_C & J_D \\ \hline \end{array}^{-1} \begin{array}{|c|} \hline \Delta P \\ \hline \Delta Q \\ \hline \end{array}$$

$$\dots (5.14)$$

Hence, $a$ and $b$ are determined and the new values, $a_k^{p+1} = a_k^p + \Delta a_k^p$ and $b_k^{p+1} = b_k^p + \Delta b_k^p$ obtained. The process is repeated ($p = p + 1$) until $\Delta P$ and $\Delta Q$ are less than a prescribed tolerance.

The Newton-Raphson method is reported to have better convergence characteristics and for many systems to be faster than the Gauss-Seidel, it has a much larger time per iteration but requires very few iterations (four is general) whereas the Gauss-Seidel requires at least thirty iterations, the number increasing with the size of system.

Acceleration factors may be used for the Newton-Raphson method.

### 5.5 Example of a Complex Load Flow

The line diagram of a system is shown in figure 5.12 along with details of line and transformer impedances and loads. This represents a slightly simplified arrangement of the network used in reference 6. All quantities are expressed as per unit and the nodes are numbered as indicated. They are not ordered consecutively to show in the solution that the numbering system is not important, although in more sophisticated studies on larger systems it has been shown that certain orderings of nodes can produce faster convergence and solutions.

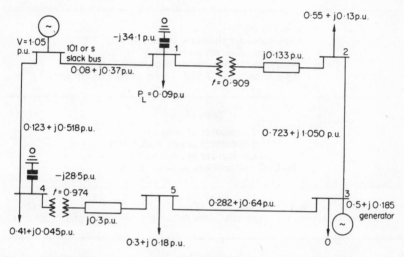

Figure 5.12   System for complex load-flow using iterative method.

The solution of this problem has been carried out by digital computer and a brief description of the arrangement of essential data will be given and the first iteration performed on a desk calculating machine. This programme was developed for instructional purposes and is not as refined or sophisticated as commercial programmes; it is hoped however that the beginner will find it easier to understand than the latter. In view of the nature of equation (5.7) the basic system data will be modified to allow less computation during the actual solution of the equations; this will be obvious as the method is described. The system admittance-matrix [Y] is stored by specifying the following: order of matrix ($n$), number of off-diagonal elements, number of off-diagonal elements per row, column numbers of the off-diagonal elements in each row, a list of off-diagonal elements ) $Y_{ij}/Y_{ii}$). This input data is arranged in two tables in the following manner.

Table A

$l$ = number of links
$n$ = number of non-slack nodes.

| $i$ | $j$ | $R$ | $X$ | $B$ | $t$ |
|---|---|---|---|---|---|
| Sending end note | Receiving end note | Resistance per unit | Reactance per unit | Susceptance per unit | Off-nominal transformer ratio |

$n$ = 1 to 99.
slack nodes are numbered starting at 101,
earth connexions are numbered 0.
When representing transformers, node $i$ must be the
tap-side node and $R$ and $X$ refer to the non-tap side

Table B

$l$ = number of links
$n$ = number of non-slack nodes
$s$ = number of slack nodes
(suffix $G \rightarrow$ Generation, suffix $L \rightarrow$ Loads)

| Node number | $V_{\text{real}}$ per unit | $V_{\text{imag}}$ per unit | $P_G$ per unit | $Q_G$ per unit | $P_L$ per unit | $Q_L$ per unit |
|---|---|---|---|---|---|---|

$P$ and $Q$ are considered positive for watts and lagging vars generated or
supplied and negative when received. It should be noted that the $Y_{ij}$'s
are in fact negative although they are shown as positive in the tables;
this is useful as the $Y_{ij}$ terms appear as negative in the nodal equations.
The connexions to the slack busbar ($V_s$) are shown separately, no equation
is necessary for the slack nodes as the voltages are fully specified. The
input data is shown in table 5.1 and identification of the various quantities
will be made easier by reference to the following system matrix:

$$\begin{bmatrix} Y_{11} & Y_{12} & Y_{13} & Y_{14} & Y_{15} & Y_{1s} \\ Y_{21} & Y_{22} & Y_{23} & Y_{24} & Y_{25} & Y_{2s} \\ Y_{31} & Y_{32} & Y_{33} & Y_{34} & Y_{35} & Y_{3s} \\ Y_{41} & Y_{42} & Y_{43} & Y_{44} & Y_{45} & Y_{4s} \\ Y_{51} & Y_{52} & Y_{53} & Y_{54} & Y_{55} & Y_{5s} \end{bmatrix}$$

The initial part of the computer flow-diagram is shown in figure 5.13.

Before proceeding with the first iteration the calculation of the admittances associated with one of the transformers will be shown. Consider
the transformer 1–2 of impedance j0·133 per unit and having $t$ = 0·909.

Table 5.1

```
+5 - - - - - - - - - - - - - - - - - - n
+8 - - - - - - - - - - - - - - - - - - l
```

$$
\left.\begin{array}{l}
+1 \\
+2 \\
+2 \\
+1 \\
+2
\end{array}\right\} \text{Off diagonal elements per row}
$$

$$
\left.\begin{array}{ll}
+2 & \\
+3 & +1 \\
+5 & +2 \\
+5 & \\
+4 & +3
\end{array}\right\} \text{Column numbers}
$$

$$
\left.\begin{array}{lll}
+0{\cdot}708238 & -0{\cdot}033932 & Y_{12}/Y_{11} \\
+0{\cdot}081853 & +0{\cdot}050025 & Y_{23}/Y_{22} \\
+1{\cdot}010063 & -0{\cdot}055033 & Y_{21}/Y_{22} \\
+0{\cdot}646931 & -0{\cdot}043097 & Y_{35}/Y_{33} \\
+0{\cdot}353069 & +0{\cdot}043097 & Y_{32}/Y_{33} \\
+0{\cdot}640699 & -0{\cdot}052397 & Y_{45}/Y_{44} \\
+0{\cdot}726081 & -0{\cdot}090184 & Y_{54}/Y_{55} \\
+0{\cdot}292797 & +0{\cdot}087839 & Y_{53}/Y_{55}
\end{array}\right\} \text{Off diagonal elements } Y_{ij}/Y_{ii}
$$

Generation Table 1

$$
\left.\begin{array}{ll}
+1{\cdot}000000 & +0{\cdot}000000 \\
+1{\cdot}000000 & +0{\cdot}000000 \\
+1{\cdot}000000 & +0{\cdot}000000 \\
+1{\cdot}000000 & +0{\cdot}000000 \\
+1{\cdot}000000 & +0{\cdot}000000
\end{array}\right\} \text{Voltage estimates, } v_i
$$

$$
\left.\begin{array}{ll}
-0{\cdot}000369 & +0{\cdot}007706 \\
-0{\cdot}019534 & +0{\cdot}066298 \\
+0{\cdot}179357 & -0{\cdot}162088 \\
-0{\cdot}014702 & +0{\cdot}076068 \\
-0{\cdot}046094 & +0{\cdot}058905
\end{array}\right\} P+jQ/Y_{ii}^{*}\text{'s}
$$

$$
\left.\begin{array}{ll}
+0{\cdot}234539 & +0{\cdot}039069 \\
+0 & +0 \\
+0 & +0 \\
+0{\cdot}366206 & +0{\cdot}055921 \\
+0 & +0
\end{array}\right\} V_s Y_{is}/Y_{ii}\text{'s} \quad \text{slack bus connexions}
$$

$$
\left.\begin{array}{ll}
+0{\cdot}558269 & -11{\cdot}652234 \\
+0{\cdot}444860 & -8{\cdot}164860 \\
+1{\cdot}021401 & -1{\cdot}954524 \\
+0{\cdot}433934 & -5{\cdot}306045 \\
+0{\cdot}576541 & -4{\cdot}641795
\end{array}\right\} Y_{ii}\text{'s}
$$

$$
\left.\begin{array}{lll}
+0{\cdot}223371 & +0{\cdot}037209 & +101{\cdot}000000 \\
+0{\cdot}000000 & +0{\cdot}000000 & +0{\cdot}000000 \\
+0{\cdot}000000 & +0{\cdot}000000 & +0{\cdot}000000 \\
+0{\cdot}348767 & +0{\cdot}053259 & +101{\cdot}000000 \\
+0{\cdot}000000 & +0{\cdot}000000 & +0{\cdot}000000
\end{array}\right\} Y_{is}/Y_{ii}\text{'s}+\text{slack nos.}
$$

For the off-tap side node number 2, $Y_{22}$ is formed using $Y_t$ only, i.e.

$$Y_{22} = \frac{1}{0 \cdot 723 + j1 \cdot 05} + \frac{1}{j0 \cdot 133}$$

$$= 0 \cdot 444860 - j8 \cdot 164860$$

$$Y_{21} = \frac{1}{j0 \cdot 133} \cdot \frac{1}{0 \cdot 909} = -j8 \cdot 278000.$$

The formation of the admittance matrix is necessary for both the Gauss-Seidel and Newton-Raphson methods. At this point the methods differ and first the application of Gauss-Seidel method will be given with a complete computer solution and then an outline of the Newton-Raphson method applied to the problem.

For node 1,

$$Y_{11} = \frac{1}{0 \cdot 08 + j0 \cdot 37} + \frac{1}{-j34 \cdot 1} + \frac{1}{j0 \cdot 133} \cdot \frac{1}{0 \cdot 909^2}$$

$$= 0 \cdot 558269 - j11 \cdot 652234$$

$$Y_{12} = Y_{21}.$$

First iteration.

Node (1)

$$V_1^{1*} = -\left(\frac{Y_{12}}{Y_{11}}\right)^* V_2^* - \left(\frac{Y_{1s}}{Y_{11}}\right)^* V_s^* + \frac{S_1}{Y_{11}^*} \frac{1}{V_1}$$

$$= (0 \cdot 708238 - j0 \cdot 033932)^* \cdot 1^* + (0 \cdot 234539 +$$

$$j0 \cdot 039069)^* + (-0 \cdot 000369 + j0 \cdot 007706) \frac{1}{1}$$

$$\therefore \quad V_1^{1*} = 0 \cdot 942409 + j0 \cdot 002569 \text{ per unit.}$$

Node (2)

$$V_2^{1*} = -\left(\frac{Y_{21}}{Y_{22}}\right)^* V_1^{1*} - \left(\frac{Y_{23}}{Y_{22}}\right)^* V_3^* + \left(\frac{S_2}{Y_{22}^*}\right) \frac{1}{V_2}$$

$$= (1 \cdot 010063 + j0 \cdot 055033)(0 \cdot 942409 + j0 \cdot 002569) +$$

$$(0 \cdot 081853 - j0 \cdot 050025)1 + (-0 \cdot 019534 + j0 \cdot 066298) \frac{1}{1}$$

$$\therefore \quad V_2^{1*} = 1 \cdot 014073 + j0 \cdot 070671 \text{ per unit.}$$

Node (3)

$$V_3^{1*} = -\left(\frac{Y_{32}}{Y_{33}}\right)^* V_2^{1*} - \left(\frac{Y_{35}}{Y_{33}}\right)^* V_5^* + \left(\frac{S_3}{Y_{33}^*}\right) \frac{1}{V_3}$$

$$= (0 \cdot 353069 - j0 \cdot 043097)(1 \cdot 014073 + j0 \cdot 070671) +$$

$$(0 \cdot 646931 + j0 \cdot 043097)1 + (0 \cdot 179357 - j0 \cdot 162088) \frac{1}{1}$$

$$V_3^{1*} = 1 \cdot 187371 - j0 \cdot 137743 \text{ per unit.}$$

Node (4)
$$V_4^{1*} = -\left(\frac{Y_{45}}{Y_{44}}\right)^* V_5^* - \left(\frac{Y_{4s}}{Y_{44}}\right)^* V_s^* + \left(\frac{S}{Y_{44}^*}\right)\frac{1}{V_4}$$

$$= (0 \cdot 640699 + j0 \cdot 052397)(1) + (0 \cdot 366206 - j0 \cdot 055921)$$
$$+ (-0 \cdot 014702 + j0 \cdot 076068)\frac{1}{1}$$

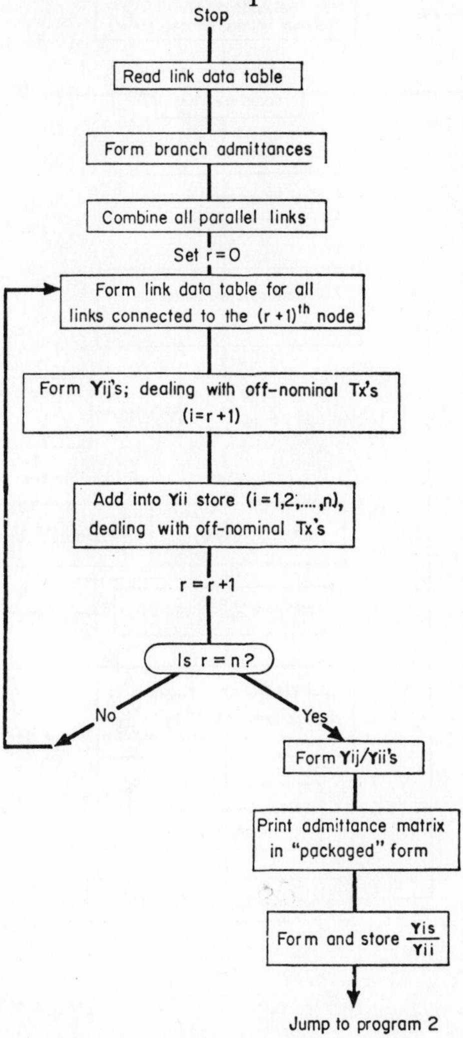

Stop

Read link data table

Form branch admittances

Combine all parallel links

Set r = 0

Form link data table for all links connected to the $(r+1)^{th}$ node

Form $Y_{ij}$'s; dealing with off-nominal Tx's
$(i = r + 1)$

Add into $Y_{ii}$ store $(i = 1,2, ...., n)$, dealing with off-nominal Tx's

r = r + 1

Is r = n?

No     Yes

Form $Y_{ij}/Y_{ii}$'s

Print admittance matrix in "packaged" form

Form and store $\dfrac{Y_{is}}{Y_{ii}}$

Jump to program 2

Figure 5.13a    Initial part of flow diagram for data processing of load-flow programme.

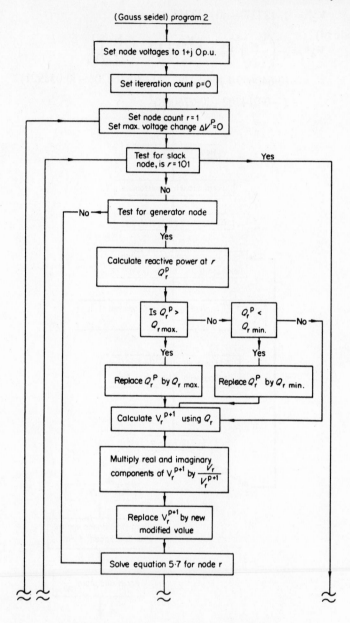

Figure 5.13b   *(continued opposite)*

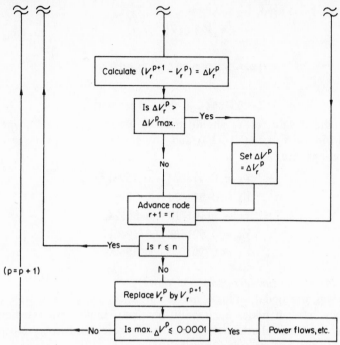

Figure 5.13b    Remainder of flow diagram for Gauss-Seidel method.

$$\mathbf{V}_4^{1*} = 0\cdot992202 + j0\cdot072543 \text{ per unit.}$$

Node (5)

$$\mathbf{V}_5^{1*} = -\left(\frac{\mathbf{Y}_{54}}{\mathbf{Y}_{55}}\right)^* \mathbf{V}_4^{1*} - \left(\frac{\mathbf{Y}_{53}}{\mathbf{Y}_{55}}\right)^* \mathbf{V}_3^{1*} + \left(\frac{\mathbf{S}}{\mathbf{Y}_{55}^*}\right)\frac{1}{\mathbf{V}_5}$$

$$= (0\cdot726081 + j0\cdot090184)(1\cdot010513 + j0\cdot069748) +$$

$$(0\cdot292797 - j0\cdot087839)(1\cdot187371 - j0\cdot137743) +$$

$$(-0\cdot046094 + j0\cdot058905)\frac{1}{1}$$

$\therefore \quad \mathbf{V}_5^{1*} = 1\cdot003342 - j0\cdot056430$ per unit.

$\mathbf{V}_s$ or $\mathbf{V}_{101} = 1\cdot05000 + j0\cdot000000$ per unit.

It will be noticed that the latest value of each nodal voltage is used. In this iteration no acceleration factor has been used (i.e. the factor = 1); if a factor of 1·6 is used, $V_1^1$ becomes

$$V_1 + 1·6(V_1^1 - V_1),$$

i.e.

$$1 + 1·6(-0·057591 - j0·002569)$$

or

$$0·907854 - j0·004110 \text{ per unit.}$$

This modified value of $V_1^1$ should then be used to evaluate $V_2^1$ when $V_2^1$ is modified to $V_2 + 1·6(V_2^1 - V_2)$ and so on. The busbar voltages after 30 iterations are,

$$V_1 = 0·918345 - j0·159312$$
$$V_2 = 0·978674 - j0·221811$$
$$V_3 = 1·101718 - j0·065242$$
$$V_4 = 0·901468 - j0·194617$$
$$V_5 = 0·903003 - j0·196604$$

*Evaluation of line currents and power flows*

Knowing the nodal voltages and admittances the current, power and var flows between nodes are readily obtained. Links with transformers however need special attention. Consider a transformer with its impedance referred to the non-tap side (figure 5.10),

$$I_i = \frac{I_j}{t} \text{ and } V_i = tV_x$$

$$\therefore \quad I_j = (V_x - V_j)Y_t$$

$$= \left(\frac{V_i}{t} - V_j\right)Y_t$$

Power at $j$,

$$P_j = V_j\left[\left(\frac{V_i}{t} - V_j\right)Y_t\right]^*$$

Also,

$$I_i = \left(\frac{V_i}{t^2} - \frac{V_j}{t}\right)Y_t$$

$$\therefore \quad P_i = V_i\left[\left(\frac{V_i}{t^2} - \frac{V_j}{t}\right)Y_t\right]^*$$

$\therefore$ Power transferred $i$ to $j$

$$= P_i - P_j.$$

*Application of the Newton-Raphson method to the system in figure* 5.12

The elements of the admittance matrix are determined as in the previous method, and it should be noted that for off-diagonal elements both $G$ and $B$ will be $(-1)$ times the values derived from the network.

Let all the busbar voltages be assigned a voltage of $(1 + j0)$ p.u.

*At node* 3 $\qquad P_3 = 0 \cdot 5, \qquad Q_3 = 0 \cdot 185$ (generated)

$$\mathbf{Y}_{32} = 0 \cdot 444860 - j0 \cdot 646063 \text{ p.u.}$$
$$\mathbf{Y}_{35} = 0 \cdot 576541 - j1 \cdot 308461 \text{ p.u.}$$
$$\mathbf{Y}_{33} = 1 \cdot 021401 - j1 \cdot 954524 \text{ p.u.}$$

$$a_3 = 1, b_3 = 0; \quad a_2 = 1, b_2 = 0; \quad a_5 = 1, b_5 = 0$$

First Iteration $(p = 0)$

$$
\begin{aligned}
P_3^0 = {} & 1 \times 1 \times (-0 \cdot 444860) + 1 \times 0 \times (-0 \cdot 646063) \\
& + 0 \times 0 \times (-0 \cdot 444860) - 0 \times 1 \times (-0 \cdot 646063) \\
& + 1 \times 1 \times (+1 \cdot 021401) + 1 \times 0 \times (1 \cdot 954524) \\
& + 0 \times 0 \times (1 \cdot 021401) - 0 \times 1 \times (1 \cdot 954524) \\
& + 1 \times 1 \times (-0 \cdot 576541) + 1 \times 0 \times (-1 \cdot 308461) \\
& + 0 \times 0 \times (-0 \cdot 576541) - 0 \times 1 \times (-1 \cdot 308461) \\
= {} & 0 \cdot 0 \text{ p.u.}
\end{aligned}
$$

Note that $\mathbf{Y}_{31}$, $\mathbf{Y}_{34}$ do not exist.
(The above result would be expected in the initial case as all the involved voltages are equal to 1 p.u.)

Similarly, $\qquad Q_3^0 = 0 \cdot 0$ p.u.

Therefore, $\qquad \Delta P_3^0 = 0 \cdot 5 - 0 = 0 \cdot 5$ p.u.,

and $\qquad \Delta Q_3^0 = 0 \cdot 185 - 0 = 0 \cdot 185$ p.u.

$\Delta P$ and $\Delta Q$ for the remaining non-slack nodes are similarly obtained.

$$\mathbf{I}_k^p = \frac{P_k^p - jQ_k^p}{(V_k^p)^*} \text{ and } \mathbf{I}_3^0 = \frac{0 \cdot 5 - j0 \cdot 185}{1 - j0}$$

hence $\qquad c_3^0 = 0 \cdot 5$ and $d_3^0 = -0 \cdot 185$

The elements of the Jacobian are determined next.

$$
\begin{aligned}
\frac{\partial P_3}{\partial a_3} &= a_3^0 \, G_{33} - b_3^0 \, B_{33} + c_3^0 \\
&= 1 \, (1 \cdot 021401) - 0 \cdot 0 + 0 \cdot 5 = 1 \cdot 521401
\end{aligned}
$$

$$\frac{\partial P_3}{\partial a_2} = a_3^0\, G_{32} - b_3^0\, B_{32}$$

$$= 1\,(-0\cdot444860) + 0\cdot0$$

$$\frac{\partial P_3}{\partial a_5} = -0\cdot576541$$

$$\frac{\partial P_3}{\partial b_3} = a_3^0\, B_{33} + b_3^0\, G_{33} + d_3^0$$

$$= 1\,(1\cdot954520) + 0\cdot0 + (-0\cdot185) = 1\cdot76952$$

$$\frac{\partial P_3}{\partial b_2} = a_3^0\, B_{32} + b_3^0\, G_{32} = 1\,(-0\cdot646063) + 0\cdot0$$

$$\frac{\partial P_3}{\partial b_5} = 1\,(-1\cdot308461) + 0\cdot0$$

Similarly, obtain

$$\frac{\partial Q_3}{\partial a_3},\ \frac{\partial Q_3}{\partial a_2},\ \frac{\partial Q_3}{\partial a_5};\ \frac{\partial Q_3}{\partial b_3},\ \frac{\partial Q_3}{\partial b_2}\ \text{and}\ \frac{\partial Q_3}{\partial b_5}$$

The Jacobian matrix for the first iteration is thus formed and inverted and then $\Delta a_k^0$ and $\Delta b_k^0$ ($k = 1 \ldots 5$) are evaluated. Hence, $V_k^{0+1} = V_k^0 + \Delta a^0 + j\Delta b^0$. The process is repeated until changes in real and reactive power at each bus are less than a prescribed amount, say $0\cdot01$ p.u.

*Summary*

The direct method involving matrix inversion and a final iterative procedure to deal with the restraints at the nodes has advantages for smaller networks. For large networks the completely iterative methods are preferable especially using accelerating factors. System information, such as specified generation and loads, line and transformer series and shunt admittances are fed into the computer on tape or punched cards. The data are expressed in per unit on arbitrary MVA and voltage bases. The computer formulates the self and mutual admittances for each node.

If the system is very sensitive to reactive-power flows, i.e. the voltages change considerably with change in load and network configuration, the computer programme may diverge. It is preferable to allow the reactive power outputs of generators to be initially without limits to ensure an initial convergence. Convergence having been attained the computer evaluates the real and reactive power flows in each branch of the system along with losses, absorption of vars and any other information that may be required. Programmes are available which automatically adjust the tap settings of transformers to optimum values. Also, facilities exist for the printing out only of information regarding overloaded and underloaded lines; this is very useful when carrying out a series of load flows investigating the outages of plant and lines.

The methods given in this section apply to complex load flows, i.e. real and imaginary components considered and therefore give the full and accurate solution of the network. In many cases such detailed knowledge is not required especially in the early stages of designing new transmission systems. For this type of work *Real Power Flows* are used which are much more economical in computer storage and operation time.

## 5.6 Real-Power Flows

In many network investigations especially in the initial design stages, analysis giving only power flows suffices. The calculations can be considerably reduced by neglecting resistance in the system. In a system with reactance and power flows only it has been shown that round a loop

$$\Sigma P_{ij} X_{ij} = 0 \tag{5.1}$$

Also the algebraic sum of the powers meeting at a point is zero. Hence power flow is analogous to current, and the nodal equation for node 1 of a multinode system could be expressed as follows,

$$P_1 = \frac{V_1}{X_{11}} - \frac{V_2}{X_{12}} - \frac{V_3}{X_{13}} \cdots \frac{V_k}{X_{1k}}$$

where $X_{11}$ is the self reactance of node 1 and $X_{1k}$ the mutual reactance between nodes 1 and $k$.

Generally,

$$V_i = \left[ P_i + \sum_{j=2}^{j=n} \frac{V_j}{X_{ij}} \right] \sum X_{ii}$$

The initial voltages at the nodes are assumed as for the complex load-flow. An approximate estimation of the $I^2R$ losses can be obtained by evaluating the currents, these losses of course are ignored in the actual flow calculations. Similarly the reactive power absorption or the $I^2X$ loss can be obtained.

Owing to the relationships between power and reactance in a network being analogous to those for current and resistance a simple resistive analogue can be designed for real power analysis.

If resistance $R = k_1 X$ and current $I = k_2 P$, then

$$\sum X_{ij} P_{ij} = \sum k_1 k_2 I_{ij} R_{ij} = k \sum I_{ij} R_{ij}.$$

Hence if reactances are represented by resistors and power inputs by currents, then current flows measured between the nodes of the analogous circuit will be proportional to power flows.


## 5.7 Network Analysers

It is not intended to give a detailed description of the various types of analysers but to indicate the general mode of operation of the main types.

*Resistance type*  The use of analogous resistance networks for load flow studies has already been discussed. This form of analogue may also be used to determine the magnitudes of fault currents in systems subjected to three-phase symmetrical faults. Resistance in the system is neglected and the branch and machine reactances represented by equivalent resistances in the analogous network which is energized by a suitably scaled alternating or direct voltage. The currents measured in the branches of the resistance analogue are then directly proportional to the magnitude of the fault currents in the actual system.

*Direct model analysers*  The lumped-constant equivalent circuit of the actual system is represented directly by resistors, inductors, capacitors and voltage sources. To reduce the size of the inductors for a given reactance frequencies of 400 Hz, 1592 Hz and 10,000 Hz are often used. This type comprises a large number of electrically separate units of variable resistance, reactance and capacitance together with transformers and autotransformers with taps and also voltage sources variable in magnitude and phase. These components are connected by means of a 'jacking-up' system. The autotransformer units are used for load matching and for the representation of plus and minus taps on the transformers. These models are essentially single phase in operation; common values of working voltage and current are 50 V, 50 mA.

*Transformer analogue analysers*  This is an indirect form of analyser employing the voltage-current relationships in an ideal transformer. Although it is more complex than the direct type it has the advantages that pure inductance and negative impedance can be represented. If null-balance methods are used for measuring the voltages the accuracy is very high, one part in a thousand. The impedance, voltage source and current source units can also be set to 0·001 per unit. If the metering is performed by dynamometer instruments fed via electronic amplifiers the accuracy of measurement is much lower.

On both direct and indirect analysers the following quantities can be read directly: voltage and current magnitudes, in-phase and quadrature components of voltage and current, watts and vars. Loads are represented by the equivalent admittance to neutral and are fed through autotransformers which keep the load voltage constant regardless of the system voltage. This maintains the values of $P_L$ and $Q_L$ regardless of the actual busbar voltage.

Load flows are performed by the adjustment of the generator outputs until all the nodal constraints are complied with. If negative impedances are available these may be used as generators of power and reactive power instead of voltage sources resulting in quicker and more easily balanced load flows.

### 5.8 Design Considerations

The entire chapter so far has been devoted to the analysis of existing systems, so also is the following chapter on faults. Before analysis can take place a network must·be designed and this will be based on the following considerations; standards of security, degree of utilization of plant, the existing system if any, standardized sizes of plant and voltage levels, amenity, location geographically. To meet a number of requirements several schemes may be produced and then the optimum, on grounds of cost, number of transmission circuits and other factors chosen. Loads will have to be predicted (the system will come into existence some time after the initial-design stage) and rough checks such as generation meeting load-requirements made.

Of great importance is the reinforcement of an existing network to meet the growing load. As the latter is a continual process the examination of the existing network must also be frequently performed. The delay of adding an extra link may save considerable expense even if delayed for only a year and this must be borne in mind.

The process of obtaining an optimum design is thus one of design-analysis-design-analysis and so on until the design constraints are met. Future development here, as with many design operations, e.g. electrical machines, lies in the realm of automatic computation. The data regarding load growth and the other considerations already mentioned must be stored along with the methods of design and analysis engineers have traditionally used. A flow diagram illustrating the sequence of operations is shown in table 5.2. The methods of logical design (reference 7) and linear programming are being used to express the design constraints in mathematical terms.

Table 5.2  Flow diagram for design—analysis operation

```
        ┌─────────────────────┐
        │ Constraints         │
        │ Load and Plant Data │
        └─────────────────────┘
                  │
                  ▼
        ┌─────────────────────────┐
        │ DESIGN                  │◄──────────────┐
        │ Conductor size—voltage  │               │
        │ level                   │               │
        └─────────────────────────┘               │
                  │                                │
                  ▼                                │
           ┌──────────────┐                        │
           │ First analysis│──────── change network│
           └──────────────┘                        │
                  │                                │
                  ▼                                │
           ┌──────────────┐                        │
           │ MW Power Flows│──── change network    │
           └──────────────┘                        │
                  │                                │
                  ▼                                │
        ┌──────────────────────┐                   │
        │ 3-Phase Short-Circuits│── sectioning     │
        └──────────────────────┘                   │
                  │                                │
                  ▼                                │
        ┌──────────────────┐                       │
        │ Complex Load Flows│──── change network   │
        └──────────────────┘                       │
                  │                                │
   ┌──────┬───────┼────────┬──────────┐            │
   ▼      ▼       ▼        ▼                        │
┌────────┐┌────────┐┌────────┐┌─────────┐          │
│steady- ││voltage ││transient││insulation│        │
│state   ││stability││stability││levels   │        │
│stability││        ││        ││         │         │
└────────┘└────────┘└────────┘└─────────┘          │
   │      │       │        │                        │
   └──────┴───┬───┴────────┘                        │
              ▼                                     │
      ┌──────────────┐                              │
      │ Check Security│──────────────────────────────┘
      └──────────────┘
              ▼
      ┌──────────────┐
      │ FINAL DESIGN │
      └──────────────┘
```

# References

**BOOKS**

1. *Electrical Transmission and Distribution Reference Book*, Westinghouse Electric Corp., East Pittsburgh, Pennsylvania, 1964.
2. El Abiad, A. H., and G. W. Stagg, *Computer Methods in Power System Analysis*, McGraw-Hill, New York, 1968.

**PAPERS**

3. Robert, R., 'Micromachines and Microreseaux', *C.I.G.R.É.*, paper 338, Paris (1950).
4. Laughton, M. A., and M. W. H. Davies, 'Numerical techniques in solution of power system load-flow problems', *Proc. I.E.E.*, **111**, No. 9 (1964).

5. Carré, B. A., 'The determination of optimum accelerating factor for successive over-relaxation', *Computer Jour.*, **4** (1962).
6. Ward, J. B., and H. W. Hale, 'Digital computer solution of power flow problem', *Trans. A.I.E.E.*, **75**, 111 (1956).
7. Knight, U. G. W., 'The logical design of electrical networks using linear programming methods', *Proc. I.E.E.*, *Part A*, **107** (1960).
8. Bennett, J. M., 'Digital computers and the load flow program', *Proc. I.E.E.*, **103**, Part B, Suppl. 1, 16 (1956).
9. Brown, H. E., G. K. Carter, H. H. Happ, and C. E. Person,' 'Power flow solution by impedance matrix iterative method', *Trans. I.E.E.E.*, *P.A. & S.*, **82**, 1 (1963).
10. Brown, R. J., and W. F. Tinney, 'Digital solutions for large power networks', *Trans. A.I.E.E.*, **76**, Part III, 347 (1957).
11. Cronin, J. H., and M. B. Newman, 'Digital load flow program for 1,000 bus systems', *Trans. I.E.E.E.*, *P.A. & S.*, **83**, 718 (1964).
12. Glimn, A. F., and G. W. Stagg, 'Automatic calculation of load flows', *Trans. A.I.E.E.*, **76**, Part III, 817 (1957).
13. Gupta, P. P., and M. W. Humphrey-Davies, 'Digital computers in power system analysis', *Proc. I.E.E.*, **108**, Part A, 383 (1961).
14. Van Ness, J. E., and J. H. Griffin, 'Elimination methods for load flow studies', *Trans. A.I.E.E.*, **80**, Part III, 299 (1961).
15. Carré, B. A., 'The solution of load flow problems by partitioning systems into trees', *Trans. I.E.E.E.*, PAS-87, 1931 (1968).
16. Baumann, R., 'Some new aspects of load flow calculations. Part I—Impedance matrix generation controlled by network topology', *Trans. I.E.E.E.*, *P.A. & S.*, **85**, 1164 (1966).

## Problems

5.1. A single-phase distributor has the following loads at the stated distances from the supply end: 10 kW at 30 yards, 10 kW at 0·9 p.f. lagging at 50 yards, 5 kW at 0·8 p.f. lagging at 100 yards, and 20 kW at 0·95 p.f. lagging at 150 yards. The loads may be assumed to be constant current at their nominal voltage values (240 V). If the supply voltage is 250 V and the maximum voltage drop is 5 per cent of the nominal value determine the cross-sectional area of the conductor and give the nearest commercially-available size. ($r + 0·376x = 0·7 \times 10^{-3}$, i.e. 7/0·136 copper.)

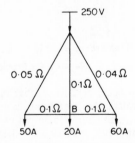

Figure 5.14   Network for Problem 5.2.

5.2. In the d.c. network shown in figure 5.14 calculate the voltage at node *B* by inverting the admittance matrix. Check the answer by Thevenin's Theorem (247·7 V).

5.3. In the interconnected network shown in figure 5.15 calculate the current in feeder BC (26·75 A).

5.4. In the network shown in figure 5.16 the loads are represented by constant impedances $1+j1$ p.u. Determine the current distributions in the network, (a) when the transformer has its nominal ratio, (b) when the transformer is tapped up 10 per cent. (Note: determine the distribution with the off-nominal voltage alone and use superposition.) (Transformer branch, (a) 0, (b) (0·0735−j0·075(p.u.)

5.5. For the system shown in figure 5.12 calculate the busbar voltages after the second iteration without the use of an accelerating factor.

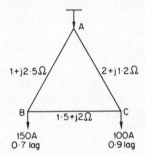

Figure 5.15   Network for Problem 5.3.

5.6. For the system shown in figure 5.12 calculate the power, reactive power and current flows in the lines for the completed load flow. (Voltages after 30 iterations are given in the text.)

5.7. For the system shown in figure 5.12 recalculate the nodal voltages and power and reactive power flows at the end of the first iteration using an accelerating factor of 1·6.

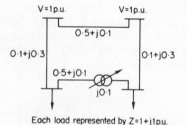

Each load represented by $Z = 1 + j1$ p.u.

Figure 5.16   Network for Problem 5.4.

5.8. In the system shown in figure 5.12 the busbar quantities are to remain the same except at busbar 3 where the generation is now to produce a constant voltage of magnitude 1·15 p.u. and a power output of 0·5 p.u. with no restriction on the var output. Determine the voltages after the first iteration without the use of an accelerating factor.

5.9. Repeat Example 5.4 using the nodal voltage-iterative method.

5.10. Repeat the Example in Appendix 2 (figure A2.7a) using an iterative method.

5.11. By suitable partitioning of the admittance matrix of the system in figure 2.7a determine the matrix of this system reduced to its simplest form and hence derive the simplified network.

5.12. For the system of figure 5.12 perform a real-power load flow (ignoring var loading and component resistance) and compare the power flows with those obtained for the complex study given in the text. Design a resistance network operating from a 50 V supply to simulate the system for real power flows only.

5.13. Enumerate the information which may be obtained from a load-flow study. Part of a power system is shown in figure 5.17. The line to neutral reactances and values of real and reactive power (in the form $P \pm jQ$) at the various stations are expressed as per unit values on a common MVA base. Resistance may be neglected. By the use of an iterative method suitable for a digital computer, calculate the voltages at the stations after the first iteration without the use of an accelerating factor.

Answer: $V_2$ 1·03333 − j0·03333 p.u.
$V_3$ 1·11666 + j0·23333 p.u.
$V_4$ 1·05556 + j0·00277 p.u.

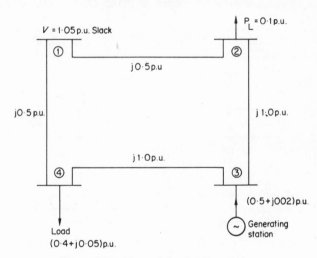

Figure 5.17 Network for Problem 5.13.

# 6

# Fault Analysis

## 6.1 Introduction

An essential part of the design of a power supply network is the calculation of the currents which flow in the components when faults of various types occur. In a fault survey, faults are applied at various points in the network and the resulting currents obtained by hand calculation, or, more likely

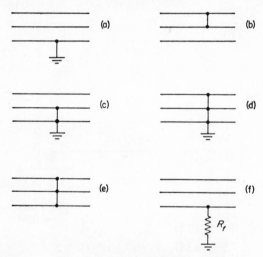

Figure 6.1    Common types of fault.

now on large networks, by analogue or digital processes. The magnitude of the fault currents give the engineer the current settings for the protection to be used and the ratings of the circuit breakers.

The types of fault commonly occurring in practice are illustrated in figure 6.1 and the most common of these is the short-circuit of a single conductor to earth. Often the path to earth contains resistance in the form of an arc as shown in figure 6.1f. Although the single-line to earth fault is the most common, calculations are frequently performed with the three-line, balanced, short-circuit (figure 6.1d and e). This is the most severe fault and also the most amenable to calculation. The causes of faults are

Table 6.1   Causes of overhead-line faults, British system 66 kV and above, 1961

|  | faults/100 miles of line |
| --- | --- |
| Lightning | 1·59 |
| Dew, fog, frost | 0·15 |
| Snow, ice | 0·01 |
| Gales | 0·24 |
| Salt spray | 0·01 |
| Total | 2 faults per 100 miles giving a total of 232 faults on system |

summarized in table 6.1 which gives the distribution of faults due to various causes on the British Central Electricity Generating Board system in 1961. Table 6.2 shows the components affected. In tropical countries the incidence of lightning is much greater than in Britain resulting in larger numbers of faults.

Table 6.2   Distribution of faults, British system

| Type | Number of faults |
| --- | --- |
| Overhead lines | 289 |
| Cables | 67 |
| Switchgear | 56 |
| Transformers | 59 |
| Total | 471 |

As well as fault current, fault MVA is frequently considered; this is obtained from the expression $\sqrt{(3)}V_L I_F \times 10^{-6}$ fault where $V_L$ is the nominal line-voltage of the faulted part. The fault MVA is often referred to as the *fault level*. The calculation of fault currents can be divided into the following two main types.

(a) Faults short-circuiting all three phases when the network remains balanced electrically. For these calculations normal single-phase equivalent circuits may be used as in ordinary load-flow calculations.

(b) Faults other than three-phase short-circuits when the network is electrically unbalanced. To facilitate these calculations a special method for dealing with unbalanced networks is used known as the method of *symmetrical components*.

The main objects of fault analysis may be enumerated as follows:

1. To determine maximum and minimum three-phase short-circuit currents.
2. To determine the unsymmetrical fault-current for single and double line-to-earth, line-to-line faults, and open circuit faults.
3. Investigation of the operation of protective relays.
4. Determination of rated rupturing capacity of breakers.
5. To determine fault-current distribution and busbar-voltage levels during faults.

### 6.2 Calculation of Three-Phase Balanced Fault Currents

The action of synchronous generators on three-phase short-circuits has been described in chapter 2. There it was seen that dependent on the time elapsing from the incidence of the fault, either the transient or the sub-transient reactance should be used to represent the generator. For specifying switchgear the value of the current flowing at the instant at which the circuit breakers open is required. It has been seen however that the initially high fault current associated with the subtransient reactance decays with the passage of time. Modern air-blast circuit breakers usually operate in 2·5 cycles of 50 Hz alternating current and are associated with extremely fast protection. Older circuit breakers and those on lower voltage networks usually associated with relatively cruder protection can take in the order of 8 cycles or more to operate. In calculations it is usual to use the subtransient reactance of generators and to ignore the effects of induction motors. The calculation of fault currents ignores the direct-current component, the magnitude of which depends on the instant in the cycle that the short-circuit occurs. If the circuit breaker opens a reasonable time after the incidence of the fault the direct-current component will have decayed considerably. With fast-acting circuit breakers the actual current to be interrupted is increased by the direct-current component and it must be taken into account. To allow for the direct-current component of the fault current the symmetrical r.m.s. value is modified by the use of multiplying factors such as the following:

8 cycle circuit breaker opening time, multiply by 1,
3 cycle circuit breaker opening time, multiply by 1·2.
2 cycle circuit breaker opening time, multiply by 1·4.

Consider an initially unloaded generator with a short-circuit across the three terminals as in figure 6.2. The generated voltage per phase is $E$ and therefore the short-circuit current is $[E/Z(\Omega)]A$, where $Z$ is either the transient or subtransient impedance. If $Z$ is expressed in per unit notation,

$$Z(\text{p.u.}) = \frac{I_{FL}Z(\Omega)}{E}$$

$$\therefore \qquad Z(\Omega) = \frac{EZ(\text{p.u.})}{I_{FL}}$$

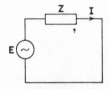

Figure 6.2    Voltage source with short-circuit and equivalent circuit.

Taking $I_{FL}$ (full load or rated current) and $E$ as base values of voltage and current,

the short-circuit current,

$$I_{\text{s.c.}} = \frac{E}{Z(\Omega)}$$

$$= \frac{E \cdot I_{FL}}{E \cdot Z(\text{p.u.})} = \frac{I_{FL}}{Z(\text{p.u.})} \qquad (6.1)$$

Also the three-phase short-circuit volt-amperes

$$= \sqrt{(3)}VI_{\text{s.c.}} = \frac{\sqrt{(3)}VI_{FL}}{Z(\text{p.u.})}$$

$$= \frac{\text{Base or full load volt-amperes}}{Z(\text{p.u.})} \qquad (6.2)$$

Hence the short-circuit level is immediately obtained if the impedance from the source of the voltage to the point of the fault is known.

*Example 6.1*    An 11·8 kV busbar is fed from three synchronous generators having the following ratings and reactances,

20 MVA, $X'$ 0·08 p.u.; 60 MVA, $X'$ 0·1 p.u.; 20 MVA, $X'$ 0·09 p.u. Calculate the fault current and MVA if a three-phase symmetrical fault occurs on the busbars. Resistance may be neglected. The voltage base will be taken as 11·8 kV and the VA base as 60 MVA.

*Solution*    The transient reactance of the 20 MVA machine on the above bases is $(60/20) \times 0{\cdot}08$, i.e. 0·24 p.u. and of the 20 MVA machine $(60/20) \times 0{\cdot}09$, i.e. 0·27 p.u. These values are shown in the equivalent circuit in figure 6.3. As the generator e.m.f.'s are equal one source may be used (figure 6.3c).

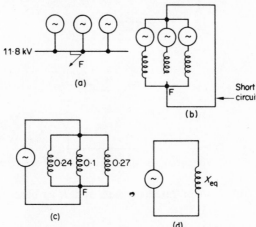

Figure 6.3    Line diagram and equivalent circuits for Example 6.1.

The equivalent reactance,

$$X_e = \frac{1}{1/0{\cdot}24 + 1/0{\cdot}27 + 1/0{\cdot}1}$$

$$= 0{\cdot}054 \text{ p.u.}$$

∴  fault MVA

$$= \frac{60}{0{\cdot}054} = 1111 \text{ MVA}$$

and fault current

$$= \frac{1111 \times 10^6}{\sqrt{(3)} \times 11{,}800}$$

$$= 54{,}300 \text{ A.}$$

*Example 6.2* In the network shown in figure 6.4 a three-phase fault occurs at point F. Calculate the fault MVA at F. The per unit values of reactance given, all refer to a base of 100 MVA. Resistance may be neglected.

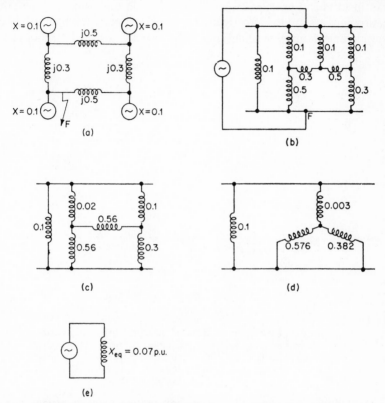

Figure 6.4 Line diagram and equivalent circuits for Example 6.2.

*Solution* The equivalent single-phase network of generator and line reactances is shown in figure 6.4b. This is replaced by the network in figure 6.4c by the use of the delta-star transformation. A further transformation is carried out on the network in figure 6.4d to give the final single equivalent reactance,

$$X_e = 0.07 \text{ p.u.}$$

The fault level at point F

$$= \frac{100}{0.07} = 1430 \text{ MVA.}$$

This value is based on the assumption of symmetrical fault current. Allowing for a multiplying factor of 1·4, the fault level is 2000 MVA.

*Current limiting reactors*    The impedances presented to fault currents by transformers and machines when faults occur on substation or generating station busbars are low. To reduce the high fault current which would do considerable damage mechanically and thermally, artificial reactances are

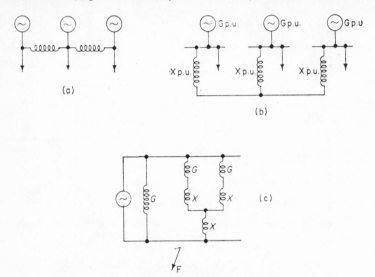

Figure 6.5   Connexion of artificial reactors; (a) Ring system; (b) Tie-bar system. Transient reactance of machines, $G$ p.u.; reactance of artificial reactors $X$ p.u.; (c) Equivalent circuit of tie bar system.

connected in the network. These current limiting reactors usually consist of insulated copper strip embedded in concrete formers; this is necessary to withstand the high mechanical forces produced by the current in neighbouring conductors. The position in the circuit occupied by the reactor is a matter peculiar to individual designs and installations. Two arrangements which have enjoyed some popularity for generating-station busbars are known as the ring and the tie-bar systems. These are shown in figure 6.5.

*Example 6.3*    The three-phase fault current in the busbar of a three-machine generating station is reduced by the installation of current limiting reactors in the tie-bar system. The generators are identical and each has a transient reactance of $G$ per unit, the reactors are of reactance $X$ per unit. Determine the fault MVA for a fault on any busbar. The per-unit values are all on a base of $C$ MVA.

*Solution*  The arrangement of machines and the equivalent single-phase circuit is shown in figure 6.5. It is obvious from the equivalent circuit that the fault current will be the same regardless of the busbar where the fault occurs. The equivalent reactance from voltage source to fault is $X_e$ and

$$\frac{1}{X_e} = \frac{1}{G} + \frac{1}{((G+X)/2) + X}$$

$$= \frac{G/2 + 3X/2 + G}{G(G/2 + 3X/2)}$$

$$\therefore \qquad X_e = \frac{G(G/2 + 3X/2)}{3G/2 + 3X/2}$$

$$= \frac{1}{3} \cdot \frac{G(G + 3X)}{G + X}$$

The fault MVA

$$= \frac{C}{X_e}$$

$$= \frac{3C(G + X)}{G(G + 3X)}$$

Consider the case when there are $n$ generators and $n$ reactors, now

$$\frac{1}{X_e} = \frac{1}{G} + \frac{1}{(G + X/n - 1) + X}$$

$$\therefore \qquad X_e = \frac{G(G + nX)}{(n-1)(G + (G + nX/n - 1))} = G\left(\frac{G + nX}{nG + nX}\right)$$

$\therefore$ short circuit current

$$= \frac{C}{G} \cdot \left(\frac{nG + nX}{G + nX}\right)$$

when $n$ is infinite,

$$I_{s.c.} = \left(\frac{G + X}{X}\right) \cdot \frac{C}{G}$$

If the circuit breakers are initially specified to cope with this current, the busbar system may be extended indefinitely in the future without replacing the existing switches.

*Example 6.4*   Two synchronous generators rated at 20 MVA and 40 MVA and of transient reactance 0·1 and 0·08 p.u. respectively are connected to a common busbar from which a feeder is supplied through a 20 MVA transformer of 0·1 p.u. reactance. It is intended to reinforce the supply to the busbar by the connexion of a grid transformer rated at 50 MVA, 0·1 p.u. reactance.

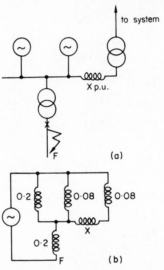

Figure 6.6   Line diagram and equivalent circuits for Example 6.4.

The breaking capacity of a circuit breaker on the feeder side of the transformer is 160 MVA, which is the limit for three-phase symmetrical faults. Indicate the manner in which the grid transformer should be connected and calculate the parameters of any apparatus used. State clearly any assumptions made including any relating to the time of operation of the circuit breaker and protective gear.

*Solution*   An MVA base of 40 is used. In the equivalent single-phase circuit shown in figure 6·6 all the p.u. values are expressed on a 40 MVA base. The transformer will have to be joined to the circuit via a reactor of $X$ p.u. reactance so as to limit the fault MVA to 160. This assumes that the circuit breaker takes such time to operate that the direct-current component of the fault current is negligible.

The equivalent reactance

$$= 0.2 + \frac{0.057(0.08 + X)}{0.057 + 0.08 + X}$$

$$= \frac{\text{Base MVA}}{\text{Fault MVA}} = \frac{40}{160}$$

from which $X = 0.33$ p.u. on a 40 MVA base.

### 6.3 Method of Symmetrical Components

This method formulates a system of three, separate phasor systems which when superposed give the true unbalanced conditions in the circuit. It should be stressed that the systems to be discussed are essentially artificial and used merely as an aid to calculation. The various sequence-component voltages and currents do not exist as physical entities in the network.

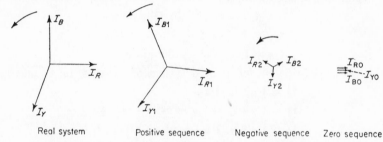

Real system      Positive sequence      Negative sequence     Zero sequence

Figure 6.7    Real system and corresponding symmetrical components.

The method postulates that a three-phase unbalanced system of voltages and currents may be represented by the following three, separate systems of phasors,

(a) a balanced three-phase system in the normal a – b – c (red-yellow-blue) sequence, called the *positive phase-sequence* system,

(b) a balanced three-phase system of reversed sequence, i.e. a — c — b (red-blue-yellow), called the *negative phase-sequence* system,

(c) three phasors equal in magnitude and phase revolving in the positive phase rotation, called the *zero phase-sequence* system.

In figure 6.7 an unbalanced system of currents is shown with the corresponding system of symmetrical components. If each of the red-phase phasors are added, i.e. $I_{1R} + I_{2R} + I_{R0}$ the resultant phasor will be $I_R$ in magnitude and direction; similar reasoning holds for the other two phases.

To express the phasors algebraically, use is made of the complex operator 'a' (sometimes denoted by $\lambda$ or h) which denotes a phase-shift operation of $+120°$ and a multiplication of unit magnitude; i.e

$V\underline{/\phi} \times \mathbf{a} = V\underline{/\phi} \times 1\underline{/120°} = V\underline{/\phi + 120°}$, $\mathbf{a} = \mathrm{e}^{\mathrm{j}\,2\pi/3}$ and $\mathbf{a}^3 = \mathrm{e}^{\mathrm{j}\,3 \times 2\pi/3} = 1$

Also,

$$\mathbf{a}^2 + \mathbf{a} = (-0\cdot5 - \mathrm{j}0\cdot866) + (-0\cdot5 + \mathrm{j}0\cdot866) = -1$$

$\therefore \qquad \mathbf{a}^3 + \mathbf{a}^2 + \mathbf{a} = 0$

and

$$1 + \mathbf{a} + \mathbf{a}^2 = 0.$$

For positive-sequence phasors, taking the red phasor as reference;

$$\mathbf{I}_{1R} = \mathbf{I}_{1R}\,\mathrm{e}^{\mathrm{j}0} = \text{reference phasor, (Figure 6.8).}$$

$$\mathbf{I}_{1Y} = \mathbf{I}_{1R}(-0\cdot5 - \mathrm{j}0\cdot866) = \mathbf{a}^2\mathbf{I}_{1R}$$

and

$$\mathbf{I}_{1B} = \mathbf{I}_{1R}(-0\cdot5 + \mathrm{j}0\cdot866) = \mathbf{a}\mathbf{I}_{1R}$$

For negative sequence quantities,

$$\mathbf{I}_{2R} = \mathbf{I}_{2R}(1 + \mathrm{j}0), \text{ (Figure 6.9).}$$

$$\mathbf{I}_{2Y} = \mathbf{I}_{2R}(-0\cdot5 + \mathrm{j}0\cdot866) = \mathbf{a}\mathbf{I}_{2R}$$

$$\mathbf{I}_{2B} = \mathbf{I}_{2R}(-0\cdot5 - \mathrm{j}0\cdot866) = \mathbf{a}^2\mathbf{I}_{2R}.$$

Returning to the original unbalanced system of currents, $\mathbf{I}_R$, $\mathbf{I}_Y$ and $\mathbf{I}_B$;

$$\mathbf{I}_R = \mathbf{I}_{1R} + \mathbf{I}_{2R} + \mathbf{I}_{0R}$$

$$\mathbf{I}_Y = \mathbf{I}_{1Y} + \mathbf{I}_{2Y} + \mathbf{I}_{0Y} = \mathbf{a}^2\mathbf{I}_{1R} + \mathbf{a}\mathbf{I}_{2R} + \mathbf{I}_{0R}$$

$$\mathbf{I}_B = \mathbf{I}_{1B} + \mathbf{I}_{2B} + \mathbf{I}_{0B} = \mathbf{a}\mathbf{I}_{1R} + \mathbf{a}^2\mathbf{I}_{2R} + \mathbf{I}_{0R}.$$

Hence in matrix form,

$$\begin{bmatrix} \mathbf{I}_R \\ \mathbf{I}_Y \\ \mathbf{I}_B \end{bmatrix} = \begin{bmatrix} 1 & 1 & 1 \\ 1 & \mathbf{a}^2 & \mathbf{a} \\ 1 & \mathbf{a} & \mathbf{a}^2 \end{bmatrix} \begin{bmatrix} \mathbf{I}_{0R} \\ \mathbf{I}_{1R} \\ \mathbf{I}_{2R} \end{bmatrix} \qquad (6.3)$$

Inverting the matrix,

$$\begin{bmatrix} \mathbf{I}_{0R} \\ \mathbf{I}_{1R} \\ \mathbf{I}_{2R} \end{bmatrix} = \tfrac{1}{3} \begin{bmatrix} 1 & 1 & 1 \\ 1 & \mathbf{a} & \mathbf{a}^2 \\ 1 & \mathbf{a}^2 & \mathbf{a} \end{bmatrix} \begin{bmatrix} \mathbf{I}_R \\ \mathbf{I}_Y \\ \mathbf{I}_B \end{bmatrix} \qquad (6.4)$$

The above also holds for voltages, i.e.

$$[\mathbf{E}_{\text{actual}}] = [\mathbf{T}_s][\mathbf{E}_{1,2,0}]$$

where $[\mathbf{T}_s]$ is the symmetrical component transformation matrix,

$$\begin{bmatrix} 1 & 1 & 1 \\ \mathbf{a}^2 & \mathbf{a} & 1 \\ \mathbf{a} & \mathbf{a}^2 & 1 \end{bmatrix}$$

In a three-wire system the instantaneous voltages and currents add to zero and therefore no single-phase component is required. A fourth wire or connexion to earth must be provided for single-phase currents to flow. In a three-wire system the zero phase-sequence components are replaced by zero in equations (6.3) and (6.4). Also,

$$I_{OR} = \frac{I_R + I_Y + I_B}{3} = \frac{I_N}{3}$$

where $I_N$ is the neutral current.

$$\therefore \qquad I_N = 3I_{RO} = 3I_{YO} = 3I_{BO}.$$

In the application of this method it is necessary to calculate the symmetrical components of the current in each line of the network and then combine them to obtain the actual values. The various phase-sequence values are obtained by considering a network derived from the actual network, in which only a particular sequence current flows; for example, in a zero-sequence network only zero-sequence currents and voltages exist. The positive-sequence network is identical with the real, balanced, equivalent network, i.e. it is the same as used for three-phase symmetrical short-circuit studies. The negative-sequence network is almost the same as the real one except that the values of impedance used for rotating machines are different. The zero-sequence network is considerably different from the real one.

### 6.4 Representation of Plant in the Phase-Sequence Networks

(a) *The synchronous machine* (table 2.1) The positive-sequence impedance $Z_1$ is the normal transient or subtransient value. Negative-sequence currents set up a rotating magnetic field in the opposite direction to that of the positive-sequence currents and which rotates round the rotor surface at twice the synchronous speed; hence the effective impedance ($Z_2$) is different from $Z_1$. The zero-sequence impedance $Z_0$ depends upon the nature of the connexion between the star point of the windings and the earth and the single-phase impedance of the stator windings in series. Resistors or reactors are frequently connected between the star point and earth for reasons usually connected with protective gear and the limitation of over-voltages. Normally the only voltage sources appearing in the networks are in the positive sequence one, as the generators only generate positive sequence e.m.f.s.

(b) *Lines and cables* The positive and negative sequence impedances are the normal balanced values. The zero sequence impedance depends upon the nature of the return path through the earth if no fourth wire is provided. It is also modified by the presence of an earth wire on the towers which protects the lines against lightning surges. In the absence of detailed information the following rough guide to the value of $Z_0$ may be used. For a single-circuit line $(Z_0/Z_1) = 3 \cdot 5$ with no earth wire and 2 with one. For a double-circuit line $(Z_0/Z_1) = 5 \cdot 5$. For underground cables, $(Z_0/Z_1) = 1$ to $1 \cdot 25$ for single core and 3 to 5 for three-cored cables.

Table 6.3

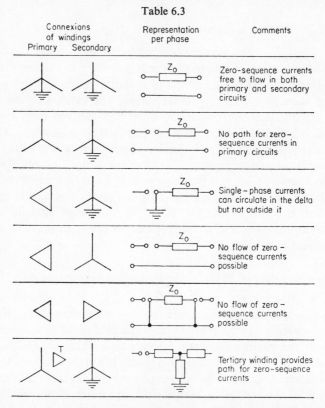

| Connexions of windings<br>Primary   Secondary | Representation per phase | Comments |
|---|---|---|
| | $Z_0$ | Zero-sequence currents free to flow in both primary and secondary circuits |
| | $Z_0$ | No path for zero-sequence currents in primary circuits |
| | $Z_0$ | Single-phase currents can circulate in the delta but not outside it |
| | $Z_0$ | No flow of zero-sequence currents possible |
| | $Z_0$ | No flow of zero-sequence currents possible |
| | | Tertiary winding provides path for zero-sequence currents |

(c) *Transformers* The positive and negative sequence impedances are the normal balanced ones. The zero-sequence connexion of transformers is however complicated and depends on the nature of the connexion of the windings. In table 6.3 is listed the zero sequence representation of transformers for various winding arrangements. Zero-sequence currents in the windings on one side of a transformer must produce the corresponding ampere-turns in the other. But three in-phase currents cannot flow in a star connexion without a connexion to earth. They can circulate round a delta winding but not in the lines outside it. Owing to the mutual impedance between the phases $Z_0 \neq Z_1$.

An example showing the nature of the three sequence networks for a small transmission link is shown in figure 6.10 and table 6.3 shows typical transformer representations for zero sequence conditions.

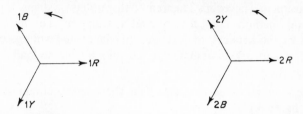

Figure 6.8  Positive-sequence phasors.

Figure 6.9  Negative-sequence phasors.

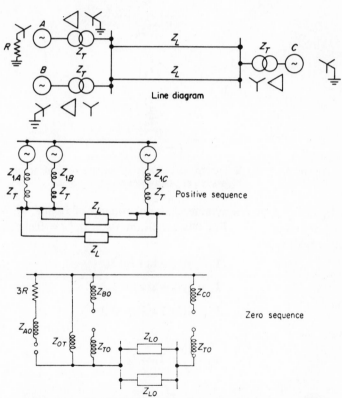

Figure 6.10  Typical transmission link and form of associated sequence networks.

Negative sequence diagram as for positive but voltage sources omitted and generator impedances are $Z_{2A}$, $Z_{2B}$ and $Z_{2C}$.

## 6.5 Types of Fault

In the following a single voltage source in series with an impedance is used to represent the power network as seen from the point of the fault. This is an extension of Thevenin's Theorem to three-phase systems. It represents the general method used for manual calculation, i.e. the successive reduction of the network to a single impedance and voltage or current source. The network is assumed to be initially on no-load before the occurrence of the fault.

(a) *Single-phase to earth fault* The three-phase circuit diagram is shown in figure 6.11.

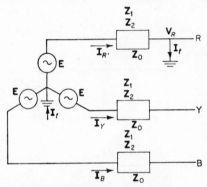

Figure 6.11   Single line to earth fault—Thevenin equivalent of system at point of fault.

Let $I_1$, $I_2$ and $I_0$ be the symmetrical components of $I_R$ and $V_1$, $V_2$ and $V_0$ the components of $V_R$. For this condition $V_R = 0$, $I_B = 0$ and $I_Y = 0$.
From equation 6.4,

$$I_0 = \tfrac{1}{3}(I_R + I_B + I_Y)$$

$$I_1 = \tfrac{1}{3}(I_R + aI_Y + a^2 I_B)$$

$$I_2 = \tfrac{1}{3}(I_R + a^2 I_Y + aI_B)$$

Hence,

$$I_0 = \frac{I_R}{3} = I_1 = I_2$$

Also,

$$V_R = E - I_1 Z_1 - I_2 Z_2 - I_0 Z_0 = 0$$

Eliminating $I_0$ and $I_2$,

$$E - I_1(Z_1 + Z_2 + Z_0) = 0$$

and

$$I_1 = \frac{E}{Z_1 + Z_2 + Z_0} \qquad (6.5)$$

The fault current,

$$I_f = I_R = 3I_1$$

$$= \frac{3E}{Z_1 + Z_2 + Z_0} \qquad (6.6)$$

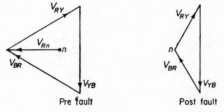

Figure 6.12   Pre and post fault phasor diagrams—single line to earth fault.

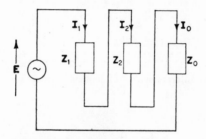

Figure 6.13   Interconnexion of positive, negative and zero sequence networks for single line-to-earth faults.

The e.m.f. of the Y-phase $= a^2E$
and

$$I_Y = I_0 + a^2I_1 + aI_2$$

$\therefore \qquad V_Y = a^2E - I_0Z_0 - a^2I_1Z_1 - aI_2Z_2$

The pre-fault and post-fault phasor diagrams are shown in figure 6.12.

It is usual to form an equivalent circuit to represent the equation (6.5) and this can be obtained from an inspection of the equations. The circuit is shown in figure 6.13 and it will be seen that $I_1 = I_2 = I_0$, and

$$I_1 = \frac{E}{Z_1 + Z_2 + Z_0}$$

(b) *Phase to phase fault* In figure 6.14, $E$ = e.m.f. per phase and the R-phase is again taken as the reference phasor. In this case, $I_R = 0$, $I_Y = -I_B$ and $V_Y = V_B$.

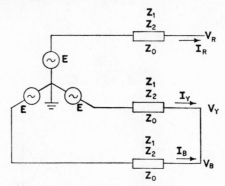

Figure 6.14  Line-to-line fault.

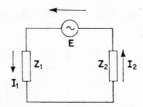

Figure 6.15  Interconnexion of sequence networks for a line to line fault.

From equation 6.4,

$$I_0 = 0$$

$$I_1 = \tfrac{1}{3}I_Y(a - a^2)$$

and

$$I_2 = \tfrac{1}{3}I_Y(a^2 - a)$$

∴

$$I_1 = -I_2$$

As $V_Y = Y_B$

$$a^2E - a^2I_1Z_1 - aI_2Z_2 = aE - aI_1Z_1 - a^2I_2Z_2$$

∴

$$E(a^2 - a) = I_1[Z_1(a^2 - a) + Z_2(a^2 - a)]$$

$$I_1 = \frac{E}{Z_1 + Z_2} \tag{6.7}$$

This can be represented by the equivalent circuit in figure 6.15 in which of course there is no zero-sequence network. If the connexion between the two lines has an impedance $\mathbf{Z}_f$ (the fault impedance) this is connected in series in the equivalent circuit.

(c) *Double phase to earth fault* (figure 6.16)

$$\mathbf{I}_R = 0, \quad \mathbf{V}_Y = \mathbf{Y}_B = 0$$

and

$$\mathbf{I}_R = \mathbf{I}_1 + \mathbf{I}_2 + \mathbf{I}_0 = 0$$

$\therefore$

$$a^2\mathbf{E} - a^2\mathbf{I}_1\mathbf{Z}_1 - a\mathbf{I}_2\mathbf{Z}_2 - \mathbf{I}_0\mathbf{Z}_0 = \mathbf{V}_Y = 0$$

and

$$a\mathbf{E} - a\mathbf{I}_1\mathbf{Z}_1 - a^2\mathbf{I}_2\mathbf{Z}_2 - \mathbf{I}_0\mathbf{Z}_0 = \mathbf{V}_B = 0$$

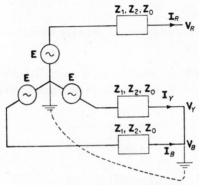

Figure 6.16  Line to line to earth fault.

Hence,

$$\mathbf{I}_1 = \frac{\mathbf{E}}{\mathbf{Z}_1 + (\mathbf{Z}_2\mathbf{Z}_0/\mathbf{Z}_2 + \mathbf{Z}_0)} \tag{6.8}$$

$$\mathbf{I}_2 = -\mathbf{I}_1\frac{\mathbf{Z}_0}{\mathbf{Z}_2 + \mathbf{Z}_0} \tag{6.9}$$

and

$$\mathbf{I}_0 = -\mathbf{I}_1 \cdot \frac{\mathbf{Z}_2}{\mathbf{Z}_2 + \mathbf{Z}_0} \tag{6.10}$$

These can be represented by the equivalent circuit shown in figure 6.17.

The inclusion of impedances in the earth path, such as the star point to earth connexion in a generator or transformer, modifies the sequence diagrams. For a phase-to-earth fault an impedance $\mathbf{Z}_g$ in the earth path is represented by an impedance of $3\mathbf{Z}_g$ in the zero-sequence network. $\mathbf{Z}_g$ can include the impedance of the fault itself usually the resistance of the arc. As $I_1 = I_2 = I_0$ and $3I_1$ flows through $Z_g$ in the physical system it is necessary to use $3Z_g$ to obtain the required effect. Hence,

$$I_f = \frac{3E}{Z_1 + 3Z_g + Z_2 + Z_0}$$

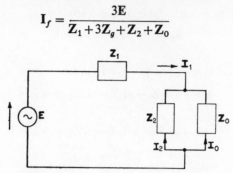

Figure 6.17   Interconnexion of sequence networks—double line to earth fault.

Again, for a double-phase to earth fault an impedance $3Z_g$ is connected as shown in figure 6.18. $Z_g$ includes both machine neutral-impedances and fault impedances.

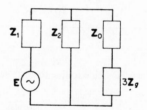

Figure 6.18   Modification of network in figure 6.17 to account for neutral impedance.

The phase shift introduced by star-delta transformers has no effect on the magnitude of the fault currents, although it will affect the voltages at various points. It is shown in chapter 2 that the positive-sequence voltages and currents are advanced by a certain angle and the negative-sequence quantities retarded by the same angle for a given connexion.

## 6.6   Fault Levels in a Typical System

In figure 6.19 a section of a typical system is shown. At each voltage level the fault level can be ascertained from the reactances given. It should be noted that the short-circuit level will change with network conditions, and there will normally be two extreme values, that with all plant connected and that with the minimum plant normally connected. The short-circuit

MVA at 275 kV busbars in Britain is normally 10,000 MVA but drops to 7000 MVA with minimum plant connected. Maximum short-circuit (three-phase) levels experienced in the British system are as follows:

275 kV, 15,000 MVA; 132 kV, 3500 MVA; 33 kV, 750/1000 MVA; 11 kV, 150/250 MVA; 415 V, 25 MVA.

As the transmission voltages increase, the short-circuit currents also increase and for the 400 kV system, circuit breakers of 35,000 MVA breaking capacity are required. In order to reduce the fault level the number

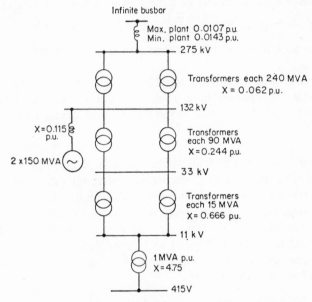

Figure 6.19 Typical transmission system. All reactances on a 100 MVA base.

of parallel paths is reduced by sectionalizing. This is usually achieved by opening the circuit breaker connecting two sections of a substation or generating station busbar. One great advantage of direct-current transmission links in parallel with the alternating-current system is that no increase in the short-circuit currents results.

*Fault calculations in large systems*  The digital and analogue methods used for load-flows and described in chapter 5 are also suitable for three-phase symmetrical fault calculations which after all are merely load flows under certain network conditions. The input information must be modified so that machines are represented by the appropriate reactances. The problem is effectively to carry out a load flow on the network with the following modifications. The generators are represented by their no-load voltages in series with the subtransient reactances. If loads are to be taken into account they are represented by the equivalent shunt admittance to neutral. This will involve some modification to the original load-flow admittance matrix, but the general form will be as in the normal load flow.

A current of 1 per unit is injected at the fault point and removed at the neutral. If the voltage at the fault becomes $V_f$ then $1/V_f$ is the fault level (i.e. short-circuit admittance) at node $f$. Also the flow along branch $kf$ into the fault is given by $\dfrac{(V_f - V_k)}{V_f} Y_{fk}$. The voltages due to the injected currents can be obtained by the same numerical techniques as used for load flows.

Three-phase fault studies are performed in conjunction with load flows. For example, if the fault level on a solid busbar is too high the busbar will be sectioned, i.e. split into two or more sections by opening switches, and new load flows required.

*Unbalanced faults*    Many fault studies performed on large systems are concerned with three-phase short-circuits. Methods have been evolved, however, for the various types of unbalanced fault. These can take the form of a load flow on an equivalent network comprising the appropriate connexion of the positive, negative and zero sequence networks with current injected at the fault point as before. With very large networks this becomes very consuming in time and computer storage and more direct methods are now available.

*Example 6.5*    A synchronous machine 'A' generating 1 p.u. voltage is connected through a star-star transformer, reactance 0·12 p.u., to two lines in parallel. The other ends of the lines are connected through a star-star transformer of reactance 0·1 p.u. to a second machine 'B' also generating 1 p.u. voltage. For both transformers $X_1 = X_2 = X_0$.

Calculate the current fed into a double-line to earth fault on the line-side terminals of the transformer fed from A.

The relevant p.u. reactances of the plant all referred to the same base are as follows:—

| | | | |
|---|---|---|---|
| 'A', | $X_1$ 0·3, | $X_2$ 0·2, | $X_0$ 0·05. |
| 'B', | $X_1$ 0·25, | $X_2$ 0·15, | $X_0$ 0·03. |

For each line $X_1 = X_2 = 0·3$, $X_0 = 0·7$.

The star points of 'A' and of the two transformers are solidly earthed.

*Solution* The positive, negative and zero sequence networks are shown in figure 6.20. All p.u. reactances are on the same base. From these diagrams the following equivalent reactances up to the point of the fault are obtained: $Z_1 = j0.23$ p.u., $Z_2 = j0.18$ p.u. and $Z_0 = j0.17$ p.u.

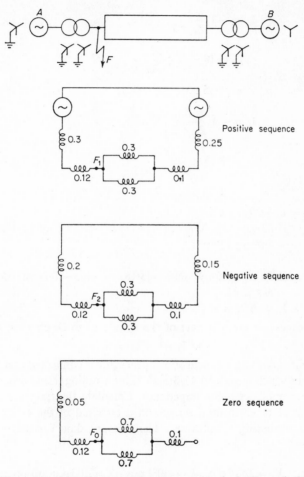

Figure 6.20   Line diagram and sequence networks for Example 6.5.

The red phase is taken as reference phasor and the blue and yellow phases assumed to be shorted at the fault point. From the equivalent circuit for a phase to phase fault,

$$\mathbf{I}_1 = \frac{1}{\mathbf{Z}_1 + (\mathbf{Z}_0\mathbf{Z}_2/\mathbf{Z}_0 + \mathbf{Z}_2)} \text{ p.u.}$$

$$= \frac{1}{j(0{\cdot}23 + (0{\cdot}17 \times 0{\cdot}18/0{\cdot}17 + 0{\cdot}18))} \text{ p.u.}$$

$$= -j3{\cdot}18 \text{ p.u.}$$

$$\mathbf{I}_2 = -\mathbf{I}_1\frac{\mathbf{Z}_0}{\mathbf{Z}_0 + \mathbf{Z}_2}$$

$$= j3{\cdot}18 \times \frac{0{\cdot}17}{0{\cdot}17 + 0{\cdot}18}$$

$$= j1{\cdot}55 \text{ p.u.}$$

$$\mathbf{I}_0 = -\mathbf{I}_1\frac{\mathbf{Z}_2}{\mathbf{Z}_0 + \mathbf{Z}_2}$$

$$= j3{\cdot}18 \times \frac{0{\cdot}18}{0{\cdot}18 + 0{\cdot}17}$$

$$= j1{\cdot}63 \text{ p.u.}$$

$$\mathbf{I}_Y = \mathbf{I}_0 + \mathbf{a}^2\mathbf{I}_{R1} + \mathbf{a}\mathbf{I}_{R2}$$
$$= j1{\cdot}63 + (-0{\cdot}5 - 0{\cdot}866j)(-j3{\cdot}18) + (-0{\cdot}5 + j0{\cdot}866)(j1{\cdot}55)$$
$$= -4{\cdot}09 + j2{\cdot}45 \text{ p.u.}$$

$$\mathbf{I}_B = j1{\cdot}63 + (-0{\cdot}5 + j0{\cdot}866)(-j3{\cdot}18) + (-0{\cdot}5 - j0{\cdot}866)(j1{\cdot}55)$$
$$= 4{\cdot}09 + j2{\cdot}45.$$
$$\mathbf{I}_Y = \mathbf{I}_B = 4{\cdot}78 \text{ p.u.}$$

The correctness of the first part of the solution can be checked as

$$\mathbf{I}_R = \mathbf{I}_1 + \mathbf{I}_2 + \mathbf{I}_0 = 0.$$

*Example 6.6*   An 11 kV synchronous generator is connected to a 11/66 kV transformer which feeds a 66/11/3·3 kV three-winding transformer through a short feeder of negligible impedance. Calculate the fault current when a single-phase to earth fault occurs on a terminal of the 11 kV winding of the three-winding transformer. The relevant data for the system are as follows:

*Generator*   $X_1 = j0{\cdot}15$ p.u., $X_2 = j0{\cdot}1$ p.u., $X_0 j0{\cdot}03$ p.u. all on a 10 MVA base; star point of winding earthed through a 3 Ω resistor.

11/66 kV *Transformer*   $X_1 = X_2 = X_0 = j0{\cdot}1$ p.u. on a 10 MVA base; 11 kV winding delta connected and the 66 kV winding star connected with the star point solidly earthed.

*Three Winding Transformer* 66 kV winding, star connected, star point solidly earthed; 11 kV winding, star connected, star point earthed through a 3 Ω resistor; 3·3 kV winding, delta connected; the three windings of an equivalent star connexion to represent the transformer have sequence

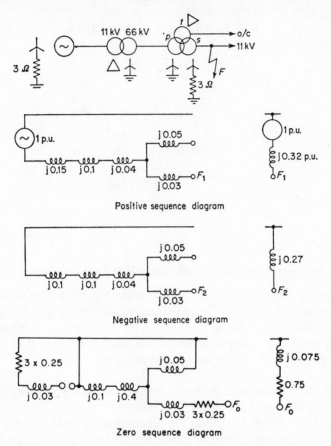

Figure 6.21  Line diagram and sequence networks for Example 6.6.

impedances,  66 kV winding $X_1 = X_2 = X_0 = $ j0·04 p.u.,  11 kV winding $X_1 = X_2 = X_0 = $ j0·03 p.u.,  3·3 kV winding $X_1 = X_2 = X_0 = $ j0·05 p.u., all on a 10 MVA base.

Resistance may be neglected throughout.

*Solution*  The line diagram and the corresponding positive, negative and zero sequence networks are shown in figure 6.21. A 10 MVA base will be used. The 3 Ω earthing resistor has the following p.u. value, $\frac{3 \times 10,000}{(11)^2 \times 1000}$ or 0·25 p.u. Much care is needed with the zero-sequence network owing to the transformer connexions. For a phase-to-earth fault the equivalent circuit shown in figure 6.13 is used, from which,

$$I_1 = I_2 = I_0 \quad \text{and} \quad I_f = I_1 + I_2 + I_0$$

$$= \frac{3 \times 1}{Z_1 + Z_2 + Z_0 + 3Z_g}$$

$$= \frac{3}{j0·32 + j0·27 + j0·075 + 0·75}$$

$$= \frac{3}{0·75 + j0·66}$$

∴ 　　　$$I_f = 3 \text{ p.u.} = \frac{3 \times 10 \times 10^6}{\sqrt{(3)} \times 11,000}$$

$$= 1575 \text{ A.}$$

## 6.7　Systematic Methods for Fault Analysis in Large Networks

The methods described so far become unwieldy when applied to large networks and a systematic approach using matrices is useful, especially when digital computers are used. The following example illustrates a method suitable for determination of balanced three-phase fault currents in a large system by means of a digital computer.

*Example* 6.7　Determine the fault current in the system shown in figure 6.4 for the balanced fault shown.

As already stated, generators are represented by their voltages behind the transient reactance and normally the system is assumed on no-load before the occurrence of the three-phase balanced fault. The voltage sources and transient reactances are converted into current sources and the admittance matrix formed (including the transient reactance admittances). The basic equation $[Y][V] = [I]$ is formed and solved with the constraint that the voltage at the fault node is zero. A current may be injected to the faulted node as explained in the previous section or the voltage made zero and the remaining nodal voltages calculated.

The system in figure 6.4 is replaced by the equivalent circuit shown in figure 6.22 in which $V_D = 0·0$ p.u.

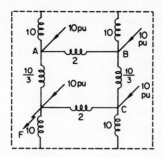

Figure 6.22 Equivalent circuit of system of Figure 6.4. All values are admittances. The generators, i.e. 1 p.u. voltage behind 10 p.u. admittance transform to 10 p.u. current sources in parallel with 10 p.u. admittance.

$$
\begin{bmatrix}
15\cdot33333 & -2\cdot00000 & & -3\cdot33333 \\
-2\cdot00000 & 15\cdot33333 & -3\cdot33333 & \\
& -3\cdot33333 & 15\cdot33333 & -2\cdot00000 \\
-3\cdot33333 & & -2\cdot00000 & 15\cdot33333
\end{bmatrix}
\begin{bmatrix}
V_A \\ V_B \\ V_C \\ 0
\end{bmatrix}
=
\begin{bmatrix}
10 \\ 10 \\ 10 \\ I_F
\end{bmatrix}
$$

As all the Ys are reactive the $j$'s will be omitted for simplicity.

For ease of calculation using a desk machine the equations are multiplied by 1·5 and also $V_D$ eliminated, i.e.

$$
\begin{bmatrix}
23 & -3 & 0 \\
-3 & 23 & -5 \\
0 & -5 & 23
\end{bmatrix}
\cdot
\begin{bmatrix}
V_A \\ V_B \\ V_C
\end{bmatrix}
=
\begin{bmatrix}
15 \\ 15 \\ 15
\end{bmatrix}
\qquad \text{from which,}
$$

$$
\begin{bmatrix}
520 & -115 \\
-5 & 23
\end{bmatrix}
\cdot
\begin{bmatrix}
V_B \\ V_C
\end{bmatrix}
=
\begin{bmatrix}
385 \\ 15
\end{bmatrix}
$$

Thus, $V_B = 0\cdot92950$, $V_C = 0\cdot85510$ and $V_A = 0\cdot77495$ p.u.

and $\quad I_F = \left(10 + \dfrac{0\cdot77495}{0\cdot3} + \dfrac{0\cdot88510}{0\cdot5}\right)$ p.u.

$\qquad\qquad = 14\cdot2985$ p.u.

Fault MVA $\quad = 14\cdot2985 \times 100 = 1429\cdot85$ MVA

In this simple circuit a direct method has been used for solving the equation, in a much larger system the Gauss-Seidel method could be used to advantage. For alternative methods and various treatments of unbalanced faults in large systems the reader is referred to reference 2.

For *unbalanced faults* an approach via the mesh method of analysis will be described. The basic unbalanced three-phase analysis is given in Appendix 3 and should be read before proceeding further.

The following equations relate system and symmetrical component quantities:

$$[E] = [T_s] [E_s] \text{ and } [I] = [T_s] [I_s],$$

where $\quad [E_s] = [E_{1,2,0}] \text{ and } [I_s] = [I_{1,2,0}]$

Also, $\quad [E_s] = [T_s]^{-1} [E] \text{ and } I_s = [T_s]^{-1} [I]$

where $\quad [T_s]^{-1} = \frac{1}{3} \begin{bmatrix} 1 & a & a^2 \\ 1 & a^2 & a \\ 1 & 1 & 1 \end{bmatrix}$ 

$$\tag{6.11}$$

In a network, $[E] = [Z_{br}] [I]$, where $[Z_{br}]$ is the branch impedance matrix.

hence, $\quad [T_s] [E_s] = [Z_{br}] [T_s] [I_s]$

and $\quad [E_s] = [T_s]^{-1} [Z_{br}] [T_s] [I_s]$

Therefore, $[E_s] = [Z_s] [I_s]$ $\hfill (6.12)$

where, $\quad [Z_s] = [T_s]^{-1} [Z_{br}] [T_s]$ $\hfill (6.13)$

It has been shown in Appendix 3 that in a three-phase system in which the impedances are assumed balanced,

$$[Z] = \begin{bmatrix} Z_p & Z_m & Z_m \\ Z_m & Z_p & Z_m \\ Z_m & Z_m & Z_p \end{bmatrix} \tag{6.14}$$

where $\quad Z_p = Z_{aa} + Z_{nn} - 2 Z_{an}$

$\qquad\quad Z_m = Z_{ab} + Z_{nn} - 2 Z_{an}$

The symmetrical component matrix

$$[Z_s] = [T_s]^{-1} [Z_{br}] [T_s]$$

$$= \begin{bmatrix} (Z_p - Z_m) & 0 & 0 \\ 0 & (Z_p - Z_m) & 0 \\ 0 & 0 & (Z_p + 2 Z_m) \end{bmatrix}$$

which may be written

$$[Z_s] = \begin{bmatrix} Z_1 & 0 & 0 \\ 0 & Z_2 & 0 \\ 0 & 0 & Z_0 \end{bmatrix} \tag{6.15}$$

where, $Z_1 = Z_2 = Z_p - Z_m = (Z_{aa} - Z_{ab})$

and $Z_0 = Z_p + 2 Z_m = (Z_{aa} + 3 Z_{nn}) + 2 (Z_{ab} - 3 Z_{an})$

This indicates that the unbalanced network may be replaced by balanced networks of impedance $Z_1$, $Z_2$ and $Z_0$, respectively, as has previously been assumed. In practice the impedances representing rotating machines are not balanced and consequently $Z_1 \neq Z_2$.

One approach is to formulate $[Z_s]$ from the network taking into account the interconnexion of the sequence networks required for a given type of fault. In many situations the only unbalance will be in the fault itself and mutual coupling is only usually of importance when a fault occurs on one of two lines in parallel. To explain this approach a system in which the only unbalance is due to the fault will be analysed. Consider the system in figure 6.23a in which a phase to earth fault exists. The manner in which the positive, negative and zero sequence networks are interconnected to represent such a fault has been derived and for the system in figure 6.23a is shown in figure 6.23b.

In figure 6.23b the generator and transformer reactances have been lumped together. Resistance is neglected and all values are in per unit on a common base. Loop currents are assigned to the network and the mesh impedance matrix is assembled either by inspection or by the use of the connexion matrix (C). Using the former method the following is obtained.

| F | $A_1$ | $B_1$ | $A_2$ | $B_2$ | $A_0$ | $B_0$ |
|---|---|---|---|---|---|---|

| | | F | $A_1$ | $B_1$ | $A_2$ | $B_2$ | $A_0$ | $B_0$ | | |
|---|---|---|---|---|---|---|---|---|---|---|
| $E_2$ | | $1 \cdot 40$ | $-0 \cdot 35$ | $-0 \cdot 15$ | $-0 \cdot 25$ | $-0 \cdot 15$ | $-0 \cdot 15$ | $-0 \cdot 35$ | | $I_F$ |
| $E_1 - E_2$ | | $-0 \cdot 35$ | $1 \cdot 07$ | $-0 \cdot 30$ | $0$ | $0$ | $0$ | $0$ | | $I_{A1}$ |
| $0$ | | $-0 \cdot 15$ | $-0 \cdot 30$ | $0 \cdot 60$ | $0$ | $0$ | $0$ | $0$ | | $I_{B1}$ |
| $0$ | $=$ | $-0 \cdot 25$ | $0$ | $0$ | $0 \cdot 82$ | $-0 \cdot 30$ | $0$ | $0$ | $\cdot$ | $I_{A2}$ |
| $0$ | | $-0 \cdot 15$ | $0$ | $0$ | $-0 \cdot 30$ | $0 \cdot 60$ | $0$ | $0$ | | $I_{B2}$ |
| $0$ | | $-0 \cdot 15$ | $0$ | $0$ | $0$ | $0$ | $1 \cdot 07$ | $-0 \cdot 70$ | | $I_{A0}$ |
| $0$ | | $-0 \cdot 35$ | $0$ | $0$ | $0$ | $0$ | $-0 \cdot 70$ | $1 \cdot 4$ | | $I_{B0}$ |

By inversion of the above matrix (e.g. by successive elimination) the loop currents $I_F$, $I_A$ . . . $I_{B0}$ are obtained and from these the sequence currents in each individual network are determined.

For the unfaulted line,

$$I_1 = I_{A1} - I_{B1}, I_2 = I_{A2} - I_{B2} \text{ and } I_0 = I_{A0} - I_{B0}$$

For machine (2), $I_1 = I_F - I_{A1}$, $I_2 = I_F - I_{A2}$

and $\qquad\qquad I_0 = I_F - I_{A0}$

Hence a connexion matrix relating branch and loop sequence currents may be assembled.

It should be noted that for a symmetrical three-phase fault the sequence networks are not interconnected and are as shown in figure 6.24.

*Example* 6.8 Determine the fault currents in the system shown in figure 6.25 in which phase (a) has a combined open circuit and earth fault.

*Solution* In this solution a branch matrix will be formed and a connexion matrix for currents used. All sequence values refer to the faulted phase.

On the 'D' side, $I_{1D} = I_{2D} = I_{0D}$. (In an earth fault the sequence components are equal)

On the 'F' side, $I_{aF} = 0$ (open circuit)

Also, $\qquad\qquad I_{0F} = \frac{1}{3} (I_B + I_C)$

$$I_{1F} = \frac{1}{3} (a I_B + a^2 I_C)$$

$$I_{2F} = \frac{1}{3} (a^2 I_B + a I_C)$$

Hence, $\qquad\qquad I_{0F} + I_{2F} + I_{1F} = I_{aF} = 0$

But $I_{0F} = 0$ as end F is not grounded $\therefore I_{2F} = - I_{1F}$

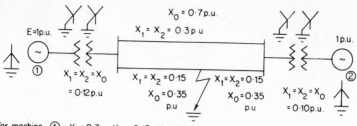

For machine ①, $X_1 = 0.3$, $X_2 = 0.15$ $X_0 = 0.10$

     " ②, $X_1 = 0.25$, $X_2 = 0.15$ $X_0 = 0.05$

(a)

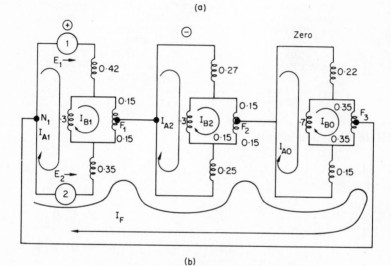

(b)

Figure 6.23a   Line diagram of faulted system
For machine (1), $X_1 = 0.3$, $X_2 = 0.15$, $X_0 = 0.10$
For machine (2), $X_1 = 0.25$, $X_2 = 0.15$, $X_0 = 0.05$
All values are p.u. on a common base.
b   Equivalent symmetrical component circuit diagram for the faulted
system of 6.23a.

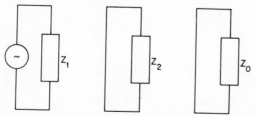

Figure 6.24  Sequence networks for a three-phase symmetrical
fault.

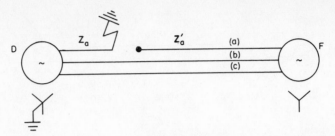

Figure 6.25   Power system with combined earth and open circuit
faults for example 6.8.

Reactances—per unit on a common base—resistance neglected.

Generator D                          Impedances of line D to fault $Z_a$
  $Z_1$   0·3 p.u.                      $Z_1$   0·1 p.u.
  $Z_2$   0·2 p.u.                      $Z_2$   0·1 p.u.
  $Z_0$   0·05 p.u.                     $Z_0$   0·2 p.u.
Generator F                          Impedances of line F to fault $Z_a$
  $Z_1$   0·25 p.u.                     $Z_1$   0·05 p.u.
  $Z_2$   0·15 p.u.                     $Z_2$   0·05 p.u.
  $Z_0$   0·03 p.u.                     $Z_0$   0·1 p.u.

Hence connexion matrix [**C**] is;

$$
\begin{bmatrix} I_{0D} \\ I_{1D} \\ I_{1F} \\ I_{2D} \\ I_{2F} \end{bmatrix}
=
\begin{bmatrix} 1 & 0 \\ 1 & 0 \\ 0 & 1 \\ 1 & 0 \\ 0 & -1 \end{bmatrix}
\begin{bmatrix} I_{0D} \\ I_{1F} \end{bmatrix}
\quad \text{and} \quad
[\mathbf{C_t}] =
\begin{bmatrix} 1 & 1 & 0 & 1 & 0 \\ 0 & 0 & 1 & 0 & -1 \end{bmatrix}
$$

$$
v = [\mathbf{C_t}][\mathbf{V}] = [\mathbf{C_t}]
\begin{bmatrix} 0 \\ E_D \\ E_F \\ 0 \\ 0 \end{bmatrix}
=
\begin{bmatrix} E_D \\ E_F \end{bmatrix}
$$

$$
\text{Branch matrix, } \mathbf{Z_{br}} =
\begin{bmatrix}
Z_{0D} & 0 & 0 & 0 & 0 \\
0 & Z_{1D} & 0 & 0 & 0 \\
0 & 0 & Z_{1F} & 0 & 0 \\
0 & 0 & 0 & Z_{2D} & 0 \\
0 & 0 & 0 & 0 & Z_{2F}
\end{bmatrix}
$$

and $[C_t]\,[Z_{br}]\,[C] = [Z] = \begin{bmatrix} Z_{0D} + Z_{1D} + Z_{2D} & 0 \\ 0 & Z_{1F} + Z_{2F} \end{bmatrix}$

As, $[v] = [Z]\,[i]$

$$\begin{bmatrix} E_D \\ E_F \end{bmatrix} = \begin{bmatrix} (Z_{0D} + Z_{1D} + Z_{2D}) & 0 \\ 0 & (Z_{1F} + Z_{2F}) \end{bmatrix} \cdot \begin{bmatrix} I_{0D} \\ I_{1F} \end{bmatrix}$$

$$\begin{bmatrix} 1\cdot4 \\ 1\cdot4 \end{bmatrix} = \begin{bmatrix} j0\cdot95 & 0 \\ 0 & j0\cdot5 \end{bmatrix} \cdot \begin{bmatrix} I_{0D} \\ I_{1F} \end{bmatrix}$$

$$I_{1D} = I_{2D} = I_{0D} = \frac{1\cdot4}{j0\cdot95} = -1\cdot475\,j \text{ p.u.}$$

Current magnitude into fault from end D $= 3 \times 1\cdot475 = 4\cdot425$ p.u.
Current into fault from end F $= I_{1F} - I_{2F} = 0$

*Note*  This method can be used for any type of fault, and although it does not use the interconnected sequence networks as such it uses the constraints which are necessary for the formation of such networks. For example, in the system shown in figure 6.23c a connexion matrix [C] can be formed knowing that for an earth-fault current $I_1 = I_2 = I_0$.

### 6.8  Power in Symmetrical Components

The total power in a three-phase network

$$= V_a\,I_a^* + V_b\,I_b^* + V_C\,I_C^*$$

where $V_a$, $V_b$ and $V_c$ are phase voltages and

$I_a$, $I_b$ and $I_c$ are line currents.

In phase (a),

$$\begin{aligned} P_a + jQ_a &= (V_{a0} + V_{a1} + V_{a2})\,(I_{a0}^* + I_{a1}^* + I_{a2}^*) \\ &= (V_{a0}\,I_{a0}^* + V_{a1}\,I_{a1}^* + V_{a2}\,I_{a2}^*) \\ &\quad + (V_{a0}\,I_{a1}^* + V_{a1}\,I_{a2}^* + V_{a2}\,I_{a0}^*) \\ &\quad + (V_{a0}\,I_{a2}^* + V_{a1}\,I_{a0}^* + V_{a2}\,I_{a1}^*) \end{aligned}$$

with similar expressions for phases (b) and (c).

In extending this to cover the total three-phase power it should be noted that $I_{b1}^* = (a^2\,I_{a1})^* = (a^2)^*\,I_{a1}^* = a\,I_{a1}^*$.
Similarly, $I_{b2}^*$, $I_{c1}^*$ and $I_{c2}^*$ may be replaced.

The total power $= 3\,(V_{a0}\,I_{a0}^* + V_{a1}\,I_{a1}^* + V_{a2}\,I_{a2}^*)$ \hfill (6.16)

i.e. 3 (the sum of the individual sequence powers in any phase).

### 6.9 Neutral Grounding

*Introduction*

From the analysis of unbalanced fault conditions it has been seen that the connexion of the transformer and generator neutrals greatly influences the fault currents and voltages. In most high-voltage systems the neutrals are solidly grounded, i.e. connected directly to the ground with the exception of generators which are grounded through a resistance to limit stator fault currents. The advantages of such grounding are as follows:

(a) Voltages to ground are limited to the phase voltage. ˙
(b) Intermittent ground faults and high voltages due to arcing faults are eliminated.
(c) Sensitive protective relays operated by ground fault currents clear these faults at an early stage.

The main advantage in operating with neutrals isolated is the possibility of maintaining a supply with a ground fault on one line which places the remaining conductors at line voltage above ground. Also interference with

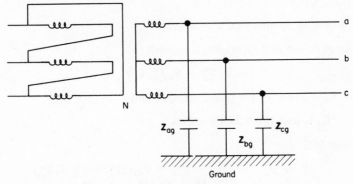

Figure 6.26 Line to ground capacitances in an ungrounded system.

telephone circuits is reduced because of the absence of zero-sequence currents. With normal balanced operation the neutrals of an ungrounded or isolated system are held at ground potential because of the presence of the system capacitance to earth. For the general case shown in figure 6.26 the following analysis applies,

$$\frac{V_{ag}}{Z_{ag}} + \frac{V_{bg}}{Z_{bg}} + \frac{V_{cg}}{Z_{cg}} = 0 \qquad (6.17)$$

Also,    $V_{ag} = V_{an} + V_{ng}$

where    $V_{an}$ = voltage of line a to neutral and

   $V_{ng}$ = voltage of neutral to ground

Similarly    $V_{bg} = V_{bn} + V_{ng}$

and    $V_{cg} = V_{cn} + V_{ng}$

Substituting in equation (6.17) and separating terms,

$$\frac{V_{an}}{Z_{ag}} + \frac{V_{bn}}{Z_{bg}} + \frac{V_{cn}}{Z_{cg}} + V_{ng}\left(\frac{1}{Z_{ag}} + \frac{1}{Z_{bg}} + \frac{1}{Z_{cg}}\right) = 0 \qquad (6.18)$$

the coefficient of $V_{ng}$ gives the ground capacitance admittance of the system.

### Arcing Faults

Consider the single phase system in figure 6.27 at the instant when the instantaneous voltages are $v$ on line (a) and $-v$ on line (b), where $v$ is the maximum instantaneous voltage. The sudden occurrence of a fault to

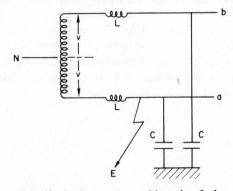

Figure 6.27   Single-phase system with arcing fault to ground.

ground causes line (b) to assume a potential of $-2v$ and (a) to become zero. Because of the presence of both $L$ and $C$ in the circuit the sudden change in voltages by $v$ produces a high frequency oscillation of peak magnitude $2v$ superimposed on the power frequency voltages (see Chapter 10) and line (a) reaches $-v$ and (b) $-3v$ as shown in figure 6.28. These oscillatory voltages attenuate quickly due to the resistance present. The current in the arc to earth on line (a) is approximately 90° ahead of the fundamental voltage and when it is zero the voltage will be at a maximum. Hence if the arc extinguishes at the first current zero the lines remain charged at $-v$ for (a) and $-3v$ for (b). The line potentials now change at power frequency until (a) reaches $-3v$ when the arc could restrike causing a voltage change of $-3v$ to 0 resulting in a transient overvoltage of $+3v$ in (a) and $+5v$ in (b). This process could continue and the voltage build up

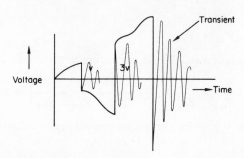

Figure 6.28    Voltage on line 'a' of Figure 6.27.

further but the resistance present usually limits the peak voltage to approximately $4v$. A similar analysis may be made for a three-phase circuit again showing that serious overvoltages may occur with arcing faults.

This condition may be overcome in an isolated neutral system by means of an *arc suppression* or Peterson Coil. The reactance of this coil which is connected between the neutral and ground is made in the range 90 to 110 per cent of the value required to neutralize the capacitance current. The phasor diagram for the network of figure 6.29a is shown in figure 6.29b if the voltage drop across the arc is neglected.

$$I_a = I_b = \sqrt{3}V\omega C$$

and
$$I_a + I_b = \sqrt{3} \times \sqrt{3}V\omega C$$

also,
$$I_L = \frac{V}{\omega L}$$

For compensation of the arc current,

$$\frac{V}{\omega L} = 3V\omega C$$

$\therefore$
$$L = \frac{1}{3\omega^2 C}$$

and
$$X_L = \frac{1}{3\omega C} \tag{6.19}$$

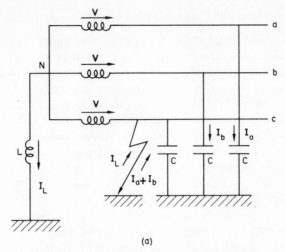

(a)

Figure 6.29a    System with arc suppression coil.

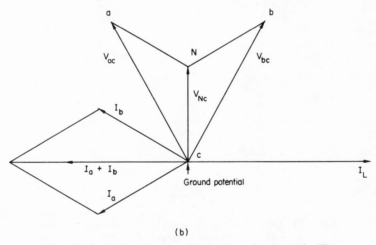

(b)

Figure 6.29b    Phasor diagram of voltages and currents in Figure 6.29a.

This result may be obtained by analysis of the ground fault by means of symmetrical components.

Generally isolated neutral systems give rise to serious arcing-fault voltages if the arc current exceeds the region 5 to 10 A which covers most systems operating above 33 kV. If such systems are to be operated with isolated neutrals arc-suppression coils should be used. Most systems at normal transmission voltages have grounded neutrals.

## 6.10 Interference with Communication Circuits

When power and telephone lines run in parallel under certain conditions voltages sufficient to cause high noise levels may be induced into the communication circuits. This may be caused by electromagnetic and electrostatic unbalance in the power lines especially if harmonics are present. The major problem, however, is due to ground faults producing large zero-sequence currents in the power line which inductively induce voltage into the neighbouring circuit. The value of induced voltage depends on the spacing, resistivity of the earth immediately below and the frequency. The induced voltage is induced into each of the communication wires so that if the latter are perfectly transposed no voltage would exist between them. However, the voltage would exist between the wire and earth. This voltage is kept low by the use of a 'drainage' coil connected between the wires and earthed at its electrical midpoint and which shunts to earth the longitudinal induced voltages but gives small attenuation at the communication frequencies.

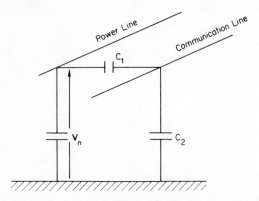

Figure 6.30  Induced voltage between power and communication circuits—induced longitudinal voltage

$$= V_n \frac{C_1}{C_1 + C_2}$$

There are two main groups of communication circuits, (a) telegraph circuits with signal frequencies up to 300 Hz and usually single line with ground return, and (b) telephone circuits with signal frequencies of 100–400 Hz and always double-line circuits. Interference from underground cable circuits is much less (10 per cent) than that from overhead lines.

## References

BOOKS

1. *Electrical Transmission and Distribution Reference Book*, Westinghouse Electric Corp., East Pittsburgh, Pennsylvania, 1964.
2. El-Abiad, A. H., and G. W. Stagg, *Computer Methods in Power Systems Analysis*, McGraw-Hill, New York, 1968.
3. Clarke, E., *Circuit Analysis of Alternating Current Power Systems*, Vol. 1, Wiley, New York, 1943.
4. Wagner, C. F., and R. D. Evans, *Symmetrical Components*, McGraw-Hill, New York, 1933.
5. Austin Stigant, S., *Master Equations and Tables for Symmetrical Component Fault Studies*, McDonald, London, 1964.
6. Kimbark, E. W., *Power System Stability*, Vol. I, Wiley, New York, 1948.


PAPERS

7. Fortescue, C. L., 'Method of Symmetrical Co-ordinates Applied to the Solution of Polyphase Networks', *Trans. A.I.E.E.*, **37**, Part II (1918).
8. Lantz, M. J., 'Digital short-circuit solution of power system networks including mutual impedance', *Trans. A.I.E.E.*, **76**, Part III, 1230 (1957).
9. Siegel, J. C., and G. W. Bills, 'Nodal representation of large complex-element networks including mutual reactances', *Trans. A.I.E.E.*, **77**, Part III, 1226 (1958).
10. Brown, H. E., C. E. Person, L. E. Kirchmayer, and G. W. Stagg, 'Digital calculation of three-phase short circuits by matrix method', *Trans. A.I.E.E.*, **79**, Part III, 1277 (1960).
11. El-Abiad, A. H., 'Digital calculation of line-to-ground short circuits by matrix method', *Trans. A.I.E.E.*, **79**, Part III, 323 (1960).
12. Taylor, G. E., G. K. Carter, and E. H. MacDonald, 'New digital computer short-circuit program for relay studies', *Trans. A.I.E.E.*, **79**, Part III, 1257 (1960).
13. El-Abiad, A. H., Ruth Guidone, and G. W. Stagg, 'Calculation of short circuits using a high-speed digital computer', *Trans. A.I.E.E.*, **80**, Part III, 702 (1961).
14. Ender, R. C., G. G. Auer, and R. A. Wylie, 'Digital calculation of sequence impedances and fault currents for radial primary distribution circuits', *Trans. A.I.E.E.*, **79**, Part III, 1264 (1960).
15. Sato, N., and W. F. Tinney, 'Techniques for exploiting the sparcity of the network admittance matrix', *Trans. I.E.E.E.*, *P.A. & S.*, **82**, 944 (1963).
16. Brown, H. E., and C. E. Person, 'Short circuit studies of large systems by the impedance matrix method', *I.E.E.E. Conference Record, P.I.C.A.*, 335 (1967).
17. Trihus, A. T., and R. Zimering, 'A digital computer program for an exhaustive fault study of a large power system network', *I.E.E.E. Conference Record, P.I.C.A.*, 343 (1967).
18. Sebo, S. A., 'Zero Sequence Current Distribution Along Transmission Lines', *Trans. I.E.E.E.*, *P.A. & S.*, **88**, 910 (1969).

**Problems**

6.1. Four 11 kV generators designated A, B, C and D each have a subtransient reactance of 0·1 p.u. and a rating of 50 MVA. They are connected in parallel by means of three 100 MVA reactors which join A to B, B to C and C to D; these reactors have per unit reactances of 0·2, 0·4 and 0·2 respectively. Calculate the voltamperes and the current flowing into a three-phase symmetrical fault on the terminals of machine B. Use a 50 MVA base.

Design a resistance fault-analyser to represent the above system using a supply of 50 V and taking 1 mA to represent a 100 MVA fault level. (940 MVA, 49,300 A.)

6.2. Two 25 MVA, 50 c/s, 11 kV generators each of subtransient reactance $X''$ 0·15 p.u. feed into a busbar designated 'A'. A 30 MVA, 11 kV generator of $X''$ 0·12 p.u. feeds into busbar B which is connected to A through a 50 MVA reactor of 30 per cent reactance. A 20 MVA, 11/66 kV transformer of reactance 10 per cent is connected to A. If a three-phase fault occurs on the high-voltage side of the transformer, calculate the current fed into the fault and the corresponding voltage at A. (1200/$\sqrt{3}$; 8·95kV.)

6.3. Two 60 MVA generators each of transient reactance 0·15 p.u. are connected to a busbar designated 'A'. Two identical machines are connected to another busbar 'B'. A feeder is supplied from A through a step-up transformer rated at 30 MVA with 10 per cent reactance.

Calculate the reactance of a reactor to connect A and B if the fault level due to a three-phase fault on the feeder side of the transformer is to be limited to 240 MVA. Calculate also the voltage on A under this condition if the generator voltage is 13 kV (line). ($X = 0·075$ p.u., $V_A = 10·4$ kV.)

6.4. A 132 kV supply feeds a line of reactance 13 Ω which is connected to a 100 MVA, 132/33 kV transformer of 0·1 p.u. reactance. The transformer feeds a 33 kV line of reactance 6 Ω which in turn is connected to an 80 MVA, 33/11 kV transformer of 0·1 p.u. reactance. This transformer supplies an 11 kV substation from which a local 11 kV feeder of 3 Ω reactance is supplied. This feeder energizes a protective overcurrent relay through 100 A/1 A current transformers. The relay has a true inverse-time characteristic and operates in 10 seconds with a coil current of 10 A.

If a three-phase fault occurs at the load end of the 11 kV feeder calculate the fault current and time of operation of the relay. (1575 A, 6·35 seconds.)

6.5. A ring-main system consists of a number of substations designated A, B, C, D and E connected by transmission lines having the following impedances per phase (ohms): AB (1·5+j2); BC (1·5+j2); CD (1+j1·5); DE (3+j4); EA (1+j1).

The system is fed at A at 33 kV from a source of negligible impedance. At each substation except A the circuit breakers are controlled by relays fed from 1500/5 A current transformers. At A the current transformer ratio is 4000/5. The characteristics of the relays are as follows:

| Current (A) | | 7 | 9 | 11 | 15 | 20 |
|---|---|---|---|---|---|---|
| Operating | Relays at A, D and C | 3·1 | 1·95 | 1·37 | 0·97 | 0·78 |
| time(s) | Relays at B and E | 4 | 2·55 | 1·8 | 1·27 | 1·02 |

Examine the sequence of operation of the protective gear for a three-phase symmetrical fault at the mid-point of line CD.

Assume that the primary current of the current transformer at A is the total fault current to the ring and that each circuit breaker opens 0·3 seconds after the closing of the trip-coil circuit. Comment on the disadvantages of this system.

6.6. The following currents were recorded under fault conditions in a three-phase system:

$$I_R = 1500 \; \underline{/45°} \; \text{A}, \qquad I_Y = 2500 \; \underline{/150°} \; \text{A}, \qquad I_B = 1000 \; \underline{/300°} \; \text{A}.$$

If the phase sequence is $R-Y-B$, calculate the values of the positive, negative and zero phase sequence components for each line. ($I_0 = (-200+j480)\text{A}$.)

6.7. A single line-to-earth fault occurs on the red phase at the load end of a 66 kV transmission line. The line is fed via a transformer by 11 kV generators connected to a common busbar. The line-side of the transformer is connected in star with the star point earthed and the generator side is in delta. The positive sequence reactances of the transformer and line are $j10·9 \; \Omega$ and $j44 \; \Omega$ respectively and the equivalent positive and negative sequence reactances of the generators, referred to the line voltage, are $j18 \; \Omega$ and $j14·5 \; \Omega$ respectively. Measured up to the fault the total effective zero-sequence reactance is $j150 \; \Omega$. Calculate the fault current in the lines if resistance may be neglected. If a two-phase to earth fault occurs between the blue and yellow lines, calculate the current in the yellow phase. (392 A and 488 A.)

6.8. A single line-to-earth fault occurs in a radial transmission system. The following sequence impedances exist between the source of supply (an infinite busbar) of voltage 1 p.u. to the point of the fault; $Z_1 = 0·3+j0·6$ p.u., $Z_2 = 0·3+j0·55$ p.u., $Z_0 = 1+j0·78$ p.u. The fault path to earth has a resistance of 0·66 p.u. Determine the fault current and the voltage at the point of the fault. ($I_f = 0·732$ p.u., $V_f = (0·43-j0·23)$ p.u.)

6.9. In the system described in Example 6·5 determine the sequence components of the current fed into a double-line to earth fault when the star point of generator B is also connected to earth. Check the result using the fact that $I_R = 0$.

6.10. In the system shown in figure 6.19 the generation connected to the 132 kV busbar is only used under maximum load conditions. Calculate the fault levels for three-phase faults at each busbar with maximum-plant infeed reactance and with and without the generation connected to the 132 kV busbars.

6.11. Develop an expression, in terms of the generated e.m.f. and the sequence impedances, for the fault current when an earth fault occurs on phase 'a' of a 3-phase generator, with an earthed star point. Show also that the voltage to earth of the sound phase 'B' at the point of fault is given by

$$V_\text{B} = \frac{-j\sqrt{3} \; E_\text{A} \; |Z_2 - aZ_0|}{Z_1 + Z_2 + Z_0}$$

Two 30 MVA, 6·6 kV synchronous generators are connected in parallel and supply a 6·6 kV feeder. One generator has its star point earthed through a resistor of 0·4 $\Omega$ and the other has its star point isolated. Determine, (1) the fault current and the power dissipated in the earthing resistor, when an earth fault occurs at the far end of the feeder on phase 'A' and (2) the voltage to earth of phase 'B'. The generator phase sequence is ABC and the relative impedances are

|                                   | Generator p.u./ph | Feeder Ω/ph |
|-----------------------------------|-------------------|-------------|
| To positive sequence currents     | j0·2              | j0·6        |
| To negative sequence currents     | j0·16             | j0·6        |
| To zero sequence currents         | j0·06             | j0·4        |

Use a base of 30 MVA.

Answer:                     5459/−52·6° A, 11·92 MW, (−1402 + j2770) V.

6.12. A 33 kV, 60 Hz overhead line has a capacitance to ground per phase of approximately 0·7 μF. Calculate the reactance of a suitable arc-suppression coil. Answer: 1315 Ω

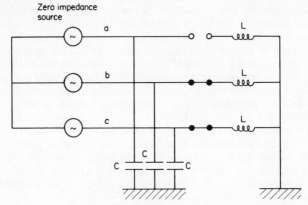

Figure 6.31   System for Problem 6.13.

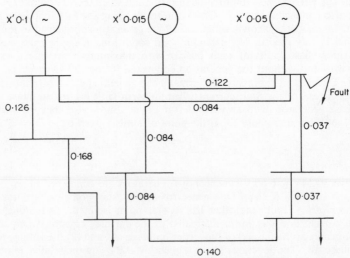

Figure 6.32   System for Problem 6.14.

6.13. The system shown in figure 6.31 has an open circuit on line (a). Determine the potential between the source neutral and ground. The phase voltage is $V$ and angular frequency $\omega$. Investigate the peculiarities arising if $\omega^2 LC < 2/3$ and $\omega^2 LC > 2/3$.

Answer: $\dfrac{V}{3\omega^2 LC - 2}$

6.14. In the system shown in figure 6.32 all reactances are in per unit. Determine the p.u. fault current for a three-phase balanced fault on the terminals of the machine of reactance 0·05 p.u.

Answer: 29 p.u.

(*Note* This system is large enough to require programming and use of a computer.)

# 7

# Stability Limits

## 7.1 Introduction

The stability of a system of interconnected dynamic components is its ability to return to normal or stable operation after having been subjected to some form of disturbance. The study of stability is one of the main concerns of the control engineer whose methods may be applied to electric power-systems.

When the rotor of a synchronous generator advances beyond a certain critical angle, the magnetic coupling between the rotor (and hence the turbine) and the stator fails. The rotor, no longer held in synchronism with the rotating field of the stator currents, rotates relative to the field and pole slipping occurs. Each time the poles traverse the angular region where stability obtains, synchronizing forces attempt to pull the rotor into synchronism. It is general practice to disconnect the machine from the system if it commences to slip poles. However a generator having lost synchronism may operate successfully as an induction generator for some time and then be re-synchronized; the possibility of allowing controlled pole-slipping for limited periods is thus being investigated. It should be remembered however that an induction generator takes its excitation requirements from the network which must be capable of supplying the requisite reactive power.

There are two forms of instability in power systems; the loss of synchronism between synchronous machines, and the stalling of asynchronous loads. Synchronous stability may be divided into two regimes, steady-state and transient. Reference to the former has already been made when discussing synchronous-machine characteristics. It is basically the ability of the power system when operating under given load conditions to retain synchronism when subject to *small* disturbances such as the continual changes in load or generation and the switching out of lines. It is most likely to result from the changes in source-to-load impedance resulting from changes in the network configuration.

Transient stability is concerned with sudden and large changes in the network condition such as brought about by faults. The maximum power transmittable, the stability limit, is less than that for the corresponding steady-state condition.

The stability of an asynchronous load is controlled by the voltage across it; if this becomes lower than a critical value, induction-motors may become unstable and stall. This is, in effect, the voltage instability problem already mentioned. In a power system it is possible for either synchronous or load instability to occur. The former is more probable and hence has been given much more attention.

## 7.2 Equation of Motion of a Rotating Machine

Before the equation is considered a revision of the definitions of certain quantities is given. The kinetic energy absorbed by a rotating mass $= \frac{1}{2}I\omega^2$ Joules. The angular momentum, $M = I\omega$ Joule-seconds per radian where $\omega$ is the synchronous speed of the rotor (radians/sec) and $I$ is the moment of inertia (kilogram-metre$^2$). The inertia constant ($H$) is defined as the stored energy at synchronous speed per volt-ampere of the rating of the machine. In power systems the unit of energy is taken as the kilojoule or megajoule. If $G$ MVA is the rating of the machine, then

$$GH = \text{stored energy (Megajoules)}$$
$$= \tfrac{1}{2}I\omega^2 = \tfrac{1}{2}M\omega \text{ and as } \omega = 180/\pi \text{ (pole pairs) or } 360f$$
electrical degrees per second,

$$GH = \tfrac{1}{2}M\,(360f)$$

$\therefore$     $M = GH/180f$ Megajoule-seconds/electrical degree.

The inertia constant $H$ for steam turbogenerators decreases from 10 kW-sec/kVA for machines up to 30 MVA to values in the order of 4 kW-sec/kVA for large machines, the value decreasing as the capacity increases. For salient-pole water-wheel machines $H$ depends on the number of poles; for machines in the range 200–400 r.p.m. the value increases from about 2 kW-sec/kVA at 10 MVA rating to 3·5 at 60 MVA. A mean value for synchronous motors is 2 kW-sec/kVA. The nett accelerating torque on the rotor of a machine, $\Delta T =$ mechanical torque input $-$ electrical torque output $= I d^2\delta/dt^2$

$$\therefore \quad \frac{d^2\delta}{dt^2} = \frac{\Delta T}{I} = \frac{(\Delta T\omega)\omega}{2 \times I\omega^2/2}$$

$$= \frac{\Delta P \cdot \omega}{2 \times \text{kinetic energy}}$$

where $\Delta P =$ nett power corresponding to $\Delta T$, i.e. $P_{mech} - P_{elect}$.

$$\therefore \qquad \frac{d^2\delta}{dt^2} = \frac{\Delta P}{M} \qquad (7.1)$$

In equation (7.1) a negative change in power output results in an increase in $\delta$. Sometimes $\Delta P$ is considered as the change in *electrical power output* and increase in $\Delta P_{el.}$ results in increase in angle $\delta$. As the power input is assumed constant the equation of motion now becomes,

$$\frac{d^2\delta}{dt^2} = -\frac{\Delta P}{M} \quad \text{or} \quad M\frac{d^2\delta}{dt^2} + \Delta P = 0.$$

### 7.3 * Steady-State Stability—Theoretical Considerations

The power system forms a group of interconnected electromechanical elements the motion of which may be represented by the appropriate differential equations. With large disturbances in the system the equations are non-linear, but with small changes the equations may be linearized

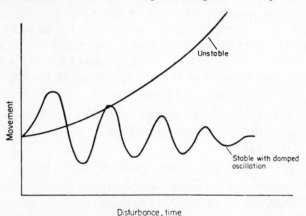

Figure 7.1 Types of response to a disturbance on a system.

with little loss of accuracy. The differential equations having been determined the characteristic equation of the system is formed from which information regarding stability is obtained. The solution of the differential equation of the motion is of the form

$$\delta = k_1 e^{a_1 t} + k_2 e^{a_2 t} + \ldots k_n e^{a_n t}$$

where $k_1, k_2 \ldots k_n$ are constants of integration and $a_1, a_2 \ldots a_n$ are the roots of the characteristic equation. If any of the roots have positive real terms then the quantity $\delta$ increases continuously with time and the original

* This section may be omitted by readers without a basic knowledge of control theory.

steady condition is not re-established. The criterion for stability is therefore that all the real parts of the roots of the characteristic equation be negative; imaginary parts indicate the presence of oscillation. Figure 7.1 shows the various types of motion. The determination of the roots is often difficult and tedious and indirect methods for predicting stability have been established, for example the Hurwitz–Routh criterion in which stability is predicted without the actual solution of the characteristic equation. No information regarding the degree of stability or instability is obtained, only that the system is, or is not, stable. One advantage of this approach is that the characteristics of the control loops associated with governors and automatic voltage regulators may be incorporated in the general treatment.

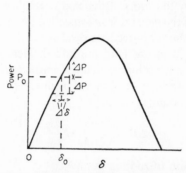

Figure 7.2 Small disturbance—initial operation on power-angle curve at $P_0$, $\delta_0$. Linear movement assumed about $P_0$, $\delta_0$.

For a generator connected to an infinite busbar through a network of zero resistance, with operation at $P_0$ and $\delta_0$ (figure 7.2),

$$M\frac{\mathrm{d}^2\Delta\delta}{\mathrm{d}t^2} = -\Delta P = -\Delta\delta\left(\frac{\partial P}{\partial\delta}\right)_0,$$

where change in $P$ causing increase in $\delta$ is positive and refers to small changes in the load angle $\delta$ such that linearity may be assumed.

$$\therefore \qquad Mp^2\Delta\delta + \left(\frac{\partial P}{\partial\delta}\right)_0 \Delta\delta = 0 \qquad (7.2)$$

where

$$p \equiv \mathrm{d}/\mathrm{d}t.$$

$Mp^2 + (\partial P/\partial\delta)_0 = 0$ is the characteristic equation which has two roots $\pm\sqrt{\dfrac{-(\partial P/\partial\delta)_0}{M}}$. When $(\partial P/\partial\delta)_0$ is positive both roots are imaginary and the motion is oscillatory and undamped; when $(\partial P/\partial\delta)_0$ is negative both roots are real, and positive and negative respectively and stability is lost. At $\delta = 90°$, $(\partial P/\partial \delta)_{90} = 0$ and the system is at the limit. If damping is accounted for the equation becomes,

$$Mp^2\Delta\delta + K_d p\Delta\delta + \left(\frac{\partial P}{\partial \delta}\right)\Delta\delta = 0 \qquad (7.3)$$

and the characteristic equation is,

$$Mp^2 + K_d p + \left(\frac{\partial P}{\partial \delta}\right) = 0 \qquad (7.4)$$

where $K_d$ is the damping coefficient.

Again if $(\partial P/\partial \delta)$ is negative, stability is lost. The frequency of the oscillation is given by the roots of the characteristic equation.

If the excitation of the generator is controlled by a fast-acting automatic voltage regulator without appreciable dead zone, the excitation voltage $E$ is increased as increments of load are added. Hence the actual power-angle curve pertaining is no longer that for constant $E$ (refer to chapter 2) and the change of power may be obtained by linearizing the $P$–$V$ characteristic at the operating point (1), when

$$\Delta P = \left(\frac{\partial P}{\partial E}\right)_1 \Delta E$$

The complete equation of motion is now,

$$Mp^2\Delta\delta + K_d p\Delta\delta + \left(\frac{\partial P}{\partial \delta}\right)_1 \Delta\delta + \left(\frac{\partial P}{\partial E}\right)_1 \Delta E = 0 \qquad (7.5)$$

Without automatic voltage control the stability limit is reached when $\delta = 90°$; with control the criterion is obtained from the characteristic equation of (7.5).

*Example 7.1* A synchronous generator of reactance 1·5 per unit is connected to an infinite busbar system ($V = 1$ per unit) through a line and transformers of total reactance 0·5 per unit. The no-load voltage of the generator is 1·1 per unit and the inertia constant $H = 5$ MW-sec per MVA. All per unit values are expressed on the same base; resistance and damping may be neglected. Calculate the frequency of the oscillations set up when the generator operates at a load angle of 60° and is subjected to a small disturbance. The system frequency is 50 Hz.

*Solution* The nature of the movement is governed by the sign of the quantity under the root sign in the equation for $p_1$ and $p_2$ (7.4). This changes when $k_d^2 = 4M(\partial P/\partial \delta)_1$; in this case $k_d = 0$. The roots of the characteristic equations give the frequency of oscillation; when $\delta_0 = 60°$,

$$\left(\frac{\partial P}{\partial \delta}\right)_{60°} = \frac{1 \cdot 1 \times 1}{2} \cos 60$$

$$= 0 \cdot 275$$

$$p_1 \text{ and } p_2 = \pm j \sqrt{\left[\left(\frac{\partial P}{\partial \delta}\right) \cdot \frac{1}{M}\right]}$$

$$= \pm j \sqrt{\left[\frac{0 \cdot 275}{5 \times (1/\pi \times 50)}\right]}$$

$$= \pm j \sqrt{8 \cdot 62}$$

$\therefore$  frequency of oscillation

$$= 2 \cdot 94 \text{ rad/sec} = \frac{2 \cdot 94}{2\pi} \text{ Hz}$$

$$= 0 \cdot 468 \text{ Hz}$$

and the periodic time

$$= \frac{1}{0 \cdot 468} = 2 \cdot 14 \text{ sec.}$$

*The stability of a two machine system*    The equivalent circuit will be as shown in figure 2.6 with $M_1$ and $M_2$ referring to machines 1 and 2 respectively.

The equations of motion are (for small changes),

$$M_1 p^2 \Delta\delta_1 + \left(\frac{\partial P_1}{\partial \delta_{12}}\right)\Delta\delta_{12} = 0$$

and

$$M_2 p^2 \Delta\delta_2 + \left(\frac{\partial P_2}{\partial \delta_{12}}\right)\Delta\delta_{12} = 0$$

As

$$\Delta\delta_1 - \Delta\delta_2 = \Delta\delta_{12},$$

$$p^2 \Delta\delta_{12} + \left[\frac{(\partial P_1/\partial \delta_{12})}{M_1} - \frac{(\partial P_2/\partial \delta_{12})}{M_2}\right]\Delta\delta_{12} = 0$$

the characteristic equation has two roots,

$$p_1, p_2 = \pm j \sqrt{\left[\frac{(\partial P_1/\partial \delta_{12})}{M_1} - \frac{(\partial P_2/\partial \delta_{12})}{M_2}\right]}$$

Stability is assured if the quantity under the square root is positive. Hence the stability limit for small disturbances is not the same as the maximum power limit discussed in chapter 2 and is in fact always larger. The difference however is never large and when one machine is effectively an infinite busbar the two limits coincide.

*Effect of governor action*  In the above analysis the oscillations set up with small changes in load on a system have been considered and the effects of governor operation ignored. After a certain time has elapsed the governor

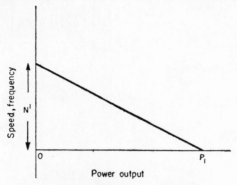

Figure 7.3    Idealized governor characteristic.

control characteristics commence to influence the powers and oscillations and this will now be considered. The basic nature of governor control has been described in chapter 3 and in figure 7.3 the idealized speed-load characteristic is shown. The treatment below is based on the method given by Rudenburg (reference 8).

Consider a governor system with a time-delay constant $\tau_g$ and a speed droop of $N_1$ rads/sec from no-load to full load ($P_1$). $\delta$ is the change in rotor operating angle, and

$$\frac{d\delta_1}{dt} - \frac{d\delta_2}{dt} = \frac{d(\delta_1 - \delta_2)}{dt} = \frac{d\delta}{dt}$$

where $d\delta/dt$ is the speed change due to $\Delta P$ and $(d\delta/dt)(1/\omega)$ is the per unit value or slip where $\omega$ is the synchronous speed.

The governor causes the steam input to the turbine to change with time according to the speed deviation, i.e.

$$\frac{d\Delta P}{dt} = \frac{P_1}{N_1\tau_g}\frac{d\delta}{dt}\frac{1}{\omega}$$

($\tau_g$ lies between 0·5 and 10 seconds).

The pilot-valve movement is influenced by the supplementary return (figure 3.1) derived from the main servo-motor position, i.e. influenced by the power.

$$\therefore \qquad \frac{d\Delta P}{dt} = \frac{P_1}{N_1 \tau_g} \frac{d\delta}{dt} \frac{1}{\omega} - \frac{\Delta P}{\tau_g}$$

or

$$\tau_g \frac{d(\Delta P)}{dt} + \Delta P = \frac{P_1}{N_1} \cdot \frac{d\delta}{dt} \cdot \frac{1}{\omega} \qquad (7.6)$$

For a machine connected to an infinite busbar

$$\frac{dP}{d\delta} = \text{synchronizing power coefficient}$$

$$= k_s$$

and

$$M \frac{d^2\delta}{dt^2} + k_s\delta + \Delta P = 0$$

from which,

$$\frac{d(\Delta P)}{dt} = -M \frac{d^3\delta}{dt^3} - k_s \frac{d\delta}{dt} \qquad (7.7)$$

Also

$$\Delta P = -k_s\delta - M \frac{d^2\delta}{dt^2} \qquad (7.8)$$

From 7.8, 7.7 and 7.6

$$\frac{d^3\delta}{dt^3} + \frac{d^2\delta}{dt^2} \cdot \frac{1}{\tau_g} + \left( \frac{k_s}{M} + \frac{P_1}{\omega N_1 \tau_g M} \right) \frac{d\delta}{dt} + \frac{k_s \cdot \delta}{M\tau_g} = 0 \qquad (7.9)$$

The condition for no governor action is represented by $\tau_g \to \infty$ when equation (7.9) becomes,

$$\frac{d^3\delta}{dt^3} + \frac{k_s}{M} \cdot \frac{d\delta}{dt} = 0$$

or

$$\frac{d^2\delta}{dt^2} + \frac{k_s}{M} = 0,$$

i.e. equation (7.2).

The stability of this system may be determined by the Hurwitz–Routh criterion. If the characteristic equation of a system can be expressed in the form,

$$a_0 p^n + a_1 p^{n-1} + a_2 p^{n-2} + a_3 p^{n-3} + \ldots + a_{n-1} p + a_n = 0,$$

then the determinant

$$\begin{vmatrix} a_1 & a_0 & 0 & 0 & \ldots & 0 & 0 & 0 \\ a_3 & a_2 & a_1 & a_0 & \ldots & . & . & . \\ a_5 & a_4 & a_3 & a_1 & \ldots & . & . & . \\ . & . & . & . & & . & . & \\ . & . & . & . & & . & . & . \\ . & . & . & . & & a_n & a_{n-1} & a_{n-2} \\ 0 & 0 & 0 & 0 & & 0 & 0 & a_n \end{vmatrix} > 0$$

for stability, i.e. the roots of the characteristic equation have negative real parts. Removing the last row and column the remaining determinant must also be $> 0$ for stability and so on, i e.

$$\begin{vmatrix} a_1 & a_0 & 0 \\ a_3 & a_2 & a_1 \\ a_5 & a_4 & a_3 \end{vmatrix} > 0; \qquad \begin{vmatrix} a_1 & a_0 \\ a_3 & a_2 \end{vmatrix} > 0; \qquad a_1 > 0.$$

In equation (7.9),

$$a_0 = 1$$

i.e. is positive,

$$a_1 = \frac{1}{\tau_g}$$

i.e. is positive,

$$a_1 a_2 = \frac{1}{\tau_g} \left( \frac{k_s N_1 \tau_g + P_1}{\omega N_1 \tau_g M} \right)$$

$$a_3 a_0 = \frac{k_s}{M \tau_g} . 1.$$

For stability,

$$a_1 a_2 - a_3 a_0 > 0,$$

i.e.

$$P_1 + k_s \omega N_1 \tau_g > k_s \tau_g \omega N_1$$

$\therefore$             $$P_1 > k_s \tau_g N_1 (\omega - \omega), \quad \text{i.e.} > 0,$$

i.e. the system is always stable for this simple (but not always representative) governor system.

## 7.4 Steady-State Stability—Practical Considerations

The steady-state limit is the maximum power that can be transmitted in a network between sources and loads when the system is subject to small disturbances. The power system is of course constantly subjected to small changes as load variations occur. To obtain the limiting value of power small increments of load are added to the system; after each increment the generator excitations are adjusted to maintain constant terminal voltages and a load-flow is carried out. Eventually a condition of instability is reached.

The stability limits of synchronous machines have been discussed in chapter 2. It was seen that provided the generator operates within the 'safe area' of the performance chart, stability is assured; usually a 20 per cent margin of safety is allowed and the limit is extended by the use of automatic voltage regulators. Often the performance charts are not used directly and the generator equivalent-circuit employing the synchronous impedance is used. The normal operating load-angle for modern machines is in the order of 60 electrical degrees and for the limiting value of 90° this leaves 30° to cover the transmission network. In a complex system a reference point must be taken from which the load angles are measured; this is usually a point where the direction of power flow reverses.

The simplest criterion for steady-state synchronous stability is, $(\partial P/\partial\delta)>0$, i.e. the synchronizing coefficient must be positive. The use of this criterion involves the following assumptions:

(a) generators are represented by constant impedances in series with the no-load voltages;
(b) the input torques from the turbines are constant;
(c) changes in speed are ignored;
(d) electromagnetic damping in the generators is ignored;
(e) the changes in load angle $\delta$ are small.

The degree of complexity to which the analysis is taken has to be decided, for example the effects of machine inertia, governor action and automatic voltage-regulators can be included; these items however greatly increase the complexity of the calculations. The use of the criterion $(\partial P/\partial\delta) = 0$ alone gives a pessimistic or low result and hence an inbuilt factor of safety.

In a system with several generators and loads, the question as to where the increment of load is to be applied is important. A conservative method is to assume the increment applied to one machine only, determine the stability and then repeat for each of the other machines in turn. Alternatively the power outputs from all but the two generators having the largest load angles are kept constant.

For calculations made without the aid of computers it is usual to reduce the network to the simplest form which will keep intact the generator nodes. The values of load angle, power and voltage are then calculated for the given conditions, $\partial P/\partial\delta$ determined for each machine and if positive the loading is increased and the process repeated.

In a system consisting of a generator supplying a load through a network of lines and transformers of effective reactance $X_T$ the value of $(\mathrm{d}P/\mathrm{d}\delta) = (EV/X_T)\cos\delta$, where $E$ and $V$ are the supply and receiving end voltages and $\delta$ the total angle between the generator rotor and the phasor of $V$.

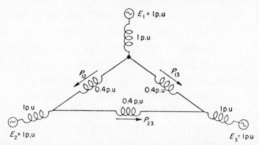

Figure 7.4   Three-machine system for Example 7.2. The output of machine (2) is constant. All values are per unit on the same voltage and MVA bases. $P_{12}$, $P_{13}$ and $P_{23}$ are power flows in the lines.

The power transmitted is obviously increased with higher system voltages and lower reactances and it may be readily shown that line capacitance increases the stability limit. The determination of $\mathrm{d}P/\mathrm{d}\delta$ is not very difficult if the voltages at the loads can be assumed constant or if the loads can be represented by impedances. Use can be made of the $P{-}V$, $Q{-}V$ characteristics of the load if the voltages change appreciably with the redistribution of the power in the network; this process, however, is extremely tedious.

*Example 7.2*   A system consisting of three interconnected synchronous machines is shown in the line diagram of figure 7.4. It is required to investigate the steady-state stability for the following operating load angles (in electrical degrees).

$$\delta_{12} = 90°, \quad \delta_{13} = 30°, \quad \delta_{23} = 60°.$$

The reactance of each machine is j per unit and the generated e.m.f. or no-load voltage is 1 per unit. The resistance of the machines and lines is negligible.

*Solution*   It is necessary to determine first the transfer reactances of the network. As the network is entirely composed of reactance for clarity the j's will be omitted. The system can be represented by the equivalent circuit shown in figure 7.5a when machine (1) is operative and the remaining machines are represented by their reactances only. Figure 7.5a can be reduced to figure 7.5b by applying the delta-star transformation on the mesh of 0·4 per unit reactances. The reactance of each branch of the star is

$$\frac{0{\cdot}4 \times 0{\cdot}4}{0{\cdot}4 + 0{\cdot}4 + 0{\cdot}4}$$

i.e.

$$0{\cdot}133 \text{ per unit.}$$

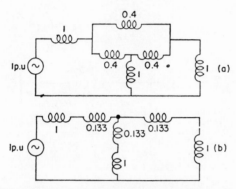

Figure 7.5a   Application of superposition to obtain the transfer reactances of the system in figure 7.4.

Figure 7.5b   Network of 7.5a reduced by transformation.

The current from machine (1) to this circuit

$$= \frac{1}{1 + 0{\cdot}133 + 0{\cdot}565}$$

$$= 0{\cdot}59 \text{ per unit.}$$

The current in the reactance of machine (2) due to (1),

$$I_{12} = 0{\cdot}295 \text{ per unit.}$$

Similarly

$$I_{13} = 0{\cdot}295 \text{ per unit.}$$

Hence,

$$X_{12} = \text{transfer reactance from (1) to (2)}$$

$$= 3{\cdot}39 \text{ per unit}$$

and similarly

$$X_{13} = 3{\cdot}39 \text{ per unit.}$$

It is evident from inspection of the circuit that,

$$X_{23} = 3{\cdot}39 \text{ per unit.}$$

$$\frac{\mathrm{d}P_{12}}{\mathrm{d}\delta_{12}} = \frac{E_1 E_2}{X_{12}} \cos \delta_{12},$$

similarly for $\delta_{23}$ and $\delta_{31}$.

The output of (1)

$$P_1 = P_{12} + P_{13}$$

and

$$\mathrm{d}P_1 = \frac{\partial P_{12}}{\partial \delta_{12}}\mathrm{d}\delta_{12} + \frac{\partial P_{13}}{\partial \delta_{13}}\mathrm{d}\delta_{13}$$

$$= \frac{E_1 E_2}{X_{12}} \cos \delta_{12} \cdot \mathrm{d}\delta_{12} + \frac{E_1 E_3}{X_{13}} \cos \delta_{13}\mathrm{d}\delta_{13}.$$

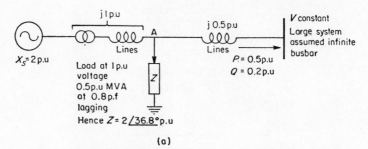

(a)

Figure 7.6a   Line diagram of system for Example 7.3.

Similar expressions are obtained for $\mathrm{d}P_2$ and $\mathrm{d}P_3$; $P_2$ is kept constant and hence $\mathrm{d}P_2 = 0$.

$$\delta_{12} = 90^\circ, \quad \delta_{13} = 30^\circ, \quad \delta_{23} = 60^\circ$$

$$\mathrm{d}P_1 = \frac{1 \times 1}{3{\cdot}39} \cos 90^\circ\, \mathrm{d}\delta_{12} + \frac{1 \times 1}{3{\cdot}39} \cos 30^\circ\, \mathrm{d}\delta_{13}$$

$$0 = \frac{1 \times 1}{3{\cdot}39} \cos 90^\circ\, \mathrm{d}\delta_{12} + \frac{1 \times 1}{3{\cdot}39} \cos 60^\circ\, \mathrm{d}\delta_{23}$$

$$\mathrm{d}P_3 = \frac{1 \times 1}{3{\cdot}39} \cos 30^\circ\, \mathrm{d}\delta_{13} + \frac{1 \times 1}{3{\cdot}39} \cos 60^\circ\, \mathrm{d}\delta_{32}$$

From these equations,

$$\frac{\mathrm{d}P_3}{\mathrm{d}\delta_{31}} = \frac{1{\cdot}366}{3{\cdot}39} = 0{\cdot}4.$$

Also,

$$\frac{dP_1}{dP_{13}} = 0.256.$$

Hence the system is stable.

*Example 7.3* For the system shown in figure 7.6 investigate the steady-state stability.

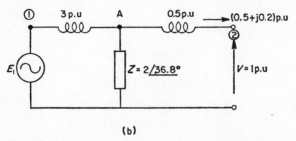

**(b)**

Figure 7.6b Equivalent circuit for Example 7.3.

*Solution* First it is necessary to determine the transfer impedance.

$$Y_{11} = \frac{1}{j3 + \dfrac{j0{\cdot}5(1{\cdot}6 + j1{\cdot}2)}{1{\cdot}6 + j1{\cdot}2 + j0{\cdot}5}} = \frac{1}{3{\cdot}423\underline{/88{\cdot}8^\circ}} = 0{\cdot}293\underline{/-88{\cdot}8^\circ} \text{ per unit}$$

$$Y_{12} = \frac{1}{j3 + j0{\cdot}5 + \dfrac{j3 \times j0{\cdot}5}{2\underline{/36{\cdot}8^\circ}}} = \frac{1}{4\underline{/98{\cdot}7^\circ}} = 0{\cdot}25\underline{/-98{\cdot}7^\circ} \text{ per unit}$$

Hence,

$$\alpha_{11} = 90 - 88{\cdot}8^\circ = 1{\cdot}2^\circ$$

$$\alpha_{12} = 90 - 98{\cdot}7^\circ = -8{\cdot}7^\circ.$$

The voltage at the tapping point A,

$$V_A = \sqrt{\left[\left(V + \frac{QX}{V}\right)^2 + \left(\frac{PX}{V}\right)^2\right]}$$

$$= \sqrt{\left[\left(1 + \frac{0{\cdot}2 \times 0{\cdot}5}{1}\right)^2 + \left(\frac{0{\cdot}5 \times 0{\cdot}2}{1}\right)^2\right]}$$

$$= 1{\cdot}105 \text{ per unit at } 5{\cdot}75^\circ \text{ to } V.$$

$$I_R^2 = \frac{P^2 + Q^2}{V^2}$$

and the reactive power absorbed by the line from A to the infinite bus

$$= I_R^2 X = \left(\frac{0\cdot5^2 + 0\cdot2^2}{1^2}\right) 0\cdot5 \text{ per unit}$$

$$= 0\cdot145 \text{ per unit.}$$

If the load at A is represented by an impedance then the actual load taken at 1·105 per unit voltage

$$= \left(\frac{1\cdot105}{1}\right)^2 0\cdot5 \underline{/36\cdot8^\circ} \text{ per unit}$$

$$= 0\cdot61 \underline{/36\cdot8^\circ} \text{ per unit}$$

$$= 0\cdot49 + j0\cdot366 \text{ per unit.}$$

Total load carried by link from generator to A

$$= 0\cdot5 + 0\cdot49 + j(0\cdot2 + 0\cdot145 + 0\cdot366)$$
$$= 0\cdot99 + j0\cdot71 \text{ per unit.}$$

Internal voltage of generator

$$= \sqrt{\left[\left(1\cdot105 + \frac{0\cdot71 \times 3}{1\cdot105}\right)^2 + \left(\frac{0\cdot99 \times 3}{1\cdot105}\right)^2\right]} = \sqrt{[9\cdot2 + 7\cdot23]}$$

$$= 4\cdot06 \underline{/41\cdot5^\circ}$$

This angle is with respect to A so that

$$\delta_{EV} = 41\cdot5 + 5\cdot75 = 47\cdot25^\circ = \delta_{12}.$$

Power from generator

$$= E^2 Y_{11} \sin \alpha_{11} + EV Y_{12} \sin (\delta_{12} - \alpha_{12})$$
$$= 4\cdot06 \times 0\cdot293 \sin 1\cdot2 + 4\cdot06 \times 1 \times 0\cdot25 \sin (\delta_{12} + 8\cdot7)$$
$$= 1\cdot09 \text{ per unit.}$$

Maximum power $= 1\cdot26$ per unit and the system is stable.

## 7.5  Transient Stability—Consideration of Rotor Angle

Transient stability is concerned with the effect of large disturbances. These are usually due to faults the most severe of which is the three-phase short-circuit which governs the transient stability limits in Britain. Elsewhere limits are based on other types of fault notably the single line to earth which is by far the most frequent in practice.

When a fault occurs at the terminals of a synchronous generator the power output of the machine is greatly reduced as it is supplying a mainly inductive circuit. However the input power to the generator from the turbine has not time to change during the short period of the fault and the rotor endeavours to gain speed to store the excess energy. If the fault persists long enough the rotor angle will increase continuously and synchronism lost. Hence the time of operation of the protection and circuit breakers is all-important.

An aspect of increasing importance is the use of *autoreclosing* circuit breakers. These open when the fault is detected and automatically reclose after a prescribed period (usually less than 1 second). If the fault persists the circuit breaker reopens and then recloses as before. This is repeated once more, when, if the fault still persists, the breaker remains open. Owing to the transitory nature of most faults often the circuit breaker successfully recloses and the rather lengthy process of investigating the fault and switching in the line is avoided. The length of the auto-reclose operation must be considered when assessing transient stability limits; in particular analysis must include the movement of the rotor over this period and not just the first swing as is often the case. If in equation (7.1) both sides are multiplied by $2(\mathrm{d}\delta/\mathrm{d}t)$,

$$\left(2\frac{\mathrm{d}\delta}{\mathrm{d}t}\right)\left(\frac{\mathrm{d}^2\delta}{\mathrm{d}t^2}\right) = \frac{\mathrm{d}}{\mathrm{d}t}\left(\frac{\mathrm{d}\delta}{\mathrm{d}t}\right)^2 = \frac{2\Delta P}{M}\left(\frac{\mathrm{d}\delta}{\mathrm{d}t}\right)$$

$$\therefore \qquad \left(\frac{\mathrm{d}\delta}{\mathrm{d}t}\right)^2 = \frac{2}{M}\int_{\delta_0}^{\delta}\Delta P\mathrm{d}\delta \qquad (7.10)$$

The rotor will swing until its angular velocity is zero in which case the machine remains stable; if $\mathrm{d}\delta/\mathrm{d}t$ does not become zero the rotor will continue to move and synchronism lost. Hence the criterion for stability is that the area between the $P-\delta$ curve and the line representing the power input must be zero. This is known as the *equal-area criterion*. It should be noted that this is based on the assumption that synchronism is retained or lost on the first swing or oscillation of the rotor, which in certain cases is subject to doubt. Physically the criterion means that the rotor must be able to return to the system all the energy gained from the turbine during the acceleration period.

A simple example of the equal-area criterion may be seen by an examination of the switching out of one of two parallel lines which connect a generator to an infinite busbar (figure 7.7). For stability to be retained the two shaded areas are equal and the rotor comes initially to rest at angle $\delta_2$, after which it oscillates until completed damped. In this particular case the initial operating power and angle could be increased to such values that the shaded area between $\delta_0$ and $\delta_1$ is equal to the shaded area between $\delta_1$ and $\delta_3$ where $\delta_3 = 180-\delta_1$; this would be the condition for maximum input power. Beyond $\delta_3$ energy would again be absorbed by the rotor from the turbine.

The power angle curves pertaining to a fault on one of two parallel lines is shown in figure 7.8. The fault is cleared in a time corresponding to $\delta_1$ and the shaded area $\delta_0$ to $\delta_1$ indicates the energy stored. The rotor swings until it reaches $\delta_2$ when the two areas are equal. In this particular case $P_0$ is the maximum operating power for a fault clearance time corresponding to $\delta_1$, and conversely $\delta_1$ is the *critical clearing angle* for $P_0$. If the angle $\delta_1$ is decreased it is possible to increase the value of $P_0$ without loss of synchronism. The general case where the clearing angle $\delta_1$ is not critical is

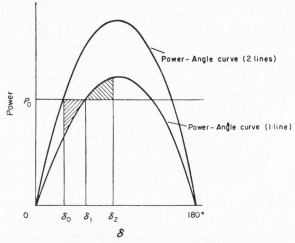

Figure 7.7    Power-angle curves for one line and two lines in parallel. Equal area criterion. Resistance neglected.

shown in figure 7.9. Here the rotor swings to $\delta_2$ where the shaded area from $\delta_0$ to $\delta_1$ is equal to the area $\delta_1$ to $\delta_2$. Critical conditions are reached when $\delta_2 = 180 - \delta_1$. The time corresponding to the critical clearing angle is called the *critical clearing time* for the particular (normally full load) value of power input. This time is of great importance to protection and switchgear engineers as it is the maximum time allowable for their equipment to operate without instability occurring. The critical clearing angle for a fault on one of two parallel lines may be determined as follows.

Applying the equal-area criterion to figure 7.8.

$$\int_{\delta_0}^{\delta_1} (P_0 - P_1 \sin \delta)\mathrm{d}\delta + \int_{\delta_1}^{\delta_2} (P_0 - P_2 \sin \delta)\mathrm{d}\delta = 0$$

$$\therefore \qquad [P_0\delta + P_1 \cos \delta]_{\delta_0}^{\delta_1} + [P_0\delta + P_2 \cos \delta]_{\delta_1}^{\delta_2} = 0$$

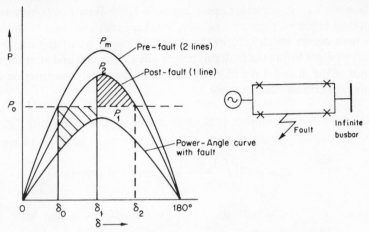

Figure 7.8 Fault on one line of two lines in parallel. Equal area criterion. Resistance neglected. $\delta_1$ is critical clearing angle for input power $P_0$.

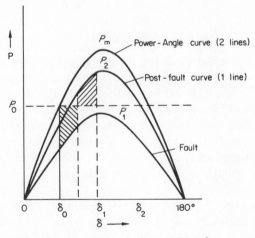

Figure 7.9 Situation as in figure 7.8 but $\delta_1$ not critical.

from which as $\delta_2 = 180 - \sin^{-1}\left(\dfrac{P_0}{P_2}\right)$

$$\cos \delta_1 = \frac{P_0(\delta_0 - \delta_2) + P_1 \cos \delta_0 - P_2 \cos \delta_2}{P_1 - P_2} \qquad (7.11)$$

Hence the critical clearing angle $\delta_1$ is determined.

*Example 7.4*   A generator operating at 50 Hz delivers 1 per unit power to
an infinite busbar through a network in which resistance may be neglected.
A fault occurs which reduces the maximum power transferable to 0·4 per
unit whereas before the fault this power was 1·8 per unit and after the clearance of the fault 1·3 per unit. By the use of the equal-area criterion determine the critical clearing angle.

*Solution*   The appropriate load angle curves are shown in figure 7.8.
$P_0 = 1$ per unit, $P_1 = 0·4$ per unit, $P_2 = 1·3$ per unit and $P_m = 1·8$ per unit.

$$\delta_0 = \sin^{-1}\left(\frac{1}{1·8}\right) = 33·8 \text{ electrical degrees}$$

$$\delta_2 = 180 - \sin^{-1}\left(\frac{1}{1·3}\right) = 180 - 50° 24' = 129° 36'.$$

Applying equation 7.11,

$$\cos \delta_1 = \frac{1(-1·6718) + 0·4 \times 0·831 + 1·3 \times 0·77}{0·4 - 1·3}$$

$$= 0·377$$

∴                     $\delta_1 = 67° 51'$ (electrical degrees).

   In a large system it is usual to divide the generators and loads into
a single equivalent generator connected to an equivalent motor. The main
criterion is that the machines should be electrically close when forming
an equivalent generator or motor. If stability with faults in various places
is investigated the position of the fault will decide the division of machines
between the equivalent generator and motor. A power system (including
generation) at the receiving end of a long line would constitutean equivalent motor if not large enough to be an infinite busbar.

*Reduction to simple system*   With a number of generators connected to
the same busbar the inertia constant ($H$) of the equivalent machine,

$$H_e = H_1\frac{S_1}{S_b} + H_2\frac{S_2}{S_b} \ldots H_n\frac{S_n}{S_b},$$

where $S_1 \ldots S_n$ = MVA of the machines, $S_b$ = base MVA. This is
obtained by equating the stored energy of the equivalent machine to the
total of the individual machines. For example, consider six identical
machines connected to the same busbar each having an $H$ of $5\dfrac{\text{MW-sec}}{\text{MVA}}$

and rated at 60 MVA. Making the base MVA equal to the combined rating of the machines the effective value of

$$H = 5 \times \frac{60}{6 \times 60} \times 6 = 5 \frac{\text{MW-sec}}{\text{MVA}}.$$

It is important to remember that the inertia of the loads must be included; normally this will be the sum of the inertias of the induction motors connected.

Two synchronous machines connected by a reactance may be reduced to one equivalent machine feeding through the reactance to an infinite busbar system. The properties of the equivalent machine are found as follows.

The equation of motion for the two-machine system is,

$$\frac{d^2\delta}{dt^2} = \frac{\Delta P_1}{M_1} - \frac{\Delta P_2}{M_2} = \left(\frac{1}{M_1} + \frac{1}{M_2}\right)(P_0 - P_m \sin \delta)$$

where $\delta$ is the relative angle between the machines. Note that,

$$\Delta P_1 = - \Delta P_2 = (P_0 - P_m \sin \delta)$$

where $P_0$ is the input power and $P_m$ the maximum transmittable power.

For a single generator of $M_e$ and the same input power connected to the infinite busbar system,

$$M_e \frac{d^2\delta}{dt^2} = P_0 - P_m \sin \delta$$

therefore

$$M_e = \frac{M_1 M_2}{M_1 + M_2} \tag{7.12}$$

This equivalent generator has the same mechanical input as the actual machines and the load angle $\delta$ it has with respect to the busbar is the angle between the rotors of the two machines.

So far the maximum powers transferable before, during and after the fault have been assumed. These values can be obtained as described in chapter 2. For the common case of a three-phase fault in the middle of one of two parallel lines as shown in figure 7.8, the mesh formed by $X_1$ and $X_2/2$ (figure 7.10a) is replaced by the equivalent star as shown in figure 7.10b. The power transmitted during the fault is $E_g V/X_e \sin \delta$ as $X_A$ and $X_B$ are purely inductive loads. The transfer reactance before and after the fault is obtained readily from the equivalent circuit. With unbalanced faults more power is transmitted during the fault period than with three-phase short-circuits and the stability limits are higher.

*Effect of automatic voltage regulators and governors* These may be represented in the equation of motion as follows,

$$M\frac{\mathrm{d}^2\delta}{\mathrm{d}t^2} + K_d\frac{\mathrm{d}\delta}{\mathrm{d}t} = (P_0 - \Delta P_0) - P_e \tag{7.13}$$

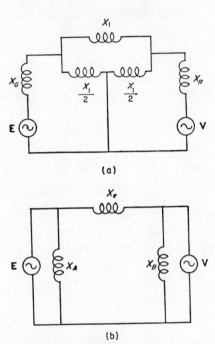

(a)

(b)

Figure 7.10a Equivalent circuit of a generator connected to an infinite busbar through two lines one of which has a three-phase fault midway along its length.

Figure 7.10b Fig. 7.10a after reduction. $X_1$ = reactance of line. $X_G$ = reactance of generator plus transformer, $X_R$ = reactance of receiving end transformer.

in which, $K_d$ is the damping coefficient,

$P_0$ is the power input,

$\Delta P_0$ is the change in input power due to governor action,

$P_e$ is the electrical power output modified by the voltage regulator.

Equation (7.13) is best solved by means of an analogue or digital computer. In most studies the governor and voltage regulator effects are ignored but their influence on stability limits have been analysed in recent investigations.

## 7.6 Transient Stability—Consideration of Time

*The swing curve* In the previous section attention has been mainly directed towards the determination of rotor angular position; in practice the corresponding times are more important. The protection engineer requires allowable times rather than angles when specifying relay settings. The solution of equation (7.1) with respect to time is performed by means of numerical methods and the resulting time-angle curve is known as the swing curve. A simple step by step method by Dahl, (reference 6) will be given in detail and references made to more sophisticated methods suitable for digital computation.

In this method the change in the angular position of the rotor over a short time interval is determined. In performing the calculations the following assumptions are made:

(a) the accelerating power $\Delta P$ at the commencement of a time interval is considered constant from the middle of the previous interval to the middle of the interval considered;

(b) the angular velocity is constant over a complete interval and is computed for the middle of this interval. These assumptions are probably better understood by reference to figure 7.11.

From figure 7.11,

$$\omega_{n-\frac{1}{2}} - \omega_{n-\frac{3}{2}} = \frac{d^2\delta}{dt^2}\Delta t = \frac{\Delta P_{n-1}}{M}\Delta t$$

The change in $\delta$ over the $(n-1)$th interval, i.e. from times $(n-1)$ to $(n-2)$

$$= \delta_{n-1} - \delta_{n-2} = \Delta\delta_{n-1} = \omega_{n-\frac{3}{2}}\Delta t$$

Over the $n$th interval,

$$\Delta\delta_n = \delta_n - \delta_{n-1} = \omega_{n-\frac{1}{2}}\Delta t$$

From the above,

$$\Delta\delta_n - \Delta\delta_{n-1} = \Delta t(\omega_{n-\frac{1}{2}} - \omega_{n-\frac{3}{2}})$$

$$= \Delta t \,.\, \Delta t \,.\, \frac{\Delta P_{n-1}}{M}$$

$$\therefore \qquad \Delta\delta_n = \Delta\delta_{n-1} + \frac{\Delta P_{n-1}}{M}(\Delta t)^2 \qquad (7.13)$$

It should be noted that $\Delta\delta_n$ and $\Delta\delta_{n-1}$ are the *changes* in angle.

Equation (7.13) is the basis of the numerical method. The time interval $\Delta t$ used should be as small as possible (the smaller $\Delta t$, however, the larger the amount of labour involved) and a value of 0·05 seconds is frequently used. Any change in the operational condition causes an abrupt change

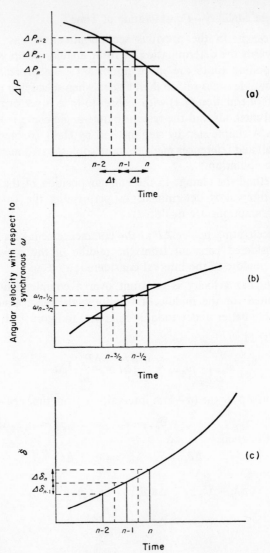

Figure 7.11a, b and c  Variation of $\Delta P$, $\omega$ and $\Delta\delta$ with time.
Illustration of step-by-step method to obtain $\delta$-time curve.

in the value of $\Delta P$. At the commencement of a fault for example ($t = 0$), the value of $\Delta P$ is initially zero and then immediately after the occurrence takes a definite value. When two values of $\Delta P$ apply the mean is used. The procedure is best illustrated by an example.

*Example 7.5*   In the system described in Example 7.4 the inertia constant of the generator plus turbine is 2·7 per unit. Obtain the swing curve for a fault clearance time of 125 milliseconds.

*Solution*   $H = 2\cdot7$ per unit, $f = 50$ c/s, $G = 1$ per unit.

$$\therefore \qquad M = \frac{HG}{180f} = 3 \times 10^{-4} \text{ per unit.}$$

A time interval $\Delta t = 0\cdot05$ seconds will be used. Hence

$$\frac{(\Delta t)^2}{M} = 8\cdot33.$$

The initial operating angle

$$\delta_0 = \sin^{-1}\left(\frac{1}{1\cdot8}\right) = 33\cdot8°.$$

Just before the fault the accelerating power $\Delta P = 0$. Immediately after the fault,

$$\Delta P = 1 - 0\cdot4 \sin \delta_0$$
$$= 0\cdot78 \text{ per unit.}$$

The first value is that for the middle of the preceding period and the second for the middle of the period under consideration. The value to be taken for $\Delta P$ at the commencement of this period is $(0\cdot780/2)$, i.e. $0\cdot39$ per unit. At $t = 0, \delta = 33\cdot8°$.

$$\Delta t_1 = 0\cdot05 \text{ seconds,} \quad \Delta P = 0\cdot39 \text{ per unit.}$$

$$\cdot \quad \therefore \qquad \Delta \delta_n = \Delta \delta_{n-1} + \frac{(\Delta t)^2}{M} \Delta P_{n-1}$$

$$\therefore \qquad \Delta \delta_n = 8\cdot33 \times 0\cdot39 = 3\cdot23°$$

$$\therefore \qquad \delta_{0\cdot05} = 33\cdot8 + 3\cdot23 = 37\cdot03°$$

$$\underline{\Delta t_2};\quad \Delta P = 1 - 0\cdot4 \sin 37\cdot03° = 0\cdot76 \text{ per unit.}$$

$$\therefore \qquad \Delta \delta_n = 3\cdot23 + 8\cdot33 \times 0\cdot76 = 9\cdot56°$$

$$\therefore \qquad \delta_{0\cdot1} = 37\cdot03 + 9\cdot56 = 46\cdot59°$$

$$\underline{\Delta t_3};\quad \Delta P = 1 - 0\cdot4 \sin 46\cdot59° = 0\cdot71 \text{ per unit}$$

$$\Delta \delta_n = 9\cdot56 + 5\cdot9 = 15\cdot46°$$

$$\delta_{0\cdot15} = 46\cdot59 + 15\cdot46 = 62\cdot05°.$$

The fault is cleared after a period of $0\cdot125$ seconds. As this discontinuity occurs in the middle of a period ($0\cdot1$ to $0\cdot15$ seconds) no special averaging is required (figure 7.12). If on the other hand the fault is cleared in $0\cdot15$ seconds an averaging of two values would be required.

From $t = 0.15$ seconds onwards,

$$P = 1 - 1.3 \sin \delta$$

$\Delta t_4;$     $\quad \Delta P = 1 - 1.3 \sin 62.05° = -0.145$ per unit

$\quad\quad\quad \Delta \delta_n = 15.46 + 8.33(-0.145) = 14.25°$

$\therefore \quad\quad \delta_{0.2} = 14.25 + 62.05 = 76.3°$

$\Delta t_5;$     $\quad \Delta P = 1 - 1.3 \sin 76.3° = -0.26$ per unit

$\quad\quad\quad \Delta \delta_n = 14.25 - (8.33 \times 0.26) = 12.09°$

and      $\quad \delta_{0.25} = 88.39°$

$\Delta t_6;$     $\quad \Delta P = 1 - 1.3 \sin 88.39 = -0.3$ per unit

$\quad\quad\quad \Delta \delta_n = 12.09 - 2.5 = 9.59°$

and      $\quad \delta_{0.3} = 97.98°.$

If this process is continued $\delta$ commences to decrease and the generator remains stable. Different curves will be obtained for other values of clearing time. It is evident from the way the calculation proceeds that for a sustained

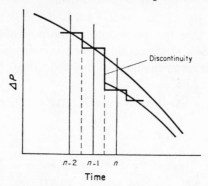

Figure 7.12   Discontinuity in $\Delta P$ in middle of a period of time.

fault $\delta$ will continuously increase and stability will be lost. The critical clearing time should be calculated for conditions which allow the least transfer of power from the generator. Circuit breakers and the associated protection operate in times dependent upon their sophistication; these times can be in the order of a few cycles of alternating voltage. A typical swing curve for a large machine over several seconds is shown in figure 7.13a. For a given fault position a faster clearing time implies a greater permissible value of input power $P_0$. A typical relationship between the critical clearing time and input power is shown in figure 7.13b; this is often referred to as the stability boundary. The critical clearing time increases with increase in the inertia constant ($H$) of turbine-generators.

Often the first swing of the machine is sufficient to indicate stability. If however the result is marginal the step by step process should be continued for several swings.

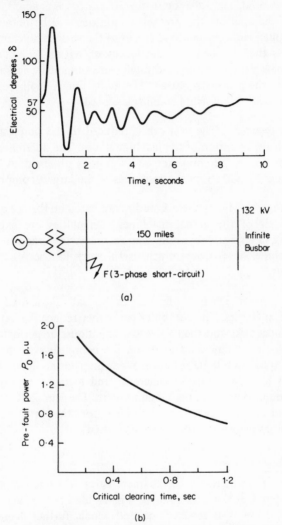

Figure 7.13a   Typical swing curve; b Typical stability boundary.

## 7.7   The Use of Computers in Transient Stability Studies

*Alternating current analysers*   The analyser is used to obtain the new value of $\Delta P$ for each value of $\delta$. In effect therefore a load flow is carried out for each new set of rotor angles, $\Delta\delta$ having been obtained in the same way as in Example 7.5. In a large multi-machine system this process is extremely tedious and the simplest possible representation of machines and loads is used, e.g. saliency in the generators is ignored. On the first balance of the analyser the prefault conditions will be obtained and the power outputs, voltages and load angles of the various machines recorded. The fault is then applied and the values of $\Delta P$ and $\delta$ obtained. From these the new values of the load angles are obtained and set up on the analyser and the process repeated. The faulty line is finally removed from the network and steady-state stability assessed.

*Analogue computers*   The network is reduced to the simplest form compatible with the retention of the machine nodes. The differential equations are then set up on the computer and solutions obtained. A system with several machines will strain the resources of the largest computer.

*Micro-machines*   Micro-réseaux studies are based on the use of electrically scaled models of the actual machines. Difficulties are experienced in reproducing in a small laboratory-size generator the time constants of the actual machine, nevertheless much useful work has been carried out on such models.

*Digital computers*   The methods used are basically those used for alternating-current analysers. The computer can either reduce the network to the machine nodes first and then solve the equations, or perform a complex load-flow for each new load angle on the complete system. The former method is quicker but information regarding power flows during rotor swinging is lost. Most of the generator and load assumptions used for analysers are also used for digital computers. The Runge-Kutta method is widely used in the digital programmes for the integration process and predictor-corrector routines are also used.

### 7.8   Stability of Loads

In chapter 4 the power-voltage characteristics of a line supplying a load were considered. It was seen that for a given load power-factor a value of transmitted power was reached, beyond which, further decreases of the load impedance produce greatly reduced voltages, i.e. voltage instability. If the load is purely static, e.g. represented by an impedance, the system will operate stably at these lower voltages. Sometimes when using alternating current network analysers this lower-voltage condition is unknowingly obtained and unexpected load-flows result. If the load contains non-static elements such as induction motors, the nature of the load characteristics is such that beyond the critical point the motors will run down to a standstill or stall. It is therefore of importance to consider the stability of composite loads which will normally include a large proportion of induction motors.

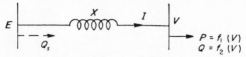

Figure 7.14 System with a load dependent on voltage as follows:
$P = f_1(V)$ and $Q = f_2(V)$; $Q_s$ = supply vars = $Q + I^2X$.
$E$ = supply voltage.

The process of voltage collapse may be seen from a study of the $V$–$Q$ characteristics of an induction motor (figure 2.51) from which it is seen that below a certain voltage the reactive power consumed increases with decrease in voltage until $(dQ/dV) = \infty$ when the voltage collapses. In

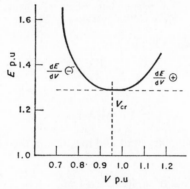

Figure 7.15 $E$–$V$ relationship for system in figure 7.14,
$V_{cr}$ = critical voltage after which instability occurs.

the power system the problem arises owing to the impedance of the connexion between the load and infinite busbar and is obviously aggravated when this impedance is high (connexion electrically weak). The usual cause of an abnormally high impedance is the loss of one line of two or more forming the connexion. It is profitable therefore to study the process in its basic form that of a load supplied through a reactance from a constant voltage source (figure 7.14). Already two criteria for load instability have been given, i.e. $(dP/ds) = 0$ and $(dQ/dV) = \infty$; from the system viewpoint voltage collapse takes place when $(dE/dV) = 0$ or $(dV/dE) = \infty$. Each value of $E$ yields a corresponding value for $V$ and the plot of $E$–$V$

is shown in figure 7.15, also the plot of $V$–$X$ for various values of $E$ is shown in figure 7.16. In these graphs the critical operating condition is clearly shown and the improvement produced by higher values of $E$ apparent, indicating the importance of the system operating voltage from the load viewpoint.

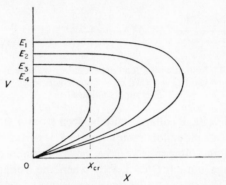

Figure 7.16 $V$–$X$ relationship. Effect of change in supply voltage $E$. $X_{cr}$ = Critical reactance of transmission link.

In the circuit shown in figure 7.14

$$E = V + \frac{QX}{V}, \quad \text{if} \quad \frac{PX}{V} \ll \frac{V^2 + QX}{V} \quad \text{(equation 2.6)}$$

$$\therefore \quad \frac{dE}{dV} = 1 + \left(\frac{dQ}{dV}XV - QX\right)\frac{1}{V^2}$$

which is zero at the stability limit and negative in the unstable region.

*At the limit,*

$$\frac{dE}{dV} = 0 \quad \text{and} \quad \frac{dQ}{dV} = \left(\frac{QX}{V^2} - 1\right)\frac{V}{X} = \frac{Q}{V} - \frac{V}{X}$$

From 2.6

$$\frac{Q}{V} = \frac{E}{X} - \frac{V}{X}$$

$$\therefore \quad \frac{dQ}{dV} = \frac{E}{X} - \frac{2V}{X} \tag{7.15}$$

*Example 7.6* A load is supplied from a 275 kV infinite busbar through a line of reactance 70 $\Omega$ phase-to-neutral. The load consists of a constant power demand of 200 MW and a reactive power demand $Q$ which is related to the load voltage $V$ by the equation:

$$(V - 0\cdot8)^2 = 0\cdot2(Q - 0\cdot8),$$

where the base quantities for $V$ and $Q$ are $275\,\text{kV}$ and $200$ MVAr respectively.

Examine the voltage stability of this system, indicating clearly any assumptions made in the analysis.

*Solution* It has been shown that,

$$E = \sqrt{\left[\left(V + \frac{QX}{V}\right)^2 + \left(\frac{PX}{V}\right)^2\right]}$$

If

$$\frac{PX}{V} < \frac{V^2 + QX}{V} \quad \text{then} \quad E = \frac{V^2 + QX}{V}$$

and

$$\frac{dE}{dV} = 1 + \left(\frac{dQ}{dV}XV - QX\right)\frac{1}{V^2} = 0.$$

In this problem, $\left.\begin{array}{l} P = 200 \text{ MW} \\ Q = 200 \text{ MVAr} \end{array}\right\}$ three-phase values

and 1 per unit

$$X = \frac{70 \times 200}{275^2} \text{ per unit}$$

$$= 0{\cdot}185 \text{ per unit}$$

$$E = V + \frac{QX}{V}$$

$\therefore$

$$1 = V + \frac{0{\cdot}185}{V}\left(\frac{(V - 0{\cdot}8)^2}{0{\cdot}2} + 0{\cdot}8\right)$$

$\therefore$

$$V^2 = V - 0{\cdot}925V^2 - 0{\cdot}59 + 1{\cdot}48V - 0{\cdot}148$$

$$1{\cdot}925V^2 - 2{\cdot}48V + 0{\cdot}74 = 0$$

$\therefore$

$$V = \frac{+2{\cdot}48 \pm \sqrt{(2{\cdot}48^2 - 4 \times 1{\cdot}925 \times 0{\cdot}74)}}{2 \times 1{\cdot}925}$$

$$= \frac{2{\cdot}48 \pm 0{\cdot}67}{3{\cdot}85}.$$

Taking the upper value, $V = 0{\cdot}818$ per unit.

$$Q \text{ at this } V = \frac{(0{\cdot}818 - 0{\cdot}8)^2}{0{\cdot}2} + 0{\cdot}8$$

$$= 0{\cdot}8016 \text{ per unit}$$

$$\frac{dE}{dV} = 1 + [10(0\cdot818 - 0\cdot8) \times 0\cdot185 \times 0\cdot818 - 0\cdot8016 \times 0\cdot185/V^2]$$

$$= 1 + \left(\frac{0\cdot027 - 0\cdot148}{0\cdot818^2}\right)$$

which is positive, i.e. the system is stable.

*Note* $PX/V \doteqdot 0\cdot16$ and $(V^2 + QX)/V \doteqdot 1$,

∴ approximation is reasonable.

When the reactance between the source and load is very high the use of tap-changing transformers is of no assistance. Large voltage drops exist in the supply lines and the 'tapping-up' of transformers increases these due to the increased supply currents. Hence the peculiar effects from tap changing noticed when conditions close to a voltage collapse have occurred in practice, i.e. tapping-up reduces the secondary voltage and vice versa. One symptom of the approach of critical conditions is sluggishness in the response of tap changing transformers.

The above form of instability is essentially a form of steady-state instability and a full assessment of the latter should not only take into account the synchronous criterion, $(dP/d\delta) \geqslant 0$, although this is the more likely to occur, but also the criterion for load instability, $(dV/dE) = \infty$.

*The use of computers*  It is difficult to detect the power transmitted in the region near to the condition of voltage collapse. A computer will detect this condition by a divergence and it is difficult to assess if this is due to the network or to other conditions. One possible method is to programme the power-voltage characteristics of the line and of the load; the intersection of these two characteristics will give the operating conditions of the system.

In steady-state stability studies the increments of load added at each step are small and this makes the results of alternating current network analysers of doubtful accuracy. Using a digital computer however, successive load flows can be performed with high accuracy in computation.

### 7.9  Further Aspects of Voltage Instability

*Faults on the feeders to induction motors*  A more common cause of the stalling of induction motors (or the low-voltage releases operating and removing them from the supply) occurs when the supply voltage is either zero or very low for a brief period due to a fault on the supply system. When the supply voltage is restored the induction motors accelerate and endeavour to attain their previous operating condition. In accelerating, however, a large current is taken and this, plus the fact that the system impedance has increased due to the loss of a line, results in a greatly depressed voltage at the motor terminals. If this voltage is too low the machines will stall or cut out of circuit.

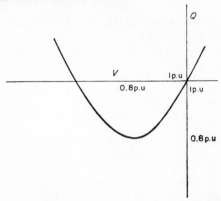

Figure 7.17   Example 7.6. Reactive power-voltage characteristic.

*Steady-state instability due to voltage regulators*   Consider a generator supplying an infinite busbar through two lines one of which is suddenly removed. The load angle of the generator is instantaneously unchanged and therefore the power output decreases due to the increased system reactance, thus causing the generator voltage to rise. The automatic voltage regulator of the generator then weakens its field to maintain constant voltage, i.e. decreases the internal e.m.f., and synchronous instability may result.

### References

**BOOKS**

1. Crary, S. B., *Power System Stability*, Vol. I (*Steady State Stability*), Vol. II (*Transient Stability*), Wiley, New York, 1945.
2. Kimbark, E. W., *Power System Stability*, Vols. 1, 2 and 3, Wiley, New York, 1948.
3. Stagg, C. W., and A. H. El-Abiad, *Computer Methods in Power System Analysis*, McGraw-Hill, New York, 1968.
4. *Electrical Transmission and Distribution Reference Book*, Westinghouse Electric Corp., East Pittsburgh, Pennsylvania, 1964.
5. Mortlock, J. R., and M. W. Humphrey Davies, *Power System Analysis*, Chapman and Hall, London, 1952.
6. Dahl, O. G. C., *Electric Power Circuits*, Vol. II. McGraw-Hill, New York, 1938.
7. Venikov, V. A., *Transient Phenomena in Electrical Power Systems* (translated from the Russian), Pergamon, London, 1964.

8. Rudenberg, R., *Transient Performance of Electric Power Systems*, M.I.T. Press, 1969
9. Chestnut, H., and R. W. Mayer, *Servomechanisms and Regulating System Design*, Wiley, New York, 1951.

PAPERS

10. Miles, J. G., 'Analysis of overall stability of multimachine power systems', *Proc. I.E.E.*, **109A**, 203 (1961).
11. Lane, C. M., R. W. Long, and J. N. Powers, 'Transient-stability studies, II', *Trans. A.I.E.E.*, **78**, Pt. III, 1291 (1959).
12. Taylor, D. G., 'Analysis of synchronous machines connected to power-system networks', *Proc. I.E.E.*, **110C**, 606 (1963).
13. Humpage, W. D., and B. Stott, 'Effect of autoreclosing circuit-breakers on transient stability in e.h.v. transmission systems', *Proc. I.E.E.*, **111** (7), 1287 (1964).
14. Humpage, W. D., and B. Stott, 'Predictor-correct methods of numerical integration in digital-computer analyses of power-system transient stability,' *Proc. I.E.E.*, **112**, 1557 (1965).
15. Scott, E. C., W. Canon, A. Chorlton, and J. H. Banks, 'Multigenerator transient-stability performance under fault conditions', *Proc. I.E.E.*, **110**, 1051, (1963).
16. Sicy, J., 'The influence of regulating transformers for the excitation of alternators on system voltages and stability', *C.I.G.R.E.*, No. 323 (1962).
17. Stagg, G. W., A. F. Gabrielle, D. R. Moore, and J. F. Hohenstein, 'Calculation of transient stability problems using a high speed digital computer', *Trans. A.I.E.E.*, **78** (1959).
18. Johnson, D. L., and J. B. Ward, 'The solution of power system stability problems by means of digital computers', *Trans. A.I.E.E.*, **75**, Pt. III, 1321 (1956).
19. Gabbard, J. L., Jr., and J. E. Rowe, 'Digital computation of induction-motor transient stability', *Trans. A.I.E.E.*, **76**, Pt. III, 970 (1957).
20. Brereton, D. S., D. G. Lewis, and C. C. Young, 'Representation of induction-motor loads during power-system stability studies', *Trans. A.I.E.E.*, **76**, Pt. III, 451 (1957).
21. Dyrkacz, M. S., C. C. Young, and F. J. Maginniss, 'A digital transient stability program including the effects of regulator, exciter, and governor response', *Trans. A.I.E.E.*, **79**, Pt. III, 1245 (1960).
22. Brown, H. E., H. H. Happ, C. E. Person, and C. C. Young, 'Transient stability solution by an impedance matrix method', *Trans. I.E.E.E.*, *P.A. & S.*, **84**, 1204 (1965).
23. Lokay, H. E., and R. L. Bolger, 'Effect of turbine-generator representation in system stability studies', *Trans. I.E.E.E.*, *P.A. & S.*, **84**, 933 (1965).
24. Olive, D. W., 'New techniques for the calculation of dynamic stability', *Trans. I.E.E.E.*, *P.A. & S.*, **85**, 767 (1966).
25. Shipley, R. B., N. Sato, Coleman, and C. F. Watts, 'Direct calculation of power system stability using the impedance matrix', *Trans. I.E.E.E.*, *P.A. & S.*, **85**, 777 (1966).
26. Harley, R. G., and B. Adkins, 'Calculation of the angular back swing following a short circuit of a loaded alternator', *Proc. I.E.E.*, **117**, 377 (1970).

27. Dharma Rao, N., 'Routh-Hurwitz condition and Lyapunov methods for the transient-stability problem', *Proc. I.E.E.*, **116**, 533 (1969).
28. Kapoor, S. C., S. S. Kalsi, and B. Adkins, 'Improvement of alternator stability by controlled quadrature excitation', *Proc. I.E.E.*, **116**, 771 (1969).
29. Coles, H. E., 'Dynamic performances of a turboalternator using 3-term governor control and voltage regulation', *Proc. I.E.E.*, **115**, 266 (1968).

## Problems

7.1. A round rotor generator of synchronous reactance 1 p.u. is connected to a transformer of 0·1 p.u. reactance. The transformer feeds a line of reactance 0·2 p.u. which terminates in a transformer (0·1 p.u. reactance) to the $L/V$ side of which a synchronous motor is connected. The motor is of the round-rotor type and of 1 p.u. reactance. On the line-side of the generator transformer a three-phase static reactor of 1 p.u. reactance per phase is connected via a switch. Calculate the steady-state power limit with and without the reactor connected. All per unit reactances are expressed on a 10 MVA base and resistance may be neglected. The internal voltage of the generator is 1·2 p.u. and of the motor 1 p.u. (5 MW and 3·14 MW for shunt reactor.)

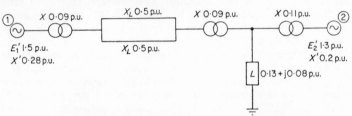

Figure 7.18    Line diagram of system in Problem 7.2

7.2. In the system shown in figure 7.18 two equivalent round-rotor generators feed into a load which may be represented by the constant impedance shown. Determine $Z_{11}$, $Z_{12}$ and $Z_{22}$ for the system and calculate $\delta_{12}$ if $P_1$ is 1·2 p.u. All per unit values are expressed on the same base and the resistance of the system (apart from the load) may be neglected. ($Z_{11} = 0·78\underline{/87·8°}$, $Z_{12} = 1·17\underline{/112°}$, $Z_{22} = 0·22\underline{/62·5°}$.)

7.3. For the system of three synchronous machines shown in figure 7.4 investigate the steady-state stability for the following operating angles: (a) $\delta_{12}$ 75°, $\delta_{13}$ 45°, $\delta_{23}$ 30°. (b) $\delta_{12}$ 85°, $\delta_{13}$ 90°, $\delta_{23}$ − 10°. ((a) stable and (b) unstable.)

7.4. A hydro-electric generator feeds a load through two 132 kV lines in parallel, each having a total reactance of 70 Ω phase-to-neutral. The load consists of induction motors operating at three-quarters of full load and taking 30 MVA. A local generating station feeds directly into the load busbars. Determine the parameters of an equivalent circuit, consisting of a single machine connected to an infinite busbar through a reactance, which represents the above system when a three-phase symmetrical fault occurs half-way along one line.

The machine data are as follows:

Hydro-electric generator: rating 60 MW at power factor 0·9 lagging;
transient reactance 0·3 p.u.;
inertia constant 3 kW-sec/kVA.

Induction motors:                composite transient reactance 3 p.u.;
                                 inertia constant 1 kW-sec/kVA.
Local generator:                 rating 30 MVA;
                                 transient reactance 0·15 p.u.;
                                 inertia constant 10 kW-sec/kVA.
($X = 9·9$ p.u., $M = 0·000121$ p.u., on a 100 MVA base).

7.5. For the system described in Example 7.4 determine the swing curves for fault clearance times of 250 and 500 milliseconds.

7.6. The $P$–$V$, $Q$–$V$ characteristics of a substation load are as follows:

| $V$ | 1·05 | 1·025 | 1 | 0·95 | 0·9 | 0·85 | 0·8 | 0·75 |
|---|---|---|---|---|---|---|---|---|
| $P$ | 1·03 | 1·017 | 1 | 0·97 | 0·94 | 0·92 | 0·98 | 0·87 |
| $Q$ | 1·09 | 1·045 | 1 | 0·93 | 0·885 | 0·86 | 0·84 | 0·85 |

The substation is supplied through a link of total reactance 0·8 p.u. and negligible resistance. With nominal load voltage $P = 1$ and $Q = 1$ p.u. By determining the supply voltage–received voltage characteristic, examine the stability of the system by the use of d$E$/d$V$. All quantities are per unit.

7.7. An induction motor load is supplied through a transformer of 0·1 p.u. reactance from a receiving end substation which is supplied from a generator through a transmission link of total reactance 0·3 p.u. Plant data are as follows:

Generator $X_s = 1·1$, $X' = 0·3$. Induction motor load $X_s = 0·2$, $R_2 = 0·03$, mechanical power output independent of speed. All per unit values are on a 60 MVA base. Examine the stability of the load when the motor takes 50 MW when the generator has, (a) no A.V.R., (b) a continuously acting A.V.R.

7.8. A large synchronous generator, of synchronous reactance 1·2 p.u., supplies a load through a link comprising a transformer of 0·1 p.u. reactance and an overhead line of initially 0·5 p.u. reactance; resistance is negligible. The voltage at the load busbar is 1 p.u. and the load is of magnitude $(0·8 + j\,0·6)$ p.u. regardless of the voltage. Assuming the internal voltage of the generator to remain unchanged, determine the value of line reactance at which instability occurs.

Answer:                      Unstable when $X = 0·5$ p.u., i.e. initial value.

7.9. A load is supplied from an infinite busbar of voltage 1 p.u. through a link of series reactance $X$ p.u. and of negligible resistance and shunt admittance. The load consists of a constant power component of 1 p.u. at 1 p.u. voltage and a per unit reactive power component $(Q)$ which varies with the received voltage $(V)$ according to the law

$$(V - 0·8)^2 = 0·2(Q - 0·8)$$

All per unit values are to common voltage and MVA bases.

Determine the value of $X$ at which the received voltage has a unique value and the corresponding magnitude of the received voltage.

Explain the significance of this result in the system described. Use approximate voltage drop equations.

Answer:                      $X = 0·25$ p.u.
                             $V = 0·67$ p.u.

# 8

# Underground Transmission

## 8.1 Introduction

A considerable amount of transmission and distribution especially in urban areas is carried out by means of underground cables. At the present time, growing public pressure to preserve the amenities of both town and countryside is forcing electricity supply authorities to consider the undergrounding of many circuits which they would prefer on economic grounds to place overhead. Even in sparsely populated regions, high-voltage bulk-transmission circuits have been placed underground where areas of outstanding natural beauty exist; in some instances disused railway tunnels are being used for this purpose. Whereas at one time a case could be put forward that an overhead line was, with some stretch of the imagination, graceful, the much higher voltages and powers now in use result in large tower structures and bundle conductors which undoubtedly mar the landscape.

Underground transmission is more expensive than the overhead alternative; at 345 kV a figure of ten times the cost of an equivalent overhead line is quoted for average suburban areas in the U.S.A., this ratio decreasing with lower voltages. A breakdown of costs for a 345 kV pipe-type cable is shown in figure 8.1. A major problem in present-day technology is the development of cables which are not only economically more attractive but physically able to carry the very large powers in use and envisaged. The major cause of the lack of current-carrying capacity in cables is the restriction on the temperature rise of the insulating material used; with overhead lines this is by no means so severe a problem. If the cable is to remain flexible there exists an upper limit to its overall diameter and hence to conductor size.

In spite of earlier hopes for plastic insulations, the combination of paper and oil still forms the most effective dielectric at the higher voltages. The difficulty of manufacturing Polythene (polyethylene) cables without the creation of voids in the insulation has limited their use to relatively low voltages, although such cables are becoming available for voltages above 100 kV. The temperature limitation of these cables is 70°c, above which the conductor tends to sink in the hot plastic, becoming offset and creating higher voltage stresses. This limit has been raised to 90°c by the use of

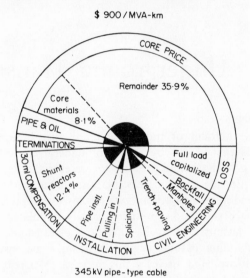

345 kV pipe-type cable

Figure 8.1   Cost components of conventional 345 kV pipe-type cable and U.S.A. practice 1968. (*Permission of the Institute of Electrical and Electronic Engineers.*)

cross-linked Polythene. An interesting development is the use of plastic tapes in combination with various insulating oils, in much the same manner as paper and oil, to avoid the formation of voids; here the problem

Table 8.1   Comparison of insulation materials

| Material | Dielectric resistance ($\Omega$ cm) | $\varepsilon_r$ | Power factor | Flame retardance | Cont. operating temp. (°c) |
|---|---|---|---|---|---|
| Chemically cross-linked polythene | $10^{19}$ | 2·3 | 0·0002 | Burns | 90+ |
| Polyvinyl chloride | $10^{14}$ | 4·5–7 | 0·08 | Self-extinguishing | 70 |
| Butyl rubber | $10^{16}$ | 3·4 | 0·008–0·015 | Burns | 85 |
| Ethylene propylene rubber | $10^{16}$ | 2·8 | 0·003–0·015 | Burns | 90 |

is one of economic viability. In low-voltage distribution the use of plastic cables is more widespread. Some basic properties of plastic insulants are summarized in Table 8.1.

A considerable development in recent years has been the widespread use of aluminium as both a conductor and sheath material. It has largely superseded lead as a sheathing material and is quickly gaining ground as a conductor. A major problem with aluminium conductors has been the making of effective joints which are in practice subject to thermal cycling and creep. Many techniques have been used and today much more confidence is placed in the available jointing techniques. The main reason for the use of aluminium is its price stability compared with copper which has severe price fluctuations. A comparison of possible conductor materials is given in Table 8.2.

Table 8.2  Conductor materials for power transmission

| Material | Resistivity $\sigma$ ($\mu\Omega$ cm) | Density $\rho$ (g/cm³) | $\rho\sigma$ | Cost per pound weight $C(\$)$ | $C\sigma\rho$ |
|---|---|---|---|---|---|
| Copper | 1·72 | 8·9 | 15·3 | 0·43 | 658 |
| Aluminium | 2·82 | 2·7 | 7·62 | 0·31 | 236 |
| Silver | 1·62 | 10·5 | 17·0 | 19·00 | 32300 |
| Sodium | 4·3 | 0·97 | 4·17 | 0·17 | 71·0 |

If $l$ = length, $A$ = area and $R$ = permitted resistance of conductor, $A = \sigma l/R$ and total cost $= C\rho Al = C\rho l^2 \sigma/R = $ Const. $\times C\rho\sigma$.

At lower voltages oil-impregnated paper-insulated cables (solid type) are used often with the three conductors contained in a single sheath. The three conductors are stranded and insulated separately and then laid up spirally together. The space between and around the conductors is packed with paper or jute to form a circular surface which is then further wrapped with insulation. This is called the *belted* type of cable (see figure 8.2) and it may have steel-wire armouring over the sheath because no induced eddy currents are produced in the armouring wires, whereas with single conductor cables severe loss and increased impedance would result. With the three-core cable high electric stresses are set up tangentially to the paper insulation surfaces in which direction the insulation strength is weakest. To overcome this each core is wrapped with a conducting layer of metallized paper which converts the cable electrically into three single-core cables with the electric stress completely in the radial direction. This form of construction was originated by Hochstadter and is known as the 'H' type. As system voltages increased above 33 kV the solid-type cable became prone to breakdown because of the voids formed (small pockets of air or gas) in the insulation when constituent parts of the cable expanded and contracted to different extents with the heat evolved on load cycles. The stress across these voids is high and local discharges occur, creating heat which chars the paper. Eventually complete breakdown results.

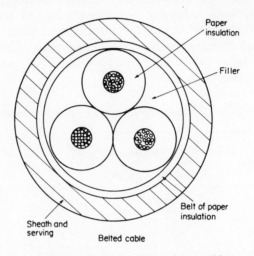

Figure 8.2   Belted type construction, 3-core cable.

In many countries, for voltages above 33 kV, the type of cable system in most common use is the oil-filled cable with paper/oil insulation, due to Emanueli[3]. In the oil-filled cable the hollow centre of the conductor is filled with insulating oil maintained under pressure by reservoirs feeding the cable along the route. As the cable heats on load, oil is driven from the cable into the reservoirs and vice versa, hence the creation of voids is avoided. In gas-pressure installations nitrogen at several atmospheres maintains a constant pressure on the inner sheath, compressing the dielectric, and so preventing the formation of voids. In the U.S.A. the paper/oil-insulated cable is often installed in a rigid pipe containing the insulating oil and a conductor oil-duct is not required. In Table 8.3 the extent of underground transmission in the U.S.A. is analysed. At the highest voltages, at present of the order of 400 kV, the oil-filled cable has a design working stress of 150 kV/cm. A 275 kV oil-filled cable and joint are shown in figure 8.3a and b and a h.v. cable sealing end in figure 8.4.

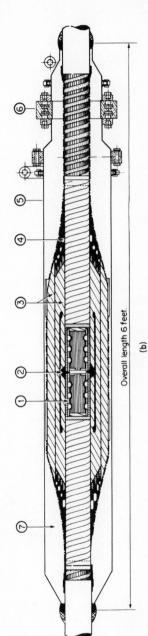

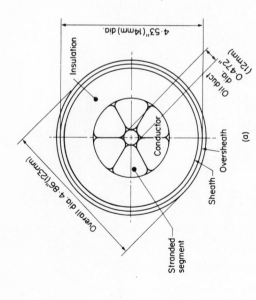

Figure 8.3a    Cross-section of a 3 in² (1935 mm²) copper conductor oil-filled cable.

Figure 8.3b    275 kV straight joint for oil-filled cable. Pirelli design for low thermal resistance. (*Permission of Institution of Electrical Engineers.*)

1 Ferrule                          4 Paper insulation
2 Stress control electrode         5 Copper sleeve
3 Filled epoxy resin casting       6 Cross-bonding insulator
                                   7 Insulating oil

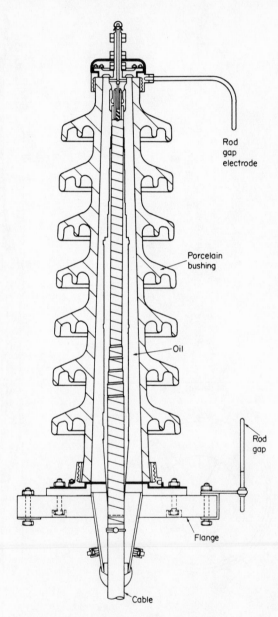

Figure 8.4   H.V. cable outdoor sealing end.

Table 8.3  Underground transmission cables in the U.S.A.

| Cable type | 69 kV | 138 kV | 230 kV | 345 kV | Total |
|---|---|---|---|---|---|
| High-pressure pipe-type circuit miles | 360 | 950 | 30 | 60 | 1400 |
| Low-pressure oil-filled circuit miles | 238 | 225 | 2 | — | 465 |
| TOTAL | 598 | 1175 | 32 | 60 | 1865 |

*Methods of laying*

The four main methods of installing cables are as follows:

(a) Direct in the soil—the cable is laid in a trench which is refilled with a backfill consisting of either the original soil or imported material of lower thermal resistivity.
(b) In ducts or troughs usually of earthenware or concrete—the cable is laid in open ducts which are sometimes filled with compound and covered with top slabs.
(c) In circular ducts or pipes through which cables are drawn—this has the advantage of further cables being installed without excavation.
(d) Where possible, cables are installed in tunnels built for other purposes.

*Current developments* include sodium conductors and a gas-filled cable consisting of a conductor supported in a rigid external pipe which is filled with a gas under pressure, usually a mixture of sulphur hexafluoride ($SF_6$) and nitrogen. One advantage of the latter is much better heat dissipation by natural convection in the gas. Perhaps more revolutionary are cables incorporating cryogenic coolants which will be described in more detail in a further section.

The possibility of transmitting large amounts of power through *wave guides* is also being considered. These would use oversized circular pipes and work in the $TE_{01}$ mode at frequencies from 3 to 10 GHz. Attenuation of the $TE_{01}$ mode in a circular wave guide decreases with increasing radii of pipe and increasing frequency. The variation of the specific attenuation ($\alpha$) per kilometre and the maximum power transmittable with radius are shown in figure 8.5. For 4 GW (4000 MW) of power to be transmitted over 1000 km with the same loss as with a 50 Hz e.h.v., a.c., line (about 1 dB) the wave guide will be at least 1·15 m at 3 GHz and 0·65 m at 10 GHz, giving power transfers of 10·6 and 3·4 GW, respectively. The relative merits of such a system have been discussed by Paul[37].

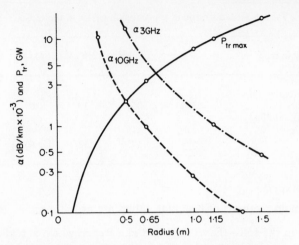

Figure 8.5   Attenuation (α) and power handling capacity ($P_{tr}$) of a circular waveguide. (Permission I.E.E.; Reference No. 37.)

Conductor cross-sectional areas are described in different units in various countries. In the U.S.A. the circular mil is used (area enclosed by a circle 0·001 in. diameter), in Britain the square inch converting in time to the square millimetre and in Europe the square millimetre. Relevant conversion factors are:

*To convert mm² to circular mils multiply by 1970*
*To convert in.² to circular mils multiply by 1·273239 × 10⁶*

## 8.2  Electrical Characteristics of Cable Systems

*Joints*

The maximum length between joints is determined by the drum size and for most cable circuits several lengths must be joined. At high voltages the design of the joint is complex and critical, especially as the paper tapes are applied by hand *in situ* and hence do not possess the electric strength of the cable tapes because of the presence of moisture. A low thermal resistance straight-through joint (connecting two lengths) is shown in figure 8.3b. Oil-filled systems need special joints to terminate the circuit ends and joints to allow oil from the reservoirs to enter the cable (feed joints). Both from an electrical and thermal standpoint, the joint represents the limiting part of a cable system.

## Cable losses

The $I^2 R_{a.c.}$ loss of the conductor represents the largest heat source. The alternating-current resistance of the conductor $R_{a.c.}$ is the direct-current resistance $R_{d.c.}$, modified to account for the skin and proximity effects. Skin effect even at power frequency is significant and increases with conductor cross-section. With large conductors, e.g. 3 in.$^2$ (1935 mm$^2$) the increase in $R_{a.c.}$ due to this cause is of the order of 20 per cent[8]. It is minimized by the use of stranded conductors which in large conductors are formed into segments which are transposed.

*Proximity* effects include the eddy currents induced in the conductor and sheath of a cable in a circuit comprising three separate single-conductor cables by the conductor fluxes of the neighbouring cables. This loss decreases with increased separation between cables. Further losses are induced into steel-wire armouring wound around some cables (especially submarine) for mechanical protection.

The dielectric loss, due to leakage and hysteresis effects in the dielectric, is usually expressed in terms of the loss angle $\delta$; $\delta = 90 - \phi_d$, where $\phi_d$ is the dielectric power-factor angle. The dielectric loss $= \omega C V^2 \cos \phi_d$ where

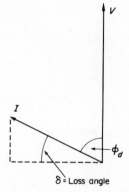

Figure 8.6   Loss angle in dielectric
Loss $= VI \cos \phi_d = VI \sin \delta$
$= \omega C V^2 \sin \delta.$

$C$ = capacitance to neutral per m, and $V$ = phase voltage. For modern cables $\cos \phi_d$ lies between 0·002 and 0·003 but this value increases rapidly with temperature above 60°c in oil-filled cables. In low-voltage cables this loss is negligible, but is very appreciable in cables at 275 kV and above. A phasor diagram illustrating the nature of the loss angle is shown in figure 8.6.

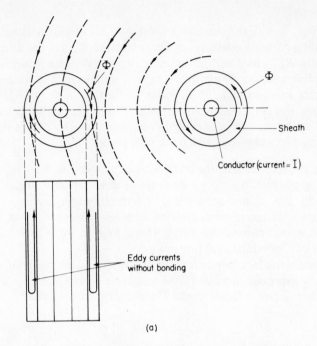

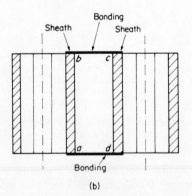

**Figure 8.7a**    Currents and fluxes in sheaths without bonding of ends of sheaths.

**Figure 8.7b**    Two cables with sheaths bonded. Spacing between centres of conductors $= s$ metres. $r$ and $r_s$ = radius of conductor (solid) and mean radius of sheath (metres).

*Currents and voltages in sheaths*

Two cables will be considered as shown in figure 8.7. If the sheaths are isolated, i.e. not grounded at the ends and the fluxes set up by one cable, cut the sheath of the other and induce eddy currents as shown in figure 8.7a. Although the loss in this condition is very small except when the cables are close or touching, voltages are set up along the sheath which may become excessive. To avoid this the sheaths are bonded at the ends forming the end-connexions indicated in figure 8.7b.

The flux through the loop abcd forms a mutual inductance $M$ between sheath and conductor equal to,

$$4 \times 10^{-7} \ln \left(\frac{s}{r_s}\right) \; H/m$$

where $s$ = spacing between sheath centres
$r_s$ = mean radius of sheath.

and for a three-phase system with equilateral spacing the effective value of $M$ is

$$\frac{2}{10^7} \ln \left(\frac{s}{r_s}\right) \; H/m$$

The self-inductance of the sheath $L_s$ is approximately equal to the mutual inductance $M$ and letting the resistance of the current path be $R_s$, the sheath current,

$$I_s = \frac{I\omega M}{\sqrt{(R_s^2 + (\omega M)^2)}}$$

and the sheath loss per phase

$$= I_s^2 R_s = R_s \left(\frac{I^2 \omega^2 M^2}{R_s^2 + \omega^2 M^2}\right) \tag{8.1}$$

Hence total effective resistance, $R_{a.c.}^* = R_{d.c.}$ (1 + skin effect factor + proximity effect factor)

$$+ \left(\frac{\omega^2 M^2}{R_s^2 + \omega^2 M^2}\right) R_s \text{ or } R_{a.c.}^* = R_{a.c.} + \left(\frac{\omega^2 M^2}{R_s^2 + \omega^2 M^2}\right) R_s$$

The equivalent circuit is shown in figure 8.8, the conductor and sheath form a coupled circuit.

Consider a metre length of the circuit, the voltage drop per metre
$= \Delta V = I (R_c + j\omega L_c) + I_s j\omega M$,
and in the sheath, $0 = I_s (R_s + j\omega L_s) + I j\omega M$.

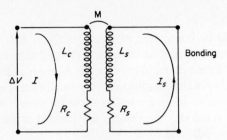

Figure 8.8   Equivalent circuit formed by conductor and bonded sheaths.

From these two equations

$$Z = \frac{\Delta V}{I} = \left( R_c + R_s \left( \frac{\omega^2 M^2}{R_s^2 + \omega^2 M^2} \right) \right) + j \left( \omega L_c - \frac{\omega^3 M^3}{R_s^2 + \omega^2 M^2} \right) \quad (8.2)$$

$$= R_{\text{effective}} + j X_{\text{effective}}$$

where, $L_s = M$.

Although the equivalent circuit shown is for a single-phase circuit, equation (8.2) applies also for a 3-phase, 3-cable system, and

$$\omega L_c = X_c = 2\omega \left( 2 \ln \frac{s}{r} + \frac{1}{2} \right) \times 10^{-7} \ \Omega/\text{m}$$

and

$$X_s = \omega M = 2\omega \left( 2 \ln \frac{s}{r_s} \right) \times 10^{-7} \ \Omega/\text{m},$$

where, $s$ = spacing between conductor and sheath centres, and $r$ and $r_s$ = radius of conductor (solid) and mean radius of sheath.

The losses occurring with sheaths short-circuited have resulted in the widespread use of a system of sheath transposition known as *crossbonding* which is illustrated in figure 8.9.  At each joint along the route the sheath

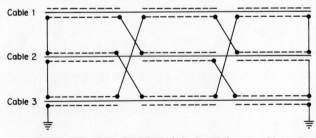

Figure 8.9   Cross bonding of single conductor cables.

is interrupted and transposed, and after three lengths have been traversed the sheaths are bonded to ground as shown. The effect is to add the sheath voltages in the adjacent sections and for balanced loads the sum will be zero. High sheath currents are avoided and induced voltages kept in reasonable proportions.

*Example 8.1* Determine the effective electrical parameters of a 3-phase, 66 kV, 50 Hz underground circuit, 65 km in length and comprising three single-core 200 mm² cables each of conductor radius 0·915 cm. The internal and external radii of the lead sheath are 2·5 and 2·8 cm, respectively. The cables are in touching equilateral formation and the sheaths are bonded to ground at several points. The conductor a.c. resistance per kilometre at 15°C is 0·0875 Ω/km and the resistivity of lead at the operating temperature may be assumed to be $23·2 \times 10^{-6}$ Ω cm.

*Solution* Conductor resistance $= 0·0875 (1 + 0·004 \times 50) = 0·105$ Ω per km. (Conductor-operating temperature assumed to be 65°C and temperature coefficient 0·004)

For the whole length, $R_{d.c.} = 0·105 \times 65 = 6·8$ Ω

Resistance of sheath

$$= \frac{23·2 \times 10^{-6} \times 65 \times 10^{5}}{\pi(2·8^2 - 2·5^2)} = 30 \ \Omega$$

In a 3-phase system the equivalent phase to neutral reactance is considered and in this case the conductor to sheath mutual inductive reactance,

$$X_m = \omega M = 2\pi \times 50 \times 2\ln\left(\frac{d}{r_s}\right) \times 10^{-7} \ \Omega/\text{m},$$

where $r_s$ = mean radius of sheath, $d$ = the axis to axis spacing (6·1 cm).

$$\therefore X_m = 2\pi \times 50 \times 2 \times \ln\left(\frac{6·1}{\frac{1}{2}(2·5 + 2·8)}\right) \times 10^{-7} \times 65{,}000$$

$$= 3·4 \ \Omega \text{ for 65 km length}$$

Effective a.c. resistance of conductor $R_{a.c.}^{*}$

$$= R_{a.c.} + \left(\frac{X_m^2 R_s}{R_s^2 + X_m^2}\right) = 6·8 + \frac{3·8^2 \times 30}{30^2 + 3·8^2} = 7·2 \ \Omega$$

Effective reactance per cable

$$= X - \left( \frac{X_m^3}{R_s^2 + X_m^2} \right)$$

(where $X$ = reactance with sheaths open circuit)

$$= \omega \times 2 \ln \left( \frac{d}{r_c} \right) \times 10^{-7} \times 65{,}000$$

$$= 314 \times 2 \ln \left( \frac{6 \cdot 1}{0 \cdot 915} \right) \times 10^{-7} \times 65{,}000$$

$$= 7 \cdot 707 \; \Omega$$

Total effective reactance

$$= 7 \cdot 75 - \frac{3 \cdot 4^3}{30^2 + 3 \cdot 4^2}$$

$$= 7 \cdot 737 \; \Omega$$

$$\frac{\text{Sheath loss}}{\text{Conductor loss}} = \left( \frac{I^2 R_s X_m^2}{R_s^2 + X_m^2} \right) \frac{1}{I^2 R}$$

$$= \left( \frac{30 \times 3 \cdot 4^2}{30^2 + 3 \cdot 4^2} \right) \frac{1}{6 \cdot 8}$$

$$= 0 \cdot 0568$$

For a current of 400 A the e.m.f. induced without bonding (per sheath)
$= X_m I = 1 \cdot 36 \, \text{kV}.$

## 8.3　System Operating Problems with Underground Cables

The determination of cable electrical parameters has been discussed in Chapter 2. It has been seen that the capacitance of cable systems is high because of their physical form. Table 8.4 shows a comparison of the charging reactive power requirements for typical overhead lines and underground cables in the United States. As the charging current flows in the cable conductor, a severe decrease in the value of load current transmittable (de-rating) occurs if the thermal rating is not to be exceeded; in the higher voltage range lengths of the order of 20 miles (32 km) create a need for drastic de-rating. A further current reduction is caused by the appreciable magnitude of dielectric losses at higher voltages.

Table 8.4 Comparison of charging MVA requirements for typical
overhead and underground a.c. transmission lines

| Voltage | Overhead 3-phase charging MVA/mile | Underground 3-phase charging MVA/mile |
|---------|-----------------------------------|---------------------------------------|
| 69      | 0·025                             | 1·9                                   |
| 138     | 0·106                             | 4·9                                   |
| 230     | 0·303                             | 8·8                                   |
| 345     | 0·852                             | 17·0                                  |
| 500     | 1·610                             | 30·3                                  |

The charging current (amperes per metre)

$$I_c = 2\pi f C V / \sqrt{3}$$

where $C$ is the capacitance (F) and $V$ the line voltage ($V$)

when $I_c = I_{rated}$, the cable length is termed critical.

Also, $C = 2\pi\epsilon_0\epsilon r / \ln(R/r)$

The total three-phase reactive power

$$= \sqrt{3}VI_c = \sqrt{3}V \, 2\pi f \left(\frac{V}{\sqrt{3}}\right) \cdot \left(\frac{2\pi\epsilon_0\epsilon r}{\ln(R/r)}\right)$$

$$= 4\pi^2 f \, V^2 \, \epsilon_0\epsilon_r \, 10^{-6} / \ln(R/r) \text{ MVAr/metre}$$

A 345-kV cable with approximately 1 in. (25·3 mm) thickness of insulation and an $\epsilon_r$ of 3·5 has a critical length of 26·5 miles (42 km), the corresponding MVAr requirement is about 17 MVAr per mile.

Apart from the reduction in transmission capability the presence of underground cables in a system presents further problems. Troubles arise when switching cables into circuit owing to their high capacitance (see Chapter 10) and also control of system voltage is difficult. Examination of the simple equivalent circuit of a cable indicates that on light or no load an appreciable voltage rise occurs at the load end (originally called the Ferranti Effect). In a system containing a reasonable proportion of h.v. cables such as in large cities, significant voltage changes occur as the load varies between its maximum and minimum value. It is difficult to control these voltage excursions by normal system equipment and it is necessary to install shunt inductive reactors in the cable circuit. These may be connected at the ends of the circuit and although technical advantage would result in connecting, say, one-third of the required amount of vars at each end and one-third at the middle, the cost of connexions at places other than the ends is relatively very high.

A further aspect of voltage control which is becoming more serious as the capacity of generators increases is the limited ability of large generators to operate at leading power factors. This requirement occurs especially at night when the load is low. It has been shown in Section 2.7 of Chapter 2 by means of a performance chart that the ability to generate leading MVArs is limited by stability considerations. In particular as the synchronous reactance becomes larger the leading MVAr capability decreases and in high-capacity modern generators the reactance is larger than in smaller machines. These problems are avoided when cables are used in direct-current transmission schemes. Against the saving in the cost of compensating reactors must be set the large cost of the d.c. conversion equipment. It will be appreciated that there is a considerable incentive to develop cable materials of low permittivity.

### 8.4  Steady-state Thermal Performance of Cable Systems

In this section it is proposed to consider the thermal characteristics of the components of a cable system from a steady state and transient viewpoint. The transient aspect can be due to short-circuits or the consideration of temperature-time characteristics when a steady load is changed under control to another value; this is electrically a steady-state condition as opposed to conditions on sudden short-circuits.

In a transmission link, the component with the lowest limit will decide the current-carrying capacity; differences in thermal time-constant may, however, have a marked influence on this. An aspect of great economical importance is the possibility of short-time overloading.

The ambient temperature depending as it does on the seasons and weather conditions influences thermal limits; this operates favourably as the low winter ambient temperatures coincide with high load demands. The thermal limitations to transmission capacity are therefore not rigidly fixed and must be determined scientifically to obtain the optimum utilization of plant.

The thermal limit on an item of plant is that value of current which produces a maximum or 'hot-spot' temperature which should not be exceeded. Care should be taken to ascertain the difference between the 'hot-spot' and mean temperatures. The critical temperature rise is usually that in the insulation adjacent to the conductor.

Depending upon the class of insulation used the maximum allowable temperature rise is determined by a mixture of scientific analysis and practical experience. Difficulty arises in the assessment of allowable temperatures as to what precisely the criterion of plant failure or length of life should be. A transformer for example will continue to function with insulation which from the point of view of many physical tests is unsound. Failure usually occurs where some undue stress is applied due to a fault condition, when high mechanical and thermal stresses are set up. Factors which influence the maximum permissible temperature rise in cables are as follows.

(a) The differential expansion between the insulation, the surrounding sheath and the conductor, sets up mechanical stresses and subsequent abrasion of the insulation. This is due to the different temperature coefficients and temperature rises of these materials.

(b) Changes in the electrical properties of the insulation especially in the dielectric loss.

(c) Changes in mechanical and chemical properties which result in electrical changes.

The following law relates the probable *life of insulation* with temperature $L = A\,e^{-m\theta'}$, where $L$ is the insulation life (years), $\theta'$ the temperature, and $A$ and $m$ constants depending on the type of insulation. From this, the rule of thumb law regarding insulation life is obtained; this states that for Class B insulation the insulation life is halved for each 10°c rise in temperature and doubled for each 10°c fall in temperature relative to a specified maximum temperature. For Class A materials the corresponding change in temperature is 8°c and for Class H, 12°c.

For an insulation life of ten years the specified maximum temperatures are as follows:

Class A, 105°c;  Class B, 130°c;  Class H, 180°c.

Recommended steady-state working temperatures of the conductors of paper/oil cables are as follows:

armoured cables buried direct in soil—65°c;

oil-filled or gas pressure cables direct in soil—85°c;

lead sheathed cables in ducts (no armouring required)—50°c.

The above values assume an earth temperature of 15°c; the range of earth temperature to be expected depends upon the depth and geographical location.

*Thermal resistance of a single-core cable*

Let the thermal resistivity of the dielectric material be $g$°c cm/W. Then considering an annulus of width d$x$ for a 1 cm length of cable (figure 8.10)

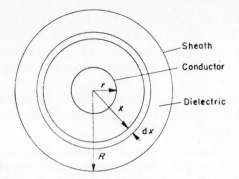

Figure 8.10   Thermal resistance of a single-core cable.

the thermal resistance

$$= g \cdot \frac{dx}{2\pi x \times 1} \, ^\circ c/W.$$

Total thermal resistance of the dielectric is

$$G_d = \int_r^R \frac{g \, dx}{2\pi x} = \frac{g \ln R/r}{2\pi} \, ^\circ c/W. \tag{8.3}$$

The thermal resistance between a core and the sheath in a three-core cable is difficult to calculate analytically. In practice a factor usually obtained by computer tests is used to modify equation (8.3).

*Thermal resistance of the cable environment*

The cables are assumed to be directly buried with the trench refilled with the original soil, i.e. they are situated in a semi-infinite homogeneous medium. The ground surface is assumed to be an isothermal plane. The thermal resistivity ($g$) of soil varies widely depending on the nature of the soil (clay, sand, etc.), and its moisture content. Values for $g$ frequently taken are 120°c cm/W and 90°c cm/W. The value used in different countries varies somewhat but is in the above region. By far the best policy when specifying a cable is to carry out a thermal resistivity survey of the proposed route. This will highlight the point of highest $g$, which value should be used in the calculation of the current rating. The mean value of $g$ taken over the route is of little use as it is the hottest section of the cable that will fail; little heat is conducted longitudinally from the sections of highest temperature.

The external thermal resistance of a single cable may be found readily by considering the image of the line heat-source in the thermal field as

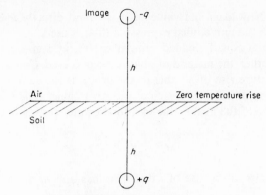

Figure 8.11  Image method for determining thermal resistance due
to surrounding soil—one cable (radius *r*).

shown in figure 8.11.  If *q* watts is the total loss in one cm of the cable,
the temperature difference between the external surface of the cable and

the ground surface is $\dfrac{q \cdot g}{2\pi}$ ln $\dfrac{2h}{r}$. *h* is normally in the order of 42 in.

(100 cm).  The external thermal resistance is therefore

$$\frac{\theta}{q} = \frac{g}{2\pi} \ln \frac{2h}{r} \; °\text{C/W per cm of cable} \qquad (8.4)$$

Modifications to (8.4) have been made in the past to allow for the difference in the actual value of *g* and that of the soil sample tested in the

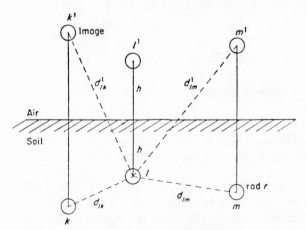

Figure 8.12  Image method for determining external thermal
resistance for group of buried cables.

laboratory. Nowadays the value of $g$ is found directly *in situ* by probe methods and the unmodified expression (8.4) is used.

With several equally loaded, similar cables, in a group, which do not touch each other, the method of images is again used. Consider figure 8.12. The temperature rise of a cable in the group is the sum of its own independent rise and that due to each of the other cables. The temperature rise at the surface of the $l$ cable due to a loss $q$ watts per cm in cable $k$ is

$$\theta_{kl} = \frac{g}{2\pi} q \ln \frac{d'_{lk}}{d_{lk}}$$

Also, the temperature rise of $l$ due to a loss $q$ in $m$

$$= \frac{qg}{2\pi} \ln \frac{d'_{lm}}{d_{lm}}$$

Hence the temperature rise of $l$ due to both $k$ and $m$

$$= \frac{qg}{2\pi} \ln \frac{d'_{lm} d'_{lk}}{d_{lm} d_{lk}}$$

The total temperature rise of cable $l$ therefore will be,

$$\frac{qg}{2\pi} \ln \frac{2h}{r} + \frac{qg}{2\pi} \ln \left( \frac{d'_{lm} d'_{lk}}{d_{lm} d_{lk}} \right)$$

For the popular configuration of three cables laid in horizontal formation with a spacing '$a$' between adjacent cable centres and a burial depth of '$h$',

$$d_{kl} = a, d'_{kl} = \sqrt{[(2h)^2 + a^2]}$$

$$d_{lm} = a, d'_{lm} = \sqrt{[(2h)^2 + a^2]}$$

$$\therefore \qquad \theta_l = \frac{qg}{2\pi} \left[ \ln \frac{2h}{r} + \ln \left( \frac{4h^2 + a^2}{a^2} \right) \right] \qquad (8.5)$$

The effective thermal resistance of the centre cable of the group is $(\theta_l/q)$ °c/W.

For cables drawn into ducts the external thermal resistance consists of the sum of the duct ground-surface resistance, the duct-wall resistance, and the cable surface to inner duct-wall resistance.

*Moisture migration* (reference 17)    An important aspect of cable environment is the increase in the resistivity of soil when it dries out. This has become important owing to the less cyclic nature of the load on some high-voltage cables now installed. When these cables form interconnexions to high .efficiency generating stations they carry full load continuously.

The moisture in the surrounding soil migrates leaving air in the interstices previously occupied by water and the thermal resistivity increases. This in turn increases the cable temperature which increases the dielectric loss and to a lesser extent the copper loss, thus causing increased moisture migration. The result is *Thermal Instability* and the electrical failure of the insulation.

*Calculation of cable steady-state temperature rises*

Methods for the calculation of the relevant thermal resistances and cable losses have been discussed and it now remains to outline the methods of calculation of the temperature rises. As with other items of electrical plant it is desirable to obtain an equivalent lumped-constant thermal circuit. This is shown for a cable of a directly buried, naturally cooled, single-core system, in figure 8.13a. It is seen that the dielectric loss is injected at the midpoint of the dielectric resistance; this is an approximation and the dielectric should be further sub-divided for more accurate representation. Usually the sheath thermal resistance is negligible. The network to represent a single-core cable with internal cooling is shown in figure 8.13(b)

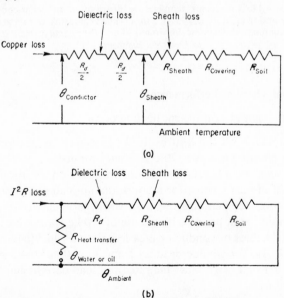

Figure 8.13a   Equivalent thermal circuit for single-core, naturally-cooled, buried cable. $R_d$ = radial thermal resistance of dielectric.
b   Equivalent circuit of single-core buried cable with internal cooling.

*Current-carrying capacity of cable joints.* Although much attention has been devoted to cable heating it has now become apparent that with artificially cooled schemes the joint will rise to the high temperature. Thermal analysis of the joint is difficult compared with cables and requires a computer; a temperature distribution is shown in figure 8.14.

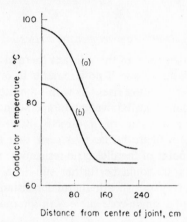

Figure 8.14  Conductor temperature distribution in water-cooled cable and straight joint, 275 kV 2 in² oil-filled cable, (a) four separate cooling pipes, (b) water jacket round joint. (*Permission of Institution of Electrical Engineers.*)

## 8.5  Transient Thermal Performance

The following regimes of transient heating occur.

(a) *Short-circuit*  All the heat evolved is assumed to be stored in the cable. A temperature limit of 120°c is often assumed.

(b) *Cyclic loading*  Calculations are often based on an idealized daily load curve of full-load current for 8 hours and no-load for 16 hours. This results in a lower temperature rise than when permanently on full load which is allowed for in practice by the use of Cyclic Rating Factors. For directly buried cables of conductor area less than 0·1 in.² (64·5 mm²) this factor is 1·09, and for conductor areas from 0·1–1·0 in.² (64·5–645 mm²) it is taken as 1·13, i.e. 1·13 times the rated full-load current can be passed on cyclic loading.

(c) *Short-time emergency loading*  An overload may be sustained for perhaps several hours without excessive temperatures due to the long thermal time-constant of the cable.

A lumped-constant thermal network may be used for transient calculations by the connexion of appropriate thermal capacitances between the nodes of the steady-state network and the ambient or reference line. For a cable if uniformly radial heat flow is assumed, i.e. the sheath is isothermal and the thermal circuit is an $R$–$C$ ladder network. The number of nodes depends on the number of annular cylinders into which the dielectric is divided. The conductor and sheath temperatures may be obtained from a simple model such as suggested by Wormer[21] and further developed by Neher[22] and shown in figure 8.15.

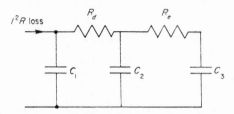

Figure 8.15 Simple model for transient thermal response of buried cable, $C_1$ = thermal capacitance of conductor + ($p$ × capacitance of insulation), $C_2$ = thermal capacitance of sheath + $(1 - p)$ × (capacitance of insulation), $C_3$ = effective capacitance of soil, $R_i$ = thermal resistance of insulation, $R_e$ = effective thermal resistance of soil, $p = 1/(2\ln(R/r)) - 1/(R^2/r^2 - 1)$ where $R$ = outside radius of cable and $r$ = conductor radius.

In transients of short-time interest the surrounding soil may be represented by a few rings with the assumption of radial flow. For an accurate simulation the temperature rise at the cable surface due to the soil alone is described by the equation

$$\theta(t) = \frac{q(t)\,g_{\text{soil}}}{4\pi}\left[\left\{-Ei\left(-\frac{d^2}{16\alpha t} + Ei\left(\frac{l^2}{\alpha t}\right)\right)\right\}\right.$$
$$\left. + \sum_{i=1}^{i=N}\left\{-Ei\left(\frac{r_{1,i}^2}{4\alpha t}\right) + Ei\left(-\frac{r_{2,i}^2}{4\alpha t}\right)\right\}\right] \qquad (8.6)$$

where $i$ = number of cable, $r_{1,i}$, $r_{2,i}$ = distances of cable $i$ and its image from reference cable,

$d$ = cable diameter, $\alpha$ = thermal diffusivity, $Ei$ = exponential integral.

If $q(t)$, the value of the heat transmitted from the cable to the soil, is known at the instant $t$, then the temperature of the cable surface is known and the cable thermal network may be solved. However, $q(t)$ varies from zero at $t = 0$ to $q$ (steady state) at $t = \infty$. A reasonably accurate formula for obtaining $q(t)$ is,

$$\frac{q(t)}{q_{\text{steady state}}} = \left(\frac{\text{Temperature across cable at } t}{\text{Steady-state temperature rise across cable}}\right)$$

(*cable surface node short-circuited*)

The simplest analysis results from a single lumped constant representation, i.e. a single $R$–$C$ network. The response of such a network to a time-varying load is illustrated in figure 8.16 in which $\tau$ is the thermal time constant ($R \times C$) and $\theta_{m1}$, $\theta_{m2}$ and $\theta_{m3}$ are steady-state temperatures.

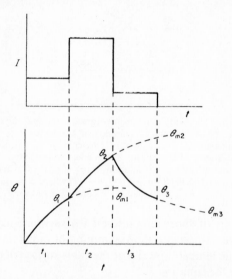

Figure 8.16   Temperature rise—time curves, varying load. $\theta_1 = \theta_m$ $(1 - e^{-t_1/\tau})$;   $(\theta_2 - \theta_1) = (\theta_{m2} - \theta_1)\,(1 - e^{-t_2/\tau})$;   $(\theta_3 - \theta_{m3}) = (\theta_2 - \theta_{m3})\,e^{-t_3/\tau}$.

Most items of power equipment possess temperature rise-time characteristics having more than one exponential term. This fact is important as often when carrying out heat runs the temperature will appear to no longer rise with the time scale being used, usually a few hours. It is quite possible, however, for the temperature to continue rising appreciably over a considerable time. In buried cables, for example, many hundred hours elapse before steady conditions are reached.

*Intermittent loading of plant*

The effects of a regularly repeated on–off load are shown in figure 8.17. In order to simplify the analysis the equipment will be assumed to have a single time-constant. From figure 8.17, on heating,

$$\theta_2 - \theta_1 = (\theta_m - \theta_1)(1 - e^{-t_1/\tau})$$

on cooling,

$$\theta_1 = \theta_2\, e^{-T_1/\tau}$$

From these,

$$\theta_2 - \theta_2\, e^{-T_1/\tau} = (\theta_m - \theta_2\, e^{-T_1/\tau})(1 - e^{-t_1/\tau})$$

$$\therefore \qquad \theta_2(1 - e^{-(T_1 + t_1)/\tau}) = \theta_m(1 - e^{-t_1/\tau})$$

From which,

$$\theta_2 = \theta_m\left(\frac{1 - e^{-t_1/\tau}}{1 - e^{-T/\tau}}\right) \qquad\qquad (8.7)$$

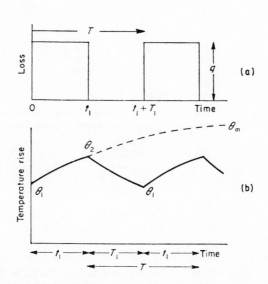

Figure 8.17a   On-off load cycle: $T = t_1 + T_1$. (b) Corresponding temperature curves. $\theta_m$ = steady-state temperature rise with q dissipated continuously.

In all the above equations the losses have been assumed constant. In practice the $I^2 R$ loss increases with increase in temperature due to the temperature coefficient of resistance of copper (0·4 per cent per °C). The losses will normally consist of a constant plus a variable portion, i.e. $q_0 + a\theta$.

### 8.6  Artificial Cooling of Underground Cables

In recent years progress has been made in schemes for reducing the external thermal resistance of cables by artificial cooling. These methods take two main forms.

(a) The pumping of water or insulating oil along a central duct formed in the cable conductor. This has been carried out using water in 11 kV plastic-insulated, single-core cables over short lengths. An engineering difficulty lies in the required gradual decrease of voltage at the cable ends where the water must enter cooling systems.

(b) Water is pumped through plastic pipes laid beside the cables as illustrated in figure 8.18a. The normal thermal resistance presented to the cable losses is shunted by the thermal resistance into the water. The water

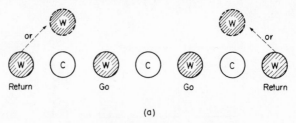

(a)

Figure 8.18a    Cooling pipes laid along buried cables—two pipes take outward flow and other two the return flow.

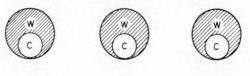

(b)

Figure 8.18b     Integral cooling.

increases in temperature along the route so that there is a limit to the individual lengths which can be cooled in this manner. In this scheme the cooling is not so efficient as in (a), as the total thermal resistance comprises the following: the soil between the cable and pipe, the pipe wall, the heat-transfer resistance into the water, and the resistance of the cable dielectric, serving, etc. In (a) the main loss, the $I^2R$ loss of the conductor, dissipates directly into the coolant impeded only by the heat-transfer resistance which is small. The heat-transfer into the coolant may be evaluated using formula (8.8). The pipe-to-soil resistance in case (b) may be calculated by the method of images assuming an infinite homogeneous medium or by field plots using appropriate techniques.

(c) A more intensive water-cooled system than the use of external pipes involves the placing of each cable in a rigid plastic pipe of approximately 20 cm diameter and pumping water through the pipe. It is more expensive than method (b) but with a 3 in.$^2$ (1935 mm$^2$) conductor cable currents of 3200 A may be achieved. This method is known as integral cooling, see figure 8.18b. In American pipe-type cables the insulating oil may be pumped thus providing enhanced heat transfer from the cable sheaths. The thermal network is shown in figure 8.18c. The necessity for artificial

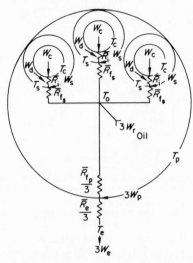

Figure 8.18c  Equivalent thermal circuit of U.S. pipe type cable system with forced cooling by passage of coil. Three cables are installed per pipe. (*Permission of Institute of Electrical and Electronic Engineers*).

$W_c$ = conductor loss
$R_i$ = insulation thermal resistance
$T_c$ = conductor temperature
$T_s$ = sheath temperature
$T_o$ = oil temperature
$T_e$ = earth temperature
$T_p$ = temperature of pipe
$R_{fs}$ = thermal resistance sheath to oil
$R_{fp}$ = heat-transfer resistance oil to pipe walls
$R_f$ = thermal resistance of pipe
$R_e$ = effective thermal resistance of soil

cooling if transmitted powers are to be increased at higher voltages is illustrated in figure 8.19. It is seen that for each mode of cooling there exists a maximum power at a certain dielectric stress after which, owing to increased dielectric losses, the power falls. With normal burial in soil of

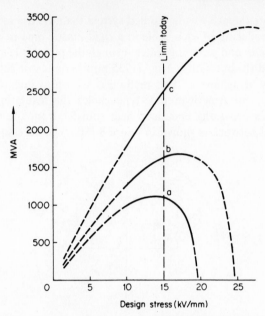

Figure 8.19   Relation between transmittable MVA and maximum working stress on cable dielectric (buried oil-filled 1935 mm² conductor cable) (a) natural cooling (normally buried) (b) use of selective backfill (normally buried) (c) artificial cooling.

thermal resistivity 120°c cm/W, the limit is reached at a stress of 150 kV/cm. The rating of cables laid in tunnels or large ducts may be increased by the provision of air flow either artificially induced or by natural ventilation.

*Calculation of essential parameters*

The heat transfer coefficient $h$ (watts/cm² per °c) between the coolant and containing pipe walls (turbulent flow and no entry effects) may be calculated from the expression[4],

$$\frac{hd}{k} = 0 \cdot 023 \left(\frac{dV\rho}{\eta}\right)^{0 \cdot 8} \left(\frac{c\eta}{k}\right)^{0 \cdot 4} \tag{8.8}$$

where
    $k$ = thermal conductivity of coolant
    $\rho$ = density of coolant
    $c$ = specific heat of coolant
    $\eta$ = viscosity of coolant
    $d$ = mean hydraulic diameter of pipe

$$\text{i.e.} \left(4 \times \frac{\text{Cross-sectional area}}{\text{Wetted perimeter}}\right)$$

For natural convection in air an empirical law applies of the form,

$$\frac{hd}{k} = 0{\cdot}41 \, (N_{\mathrm{Gr}}, N_{\mathrm{pr}})^{0{\cdot}25} \qquad (8.9)$$

where $N_{\mathrm{Gr}} = \dfrac{\beta g \theta d^3 \rho^2}{\eta^2}$ and $N_{\mathrm{pr}} = \dfrac{\eta c}{k}$

$\beta$ = volumetric coefficient of expansion of coolant
$g$ = acceleration due to gravity
$d$ = characteristic dimension, e.g. diameter of cable

*Calculation of coolant temperature rise*

In artificially cooled schemes the reference temperature is that of the coolant and the temperature rise of this is a limiting parameter; it may be calculated as follows. Consider a pipe, tunnel or duct traversed by a coolant at $W \, \mathrm{m}^3/\mathrm{s}$ entering at $\theta_0\,°\mathrm{c}$. The cables emit a loss in the steady state of $q$ Watts/m and the thermal resistance through the containing

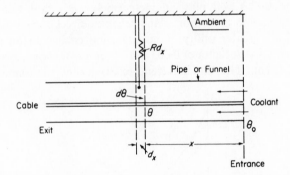

Figure 8.20 Temperature rise of cooling medium in a tunnel or duct.

walls and surrounding soil to ambient is $R°\mathrm{c}/\mathrm{W}/\mathrm{m}$. At a distance $x$ from the entrance the coolant temperature is $\theta$ (see figure 8.20) and the following equation applies:

$$\frac{\theta}{R/\mathrm{d}x} + CW\,\mathrm{d}\theta = \mathrm{d}x\,q,$$

where $C$ is the volumetric specific heat of the coolant.

Hence,

$$\frac{d}{Rq - \theta} = \frac{\mathrm{d}x}{CWR}$$

integrating and putting $CWR = \lambda$,

$$\ln\left(\frac{Rq - \theta}{A}\right) = e^{-x/\lambda}$$

where $A$ is the integration constant.

When $x = 0$, $\theta = \theta_0$, and $Rq - \theta_0 = A$. Also with zero coolant velocity the temperature rise, $\theta_m = qR°c$. So that

$$\theta = \theta_m - (\theta_m - \theta_0)\, e^{-x/\lambda} \qquad (8.10)$$

Hence the steady-state coolant temperature at any distance may be calculated. The actual temperature of the cable conductor is obtained by adding to the coolant rise the temperature drop across the cable dielectric and covering plus the drop across the cable-to-air heat transfer resistance.

### 8.7   Cables for Direct Current Transmission

Owing to the absence of periodic charging current with direct voltage, high-voltage cables will play an increasingly important role in direct-current transmission links. For inland schemes direct current will be used because of the greater utility of underground cables. In an alternating current cable a power factor of 0·003 can be represented by a loss resistance of $3 \times 10^{12}$ $\Omega$-cm. The direct-current resistivity of the same dielectric

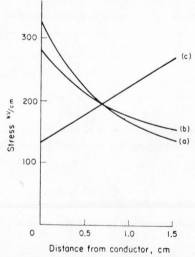

Figure 8.21   Distribution of electric stress in a direct-current cable. Curve (a) on a.c. curve (b) on d.c. (no load), curve (c) on d.c. with load, working voltage 300 kV. (*Permission of Electrical Journal.*)

would be greater than $10^{14}$ $\Omega$-cm, hence the loss in the dielectric on direct current is only about 3 per cent of that on alternating current. Whereas the electric stress distribution in alternating-current cables is determined by the dielectric capacitance, in direct current cables it is determined by the electric resistance of the dielectric. The electrical resistivity of the conventional dielectrics is very temperature dependent; for oil-impregnated paper for example the resistivity at 20°c is one hundred times that at 60°c. In direct-current cables thermal considerations not only determine the rating but also influence the electric stress distribution in the dielectric. The electrical resistivity also varies with the electric stress. Instead of the electric stress decreasing through the dielectric from the conductor to the sheath, in direct-current cables the stress increases and can be larger at the sheath, as shown in figure 8.21. This is known as stress inversion and can lead to troubles at terminations and joints where longitudinal stresses are created.

*Example 8.2* In a cable system with artificial cooling by water pipes, calculate the total resistance to heat flow presented by each pipe, i.e. heat transfer resistance and pipe wall resistance. The pipe details are as follows:

Internal diameter            0·75 in

Material polythene          $g = 570°$c cm/W

Wall thickness               0·125 in

*Coolant details* (water) at 20°c

$$\eta = 0{\cdot}994 \times 10^{-2} \text{ g/cm-sec}$$
$$k = 0{\cdot}00598 \text{ (W/cm}^2)(°\text{c/cm)}$$
$$\rho = 1 \text{ g/cm}^3$$
$$c = 4{\cdot}2 \text{ J/g per °c.}$$

Velocity of flow $V = 7{\cdot}95$ ft/sec.
Using the formula,

$$\frac{hd}{k} = 0{\cdot}023\left(\frac{dV\rho}{\eta}\right)^{0{\cdot}8}\left(\frac{c\eta}{k}\right)^{0{\cdot}4}$$

$$h\frac{0{\cdot}75 \times 2{\cdot}54}{0{\cdot}00598} = (46{,}300)^{0{\cdot}8}(6{\cdot}96)^{0{\cdot}4} \, 0{\cdot}023$$

$$= 265$$

$$\therefore \qquad h = 0{\cdot}83 \text{ W/cm}^2 \text{ per °c.}$$

This is the heat transfer coefficient into the coolant. The heat transfer conductance = area $\times h$.

Considering 1 cm length of pipe,

$$R_{H.T.} = \frac{1}{0 \cdot 83 \times \cdot 75\pi \times 2 \cdot 54}\,°c/W$$

$$= 0 \cdot 2\,°c/W.$$

The thermal resistance of the pipe wall =

$$\frac{gt}{\text{area}} = \frac{570 \times 0 \cdot 125 \times 2 \cdot 54}{0 \cdot 875 \times 2 \cdot 54 \times 1 \times \pi} = 25 \cdot 9°c/W$$

Hence the total resistance = $25 \cdot 92°c/W$. (*Note the importance of the pipe wall resistance.*)

*Example 8.3*  Three 132 kV single core cables rated at 650 A are laid at a depth of 42 in. in horizontal formation 2 in. apart. The thermal resistivity of the soil is 90°c cm/W. The cable details are as follows:

| | |
|---|---|
| Copper conductor | 0·45 in² cross-section |
| Insulation | 0·42 in radial thickness |
| | Outer radius 0·917 in |
| | $g = 500°c$ cm/W |
| Lead sheath | Thickness 0·09 in |
| | $g = 2·9°c$ cm/W |
| Covering | Thickness 0·22 in |
| | $g = 500°c$ cm/W. |

Rated-load copper loss at 85°c is 0·323 W/cm

Dielectric loss per cm is 0·0166 W/cm

Cables cross bonded—sheath loss negligible.

Calculate the temperature rise of the conductor and sheath.

*Solution*  The equivalent circuit is shown in figure 8.13. The radial thermal resistances of 1 cm length of cable are as follows:

$$R_d = \frac{500}{2\pi} \ln\left(\frac{0 \cdot 917}{0 \cdot 497}\right)°c/W$$

$$= 52 \cdot 08°c/W$$

$$R_s = \frac{2 \cdot 9}{2\pi} \ln\left(\frac{1 \cdot 007}{0 \cdot 917}\right)°c/W$$

$$= 0 \cdot 512°c/W$$

$$R_0 = \frac{500}{2\pi} \ln\left(\frac{1 \cdot 227}{1 \cdot 007}\right)°c/W$$

$$= 12 \cdot 8°c/W.$$

The thermal resistance presented to the centre cable of the group by the surrounding soil,

$$R_e = \frac{g}{2\pi} \ln\left(\frac{2h}{r}\right) + \frac{g}{2\pi} \ln\left(\frac{4h^2 + p^2}{p^2}\right)$$

where 
$h = $ depth of burial

$r = $ outer radius of cable

$p = $ spacing between cables

∴ $R_e = 145 \cdot 5\,°c/W$ per cm of cable.

The conductor temperature rise due to the $I^2R$ loss

$$= 0 \cdot 323\,(145 \cdot 5 + 70)$$

$$= 69 \cdot 5\,°c.$$

Assuming the dielectric loss to be injected at the electrical centre of $R_d$ the rise due to this is 3°C, giving a total conductor temperature of $69 \cdot 5 + 3 + 15$, i.e. 87·5°C where the ambient is 15°C.

*Example 8.4* An underground cable system is cooled by water pipes laid along the route. On normal operation with the water cooling in action the cable conductors attain a steady temperature rise of 70°C; this value increases to 120°C without water cooling. The temperature-time relation without water cooling is an exponential with a time constant of 7 hours while with water cooling the time constant is 0·65 hours. If the water cooling is out of action for 7 hours calculate the time that will elapse after the cooling is restored for the conductor temperature rise to be 80°C.

This example illustrates the important case of the failure of the water cooling system and the time available for recovery. Single time constants for the transient curves have been assumed for simplicity. The following relationships apply:

$$\theta_2 - \theta_1 = (\theta_{2m} - \theta_1)(1 - e^{-t_2/\tau_2})$$

See figure 8.16.

$$\theta_3 - \theta_{3m} = (\theta_2 - \theta_{3m})e^{-t_3/\tau_3}$$

Hence,

$$\theta_2 - 70 = (120 - 70)(1 - e^{-7/7})$$

∴ $$\theta_2 = 101 \cdot 6\,°c.$$

With water cooling on:

$$\theta_3 = 80\,°c$$

and

$$(80 - 70) = (101 \cdot 6 - 70)e^{-t_3/0 \cdot 65}$$

$$\therefore \qquad \frac{t_3}{0.65} = 1.15$$

$$t_3 = 0.75 \text{ hour.}$$

Time to fall to 80°c

$$= 0.75 \text{ hour.}$$

## 8.8 Underground Distribution Systems

Although the emphasis in this book tends to be on technological problems concerned with transmission at high voltages, the distribution of energy to individual consumers should not be thought of as of lesser importance. The capital cost of installed distribution plant is vast and improvements in technique are constantly being made. Traditionally the main transmission circuits feed 11 kV (15 kV in U.S.A.) substations from which a network of feeders, usually underground in urban areas, in turn supply local sub-stations where 415 V, 3-phase and 240 V, single-phase (208 V 3-phase and 120 V single-phase in the U.S.A.) feeders form a network supplying consumers. Arrangements at this level vary somewhat from one supply company to another, but where radial feeders are used effective switching to ensure continuity of supplies in case of faults is made. Recently new developments in distribution (mainly in the U.S.A.) have aroused great interest. These, to some extent, have been brought about by the growing public pressures to underground circuits.

*Improved cables and multiple earthing*

Although oil-impregnated paper insulation is still widely used for medium voltage cables, synthetic insulations such as elastomerics are becoming important. P.V.C. insulated and sheathed split-concentric neutral-earth cable for house services is being used by some authorities.

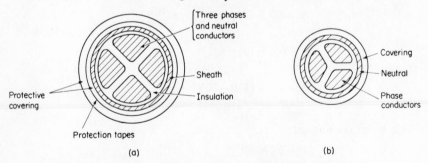

Figure 8.22   Relative size of a conventional cable with the concentric neutral type (a) Conventional (b) Concentric neutral type.

The use of the neutral for the dual purpose of neutral connexion and consumers' earth connexion requires the use of a cable with only four conductors, the neutral being in the form of a concentric conductor as shown in figure 8.22. The neutral conductor is earthed at the supply transformer, at the remote end and at intermediate points such that the resistance to earth is always less than 10 Ω. This system is known as *protective multiple earthing (p.m.e.)*. The conductors are normally of aluminium and one possible hazard is the neutral becoming discontinuous because of corrosion.

*High-voltage single-phase networks*

There is a growing tendency in the U.S.A., to use high voltage for single-phase feeds traditionally at 240 V or 120 V. A schematic diagram of such a system is shown in figure 8.23. The distribution point separates the three-phase supply into three single-phase to neutral supplies each of

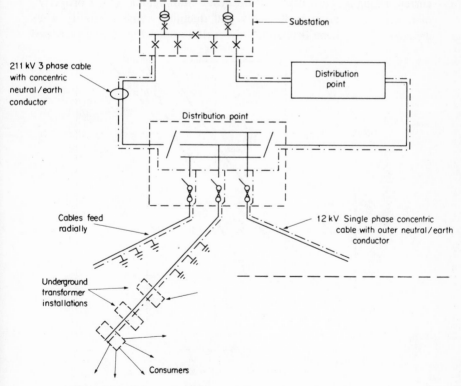

Figure 8.23   Schematic diagram of high voltage phase to neutral distribution scheme.

up to 1 MVA. These cables radiate from the distribution point supplying 12 kV/120 V transformers of 50 to 100 kVA capacity which in turn supply the customers. These transformers are frequently situated underground.

The economics of the system depends on the housing density, at 10 houses per hectare (typical of the U.S.A.) the h.v. single-phase system is considerably cheaper than traditional schemes; at 30 houses per hectare (typical of Britain) the system is marginal.


### 8.9   Sodium Conductor Cables

The parameters of sodium as a conductor material which are summarized in Table 8.2 indicate that it possesses many advantages over copper or aluminium. In addition it exists abundantly in the Earth's crust and is cheaply extracted from sodium chloride by electrolysis. For a given current rating it is less than half the cost of aluminium. Other properties of interest are: good malleability and ductility giving flexibility when encased by polythene insulation, low melting point (98°c) and increased

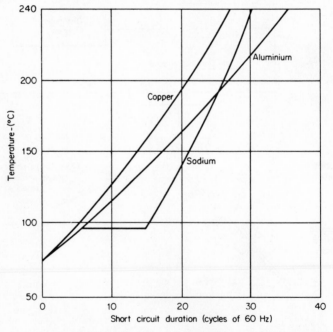

Figure 8.24   Short circuit performance of sodium cables. (*Permission of Union Carbide Corporation.*)

resistivity when molten, and very important, a violent chemical reaction when exposed to water. At the present time sodium conductor cables have been manufactured for working voltages up to 15 kV, the insulant being Polythene (polyethylene). They are produced by the extrusion of a tube of Polythene which is simultaneously filled with sodium. Virtually all of the mechanical strength of the cable is in the insulation and this is normally adequate owing to the light weight and flexibility.

The short-circuit capacity is increased owing to the latent heat of fusion when the sodium melts. Figure 8.24 compares the short-circuit performance with that of copper and aluminium. The sharp increase in sodium resistivity on melting tends to reduce any overcurrent which has caused the conductor to melt. The advantages of cost and weight are such that serious consideration should be given to the use of sodium conductors in distribution systems.

## 8.10 Cryogenic Cables (Cryocables)—Introduction

With the magnitudes of power to be transmitted steadily increasing attention is constantly being given to new materials and techniques. Ideas which theoretically are attractive frequently cannot be translated into an engineering entity because of lack of suitable materials or the fact that an existing process is cheaper. The continued use of paper/oil insulation up to the highest voltages over most of this century is a sobering lesson on the availability of materials and the importance of economics on engineering processes. The application of superconductivity to electric transmission was for many years a dream; the superconductors available lost their remarkable properties at relatively low magnetic fields, being of little use in engineering applications. In 1961 niobium and a compound of niobium and tin ($Nb_3 Sn$) became available and remained superconducting at $8.8$ $Wb/m^2$ (Tesla) with a current density of $10^5$ $A/cm^2$ and their application to electric transmission and machinery became the object of intense research activity.

A possible alternative to the superconducting cable is the resistive operation of a conductor of conventional material (probably aluminium) at cryogenic temperatures. This utilizes the decrease in resistivity with decrease in temperature (see figure 8.25) and uses a cryogenic fluid in the same manner as a conventional coolant; such a cable is known as cryogenic-resistive.

A major cost of any cryogenic scheme will be the refrigerators, the size of which will be considerable owing to their low efficiencies especially at

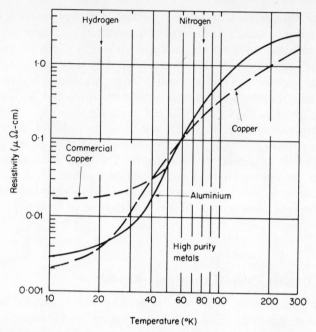

Figure 8.25    Relation between the electrical resistivity of metals with temperature.

very low temperatures. The basic parameters of coolants and refrigerators related to a conventional cable are summarized in Table 8.5.

Table 8.5 Properties of cryogenic liquids

| Cable coolant | Temperature (K) | Relative resistivity copper and aluminium | Refrigerator ratio Watts in/Watts load | |
|---|---|---|---|---|
| | | | Theoretical | Practical |
| Conventional oil | 293 | 1 | — | — |
| Liquid $N_2$ | 77 | $\frac{1}{8}$ | 3 | 6–10 |
| Liquid $H_2$ | 20 | $\frac{1}{500}$ | 14 | 50–100 |
| Liquid He | 4·2 | Super-conducting | 75 | 500–1000 |

For resistive cables the optimum coolant theoretically would appear to be liquid hydrogen but the cheapness and complete safety of nitrogen make it very attractive and the use of both these coolants is being actively investigated. Whereas in conventional cables the major problem is dissipating the heat from the cable to ambient, is cryogenic cables the reverse holds.

Although it would seem attractive to operate superconducting cables at low voltages there are practical difficulties such as conductor a.c. loss and the transformer requirements at the ends. Also high operating voltages appear to be economically advantageous. Some design aspects are common to both superconducting and resistive cables, e.g. electrical insulation, cooling systems and the cryogenic envelopes to minimize heat entering the cable from the ambient (inleak); these requirements will be discussed first and then possible designs described.

*Cryogenic envelopes*

It is necessary to keep the heat inleak to a very low value, e.g. in liquid helium systems a few $\mu$W/cm$^2$. The general form of the envelope is shown in figures 8.26 and 8.27. Superinsulation is made up of multiple reflective metallic surfaces spaced in an annular region under vacuum. It is most

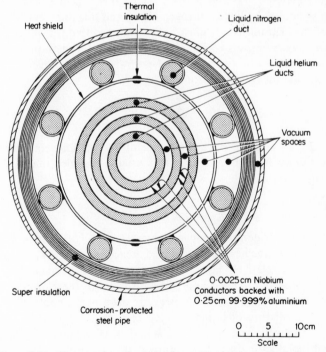

Figure 8.26   Cross section of proposed superconducting cable Rating 33 kV, 750 MVA. (*Permission of the Electrical Review.*)

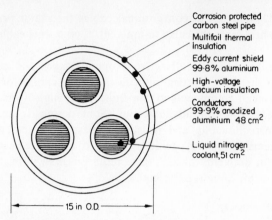

Figure 8.27     Cross section of proposed resistive cryocable. Coolant—
liquid nitrogen, electrical insulation—vacuum. (*Permission of the
Institute of Electrical and Electronic Engineers.*)

effective between 100 and 300 K; below this a heat shield is also used
which is cooled by liquid nitrogen (77 K) and this intercepts about 90 per
cent of the heat inleak. It must be remembered that it is much cheaper to
remove heat by means of liquid nitrogen than by the liquid-helium conductor coolant. For a vacuum the heat transmitted is governed mainly by
the radiation loss given by,

$$q = 5 \cdot 67 \times 10^{-12} \, (T_1^4 - T_2^4) \, \frac{\varepsilon_1 \varepsilon_2}{\varepsilon_1 + \varepsilon_2 - \varepsilon_1 \varepsilon_2} \, \text{W/cm}^2$$

where $T_1$, $T_2$ and $\varepsilon_1 \varepsilon_2$, are the temperatures (K) and emissivities of the
surfaces containing the vacuum; it is usually assumed that $\varepsilon_1$ and $\varepsilon_2$ are
both equal to 0·03.

### Cooling systems and refrigeration

The cost of refrigerators approximates to $(a + bq^{\frac{1}{2}})$, where $a$ and $b$ are
constants and $q$ the heat to be removed, hence with larger systems refrigeration costs are relatively less. Two factors control the maximum length of
cable between refrigerator and pumping stations, these are the maximum
temperature rise and maximum pressure drop of the coolant. Two-phase
flow, i.e. a mixture of liquid and gas sets up high pressure drops in the
cooling channels and should be avoided. Coolants may be pumped
through conductor ducts and returned through pipes separate from the
cable or returned through the conductor ducts of a second cable as shown

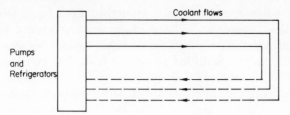

Figure 8.28   Coolant flow paths in cryogenic cables—two, three-phase cables, i.e. six conductors, coolant flows outward through one cable and returns through second.

in figure 8.28.   For single-phase flow (pressurized liquid or gas) the pressure drop

$$\Delta p = \frac{4fG^2L}{2D}$$

where, $G$ = mass flow = $\rho V$, $L$ = length, $\rho$ = density, $V$ = velocity, $D$ = diameter of duct or channel and $f$ = friction factor.   Also,

$$qL = G \frac{\pi D^2}{4} . c_p \theta$$

where, $\theta$ = temperature rise (°C), $c_p$ = specific heat, and $q$ = total loss per m.   Pumping power

$$= \Delta p \frac{G \pi D^2}{\rho 4}$$

From the above it is seen that $\Delta p \propto L^3$ whereas the heat load $\propto L$.   The total heat load, $q$, is the sum of the conductor a.c. loss, heat inleak and the heat evolved due to the turbulent flow of coolant.

*Dielectrics*

A number of possibilities exist for the electrical insulation which, because of the use of high voltages, is of paramount importance.   As vacuum is needed for thermal insulation it could be used as the dielectric as well. Much work has been performed on the breakdown characteristics of vacuum although in this case the presence of spacers to separate the annular surfaces reduces the breakdown voltage compared with that of plain vacuum gaps.   The low-temperature environment will enhance the dielectric properties of vacuum which include very low dielectric loss. The use of the cryogenic fluids themselves perhaps with a suitable filler such as paper tapes (tan $\delta$ = 0·002 with liquid N and 0·0004 with liquid H) or plastic, e.g. polypropylene, tapes is attractive. Both liquid nitrogen and hydrogen have breakdown and loss characteristics superior to those of conventional paper/oil, while those of liquid helium are slightly worse although still usable. A further advantage is the low permittivity of the cryogenic liquids compared with paper/oil giving lower capacitance and reducing the system operation voltage control problem of cable systems on

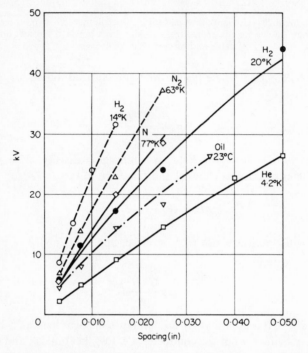

Figure 8.29 Electrical breakdown properties of cryogenic fluids.
(*Permission of the Institute of Electrical and Electronic Engineers.*)

light load. If fillers are not used then the problems associated with spacers are again encountered. Typical dielectric loss angles (tan $\delta$) are $2 \times 10^{-7}$ for hydrogen at 14 K, $2 \times 10^{-6}$ for helium at 4·2 K and $2 \times 10^{-3}$ for paper/oil at 293 K. Breakdown properties are summarized in figure 8.29.

### 8.11 Cryogenic Cables–Prototypes

*Superconducting*

Below their critical temperature $(T_c)$ superconductors exhibit perfect diamagnetism and the current flows on the conductor surface, above $T_c$

they become resistive. Metals which switch state in this manner are called Type I and include most pure metals (e.g. lead); another variety known as Type II superconductors switch to a mixed state partly superconducting and partly resistive above the original critical field and do not become completely resistive until a much higher magnetic field strength. When the Type II conductor is in the mixed state the current and flux penetrate the conductor and on alternating current a loss occurs which at 50 Hz in niobium strip of thickness 0·0025 cm and carrying 400 A/cm of circumference amounts to 10 $\mu$W/cm²; for a thickness of 0·016 cm with 800 A/cm the loss is 40 $\mu$W/cm². It is usual to express currents in terms of distance around the circumference as the current largely flows on the surface. Type II conductors comprise mainly brittle intermetallic compounds such as niobium tin ($Nb_3$ Sn), niobium–zirconium (NbZr) or niobium itself.

In the practical cable the conductor must withstand several times the rated current on faults and to meet this condition the superconductor, normally niobium, is formed on to a strip of high-purity aluminium of thickness 0·25 cm; the aluminium takes most of the fault current with the niobium resistive and also provides mechanical strength. The cross-section of a proposed superconducting cable rated at 33 kV, 750 MVA is shown in figure 8.26. As the critical temperatures of superconductors lie below 20 κ liquid helium is used and pumped through the ducts shown and a liquid-nitrogen heat-shield provided. The size of gap for the vacuum electric insulation is decided by the impulse voltage requirement of 190 kV

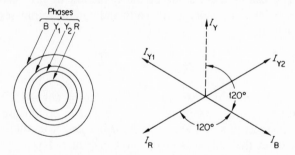

Figure 8.30 Phase relationships of conductor currents for arrangement of Figure 8.26. (*Permission of the Electrical Review.*)

(peak) and a 12·7 mm gap is assumed. The conductor diameters are determined by the current density, the inner conductor value is 375 A (r.m.s.) per cm of circumference. With the concentric conductor arrangement there are no net mechanical forces between the conductors. The phase relationship for the currents is shown in figure 8.30, the split yellow conductor takes the return current from the other two phases giving a resultant current in the correct phase relationship. The spacers between the conductor tubes are designed of helically wound fluorinated ethylene propylene (f.e.p.) strip, the outer duct carries the 'go' flow and the two inner ducts the return flow of helium, the go and return paths being thermally isolated from each other. The inlet and outlet pressures for a 10 km length are 2·2 and 2 atm and the corresponding temperatures 4·2 and 5 K. The cable is rigid and the maximum prefabricated lengths determined by transportation limits, hence the number of joints between sections will be high and the problem òf accommodating the large contraction of the conductors relative to the outer casing is formidable.

*Resistive cables*

A design using liquid nitrogen is shown in figure 8.27. The cable is rated at 345 kV and the circuit comprises two separate three-phase cables laid one above the other in a trench 50 cm wide. As there is refrigeration at one end only of the link the coolant flows through one cable and returns through the other. In the event of a failure in one cable valves are incorporated to enable the coolant to return through one of the three conductor channels of the operative cable. For a circuit of length 10 km, 1·6 MW of heat are removed and the rating is 3660 MVA.

A major problem with resistive cables is the large influence of electromagnetic effects such as those of skin and proximity on the conductor resistance; because of these some of the reduction in resistance at low temperatures is lost. The sum of the eddy and conduction losses per unit length is equal to,

$$\frac{I^2 \rho}{A} + \frac{K.A. B_m^2 (\omega d)^2}{\rho}$$

where $\rho$ = resistivity, $d$ = strand diameter, $\omega$ = angular frequency, $K$ = constant and $B_m$ = maximum flux density, $A$ = conductor area.

Hence an increase in area ($A$) and decrease in $\rho$ decreases the conduction loss but increases the eddy loss. The effect may be reduced by stranding the conductor (a suitable diameter of strand being 0·025 cm) and transposing the strands. Eddy-current losses also occur in the containing pipes although these may be reduced considerably by the use of low-resistivity shields around the conductors.

*Economics*

In cryogenic cables it has been suggested[34] that the economics be based entirely on the capital cost and that losses are of minor importance. One method of spreading the capital cost is the gradual addition of refrigeration capacity as the load builds up. Normally transmission links are installed appreciable times before their loading reaches the rated value and in this case the matching of the refrigerator capacity to the load would produce a considerable saving. An economic assessment of the various possibilities for cryogenic cable has been made by Rogers[35], some of the results are shown in Table 8.6 which relate to a 132 kV, 3000 MVA cable. The values relate to British costs in 1969.

Table 8.6  Percentage breakdown of capital costs of
3000 MVA 132 kV cryo-cables

| Cable type | Cryo-resistive | | Super-conducting |
|---|---|---|---|
| Coolant | Liquid nitrogen | Liquid hydrogen | Liquid helium |
| | % | % | % |
| Conductors and duct | 29 | 10 | 17 |
| Dielectric | 6 | 3 | 3 |
| Cryogenic envelope | 18 | 9 | 17 |
| LHe filling | — | — | 24 |
| LH$_2$ filling | — | 17 | — |
| LN filling | 2 | — | 2 |
| Duplicate refrigerators | 18 | 41 | 19 |
| Contraction joints | 3 | 2 | 2 |
| Factory assembly and inspection | 6 | 4 | 4 |
| Buildings and plant | 3 | 2 | 2 |
| Open country installation | 12 | 9 | 8 |
| Stop joints | 2 | 2 | 1 |
| Instrumentation | 1 | 1 | 1 |
| *Costs per* MVA/Km (£) | | | |
| Capital cost | 122 | 159 | 71 |
| 10 year running cost | 73 | 77 | 37 |
| Total | 195 | 236 | 108 |

(Permission IEE—Proceedings of conference on Low Temperature and Electric Power, London, 1969.)

Work in the United States has resulted in the information given in figure 8.31 in which the cost per MVA per mile for conventional and cryogenic cables is related to MVA rating. The same organization give a comparison between the transmission capability of conventional (high-pressure oil-filled) and superconducting cable systems and this is shown in Table 8.7. By the 1980s there will be many metropolitan systems (especially

Table 8.7 Comparison of a.c. superconducting and conventional H.P.O.F. cable systems

| | 69 kV | | 138 kV | | 230 kV | | 345 kV | |
|---|---|---|---|---|---|---|---|---|
| | Conv. | SC | Conv. | SC | Conv. | SC | Conv. | SC |
| Capability | | | | | | | | |
| MVA* | 120 | 423 | 230 | 1,690 | 370 | 4,710 | 500 | 10,590 |
| Amperes | 990 | 3,540 | 950 | 7,050 | 930 | 11,800 | 840 | 17,700 |
| Permissible peak losses, watts/conductor foot | 10·34 | 3·54† | 10·35 | 5·42 | 10·35 | 8·2 | 10·06 | 11·7 |
| Pipe size, inches | $6\frac{5}{8}$ | $9\frac{5}{8}$ | $8\frac{5}{8}$ | $12\frac{3}{4}$ | $10\frac{3}{4}$ | $18\frac{3}{4}$ | $10\frac{1}{4}$ | $24\frac{3}{4}$ |

*MVA for conventional cable are for 75 per cent load factor and 100 per cent load factor for superconducting cable.
†Superconducting cable losses are expressed in terms of room temperature refrigeration power required.
(Permission *Union Carbide Corp.* research funded by Edison Electric Institute)

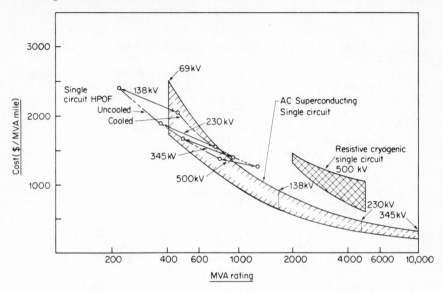

Figure 8.31 Cost comparison between superconducting cryogenic and conventional pipe cables—year 1969. (*Permission of the Union Carbide Corporation*—work funded by Edison Electric Institute as Research Project Rp78–7.)

in the U.S.A.) whose principal load centres will have a peak load of 7500 to 10,000 MW. Using conventional 345-kV cables, twenty-two cable circuits would be required to meet such a load. There is therefore an urgent requirement for much higher capacity underground transmission circuits by the 1980s.

# References

### BOOKS

1. *Underground Systems Reference Book*. Edison Electric Institute, 1957.
2. Barnes, C. C., *Electric Cables*, Pitman, London, 1964.
3. Emanueli, L., *High Voltage Cables*, Chapman and Hall, London, 1926.
4. McAdams, W. H., *Heat Transmission*, 2nd ed., McGraw-Hill, New York, 1951.
5. *Electrical Transmission and Distribution Reference Book*. Westinghouse Electric Corp., Pennsylvania, 1964.
6. Waddicor, H., *Principles of Electric Power Transmission*, 5th ed., Chapman and Hall, London, 1964.

PAPERS

7. Allen, P. H. G., 'The thermal properties of high voltage insulants', *Proc. I.E.E., C, Monograph*, No. 250 M, 1957.
8. Ball, E. H., and G. Maschio, 'The a.c. resistance of segmented conductors used in power cables', *I.E.E.E. Trans., P.A. & S.*, **87**, 1143, 1968.
9. Arnold, A. H. M., 'Eddy current losses in multi-core paper insulated lead covered cables', *Jour. I.E.E.*, **88**, 52, 1941.
10. Rhodes, D. J., and A. Wright, 'Induced voltages in the sheaths of cross bonded a.c. cables', *Proc. I.E.E.*, **113**, 99, 1966.
11. Ball, E. H., E. Occhini, and G. Luoni, 'Sheath overvoltages in H.V. cables resulting from special sheath bonding connexions', *Trans. I.E.E.E., P.A.S.*, **84**, 974, 1965.
12. Gibbons, J. A. M., P. R. Howard, and D. J. Skipper, 'Gas-pressurized lapped-polythene dielectric for EHV power-cable system', *Proc. I.E.E.*, **112**, 89, 1965.
13. Beale, H. K., 'Some economic aspects of extra high-voltage underground cables', *I.E.E.E., Proc.*, **112**, 109, 1965.
14. Williams, H. B., and S. C. Chu, 'Computer solution of E.H.V. underground cable cooling problems', *Proc. I.E.E.*, **115**, 1968.
15. Goldenberg, H., 'The calculation of continuous current rating and rating factors for transmission and distribution cables', *E.R.A.*, Rept. F/T 187, 1958.
16. Ralston, P., and G. H. West, 'The artificial cooling by water of underground cables', *C.I.G.R.É.*, Paris, Paper 215, 1960.
17. Milne, A. G., and K. Mochlinski, 'Characteristics of soil affecting cable ratings', *Proc. I.E.E.*, **111**, No. 5, May 1964.
18. Williams, A. L., 'Direct current cables', *Elec. J.*, March, 1960.
19. Weedy, B., 'Thermal transients in a high voltage cable system with natural and artificial cooling', *Proc. I.E.E.*, **109A**, 1962.
20. Lis, J., and M. J. Thelwell, 'Analogue investigation of water cooling of E.H.V. cables', *Proc. I.E.E.*, **112**, No. 2, 1965.
21. Wormer, I., 'Calculating cable temperature transients', *Trans. A.I.E.E.*, **74**, 1955.
22. Neher, J. H., 'The transient temperature rise of buried cable systems', *Trans. A.I.E.E.*, **83**, 1964.
23. 'Report on the work of Study Committee No. 2 (High Voltage Cables)', *C.I.G.R.É.*, Paper 233, Paris, 1964.
24. Weedy, B. M., and J. P. Perkins, 'Steady state thermal analysis of a 400 kV cable through joint', *Proc. I.E.E.*, **114**, 109, 1967.
25. Beale, H. K. et al., 'The application of intensive cooling techniques to oil-filled cables and accessories for heavy duty transmission circuits', *C.I.G.R.É.*, Paper 21-09, 1968.
26. Pasqualini, G., 'High power long distance cable cooling', *C.I.G.R.É.*, Paper 21-08, 1968.
27. Kitagawa, K., 'Forced cooling of power cables in Japan. Its studies and performance', *C.I.G.R.É.*, Paper 213, 1964.
28. McFee, R., 'Superconducting transformers and transmission lines', *Power Engineering*, **65**, 1961.

29. Wilkinson, K. J. R., 'Prospect of employing conductors at low temperature in power cables and in power transformers', *Proc. I.E.E.*, **113**, 1509, 1966.
30. *Underground Power Transmission*, Industry Advisory Committee Report to the Federal Power Commission, April, 1966. (U.S.A.).
31. Rogers, E. C., and D. R. Edwards, 'Design for a 750 MVA super-conducting power cable', *Electrical Review*, pp. 2–5, September 8, 1967.
32. Mathes, K. N., 'Cryogenic cable dielectrics', *Trans. I.E.E.E. Electrical Insulation*, **EL4**, No. 1, 1969.
33. Afshartous, S. B., P. Graneau, and J. Jeanmonod, 'Economic assessment of a liquid nitrogen cooled cable', *I.E.E.E., P.A.S.*, **89**, 8, 1970.
34. Graneau, P., 'Economics of underground transmission with cryogenic cables', *I.E.E.E., P.A.S.*, **89**, 1, 1970.
35. Rogers, E. C., 'A comparison of superconducting and cryo-resistive cables', *Proceeding of Conference on Low Temperature and Electric Power*, March, 1969—London, British Cryogenic Council, p. 83.
36. McAnulla, R. J., 'Sodium conductor for power cables', *I.E.E. Electronics and Power*, November, p. 434, 1968.
37. Eigenbrod, L. K., H. M. Long, and J. Notaro, *Conceptual Design and Economic Analysis of a Superconducting a.c. Power Cable System*, Paper 70 TP 168-PWR, presented at the I.E.E.E. 1970 Winter Power Meeting.
38. Paul, H., 'Power transmission of the future', *I.E.E. Electronics and Power*, p. 171, May, 1970.

**Problems**

8.1. The winding of an oil-cooled transformer, when passing full-load current, reaches 32°c after 1 hour and 41°c after a further hour, the ambient temperature being 20°c.

The transformer is energized when 'cold' and passes 75 per cent full-load current for 1 hour and then, due to an emergency, 150 per cent full load for 5 hours. Calculate the time for which the transformer must be taken off load although still energized, before full-load current can be again taken, if the normal maximum permissible temperature is that existing at steady state on full load. The copper loss on full load is 1·5 times the iron loss and the cooling time-constant is 0·75 of the heating time-constant.

Discuss the validity in practice of your solution. (1·25 hours.)

8.2. A 66 kV, single-core, lead covered cable is laid direct in soil of thermal resistivity 120°c/W per cm at a depth of 1 metre. The cable specification is as follows.

| | |
|---|---|
| Cross section of copper conductor | 260 mm² |
| Insulation: Thickness | 7·62 mm |
| Thermal resistivity | 550°c/W per cm |
| Lead sheath: Thickness | 2·03 mm |
| Diameter over sheath | 42 mm |
| Diameter over serving | 47 mm |
| Thermal restivity of serving | 400°c/W per cm |
| Conductor alternating-current resistance at 20°c | 0·0658 Ω/km |
| Copper temperature-resistance coefficient | 0·004/°c. |

Calculate the maximum current to be taken by the cable if the conductor temperature rise is not to exceed 50°C above an ambient of 15°C (670 A.)

8.3. Investigate the variation of permissible steady-current in the cable of question 8.2 with different values of soil thermal resistivity in the range 70 to 150°C/W per cm.

8.4. For the cables specified in Example 8.2 calculate the maximum temperature if the cables are laid in equilateral formation with 10 cm spacing.

8.5. Determine the equivalent thermal circuit for transient studies for the 66 kV cable in question 8.2. The following specific heats (in J/°C—gm) and densities (gm/cm³) obtain:

$$\begin{array}{lll} \text{Copper} & 0\cdot39 \text{ and} & 8\cdot93 \\ \text{Dielectric} & 1\cdot6 \text{ and} & 1\cdot05 \\ \text{Lead} & 0\cdot13 \text{ and} & 11\cdot37 \end{array}$$

8.6. A mile of 3-core, 3-phase, metal-sheathed cable gave the following results on test for capacitance:

(a) capacitance between shorted conductors and sheath, $1\cdot0\ \mu\text{F}$;

(b) capacitance between two conductors shorted with the sheath and the third conductor, $0\cdot6\ \mu\text{F}$.

With the sheath insulated, determine the capacitance

  (i) between any two conductors, and

 (ii) between any two shorted conductors and the third conductor.

(iii) Calculate the capacitance current per conductor per mile when connexion is made to 100 kV, 60 Hz busbars.

Answer: (i) $0\cdot367\ \mu\text{F}$; (ii) $0\cdot489\ \mu\text{F}$; (iii) 16 A.

8.7. A 3-phase underground circuit consists of three single-core cables each of effective conductor radius of $0\cdot75$ cm and spaced $5\cdot1$ cm apart (axis to axis) in equilateral formation. The diameter of the lead sheath is $2\cdot3$ cm and the sheath thickness $0\cdot15$ cm. The specific resistance of lead is $22\cdot0 \times 10^{-6}\ \Omega\text{cm}$ at the working temperature and the conductor resistance per mile is $0\cdot26\ \Omega$ at 65°C. For a cable length of one mile ($1\cdot6$ km) and a load of 300 A, determine, (a) the induced sheath voltage without bonding (47 V per sheath), (b) the ratio of sheath loss to conductor loss ($0\cdot029$).

8.8. An underground tunnel has a cross-section of $5\cdot4$ m² and is 250 m in length. A group of cables laid along the length of the tunnel dissipates a total of 10 W/cm. The effective thermal resistance from the air to ambient (air to tunnel wall heat-transfer and transfer through wall and soil) is 6°C cm/W. If the maximum steady-state temperature rise of the air blown through the tunnel is to be not greater than 32°C determine the air velocity required ($1\cdot3$ m/s).

8.9. Two 3-phase circuits each consist of three cables 1 km in length. The cables are cryo-resistive employing a conductor with a 2 cm² cross-sectional area duct for the passage of liquid hydrogen. Around the conductor is a nitrogen heat shield surrounded by vacuum super-insulation. The diameter over the nitrogen channel is $7\cdot45$ cm, over the hollow conductor 6 cm and over the hydrogen duct 6 cm. The cross-sectional area of the nitrogen shield channel is $3\cdot5$ cm². The channels in three of the cables form the go and the other three the return flow channels. The a.c. loss for each conductor is 168 W/km and the heat inleak through the thermal insulation may be approximated to by the expression,

$1.44 \times 10^{-8} \, dT^4$ W/km, where $d = 7.45$ cm and $T = 288$ K (ambient). The duct velocities of liquid nitrogen and hydrogen are both 40 cm/sec. Determine the temperature rise of the coolants and the approximate size of refrigerators required. The loss due to friction in the ducts may be assumed to be 18 W/km for hydrogen flow and 200 W/km for nitrogen flow. The specific heat of hydrogen is 0.424 and of nitrogen 1.76 J/cm$^3$ per °C (Hydrogen 11°C, Nitrogen 7.8°C, H$_2$ 110 kW, N$_2$ 57 kW).

8.10. Determine the optimum ratio of sheath-to-conductor radius for a single conductor cable assuming these dimensions are decided entirely by electric stress considerations (Answer: 2.718).

8.11. The paper/oil insulation of a single conductor cable contains a void (a small pocket of air or gas). If the relative permittivity of the insulation is 3.5 calculate the stress across the void in terms of the stress in the adjacent insulation. Comment on the practical implications of your result. (Answer: 3.5.)

# 9

# Direct Current Transmission

## 9.1 Introduction

The drawbacks to the use of alternating current for transmission of large powers over long distances will now be apparent to the reader. However, large series inductive reactances can be substantially reduced by series capacitors along the line and the stability limit greatly increased. For this form of transmission the decision to use either alternating current or direct current is entirely economic in nature. Critical lengths of lines have been quoted above which the use of d.c. is more economical (e.g. for 750 MW a critical distance of 350–450 miles), but these are largely dependent on the cost of the valves and associated devices which, with the increased use of direct current, will probably become cheaper.

In two cases there are strong technical reasons for the use of direct current transmission. These are as follows:

(a) For the connexion of large systems through links of small capacity. An example is the Britain–France cross-channel link where slightly different frequencies in the two large systems would produce serious problems of power transfer control in the small capacity link. A d.c. line is an asynchronous or flexible link between two rigid systems.

(b) Where high voltage underground cables are needed for reasonable transmission distances. The limitations of cables due to charging current with a.c. have been discussed and to increase lengths either artificial reactors or d.c. must be used. D.c. transmission by cables inland may be expected to take place in areas where amenity considerations restrict the use of overhead lines. The use of cables for cross channel crossings is well established. A comparison between the use of a.c. and d.c. with underground cables is interesting. Six 275 kV, 3 in², * cables in two groups of three in horizontal formation (total width of trench 5·2 m) in soil of resistivity 120°c

* $3 \text{ in}^2 = 1935 \text{ mm}^2 = 3\cdot8 \times 10^6$ cmil.

cm/W have an a.c. capacity of 1520 MVA. Two cables at $\pm 500$ kV d.c. have a capacity of 1600 MW with a trench width of only 0·68 m.

Further advantages in the use of d.c. are as follows: the corona loss in a d.c. line operating at a voltage corresponding to the peak value of the equivalent alternating voltage is substantially less than for the a.c. line. This is important, not so much because of the loss of power, but due to the resulting interference with radio and television transmissions. Generally the line loss will be smaller than for the equivalent a.c. line. Investigations have shown that the fault levels in an a.c. system with a d.c. link operating are less than with the link replaced by an a.c. equivalent. This is of great importance when the use of higher voltages and many interconnexions has greatly increased the fault MVA to be withstood by circuit breakers.

Disadvantages are as follows:

(a) The much more onerous conditions for circuit breaking when the current does not reduce to zero twice a cycle. Because of this, switching is not carried out on the d.c. link but effected by means of the terminal rectifiers and inverters. This severely hampers the creation of an interconnected d.c. system with tee-junctions. However the use of multiple links with convertor stations in series or parallel with each other and with the valves used as switches is likely.

(b) Voltage transformation has to be provided on the a.c. sides of the system.

(c) Rectifiers and inverters absorb reactive power and this must be supplied locally.

(d) D.c. converting stations are much more expensive than conventional a.c. substations.

*Converter devices* At the moment the *mercury-arc valve* is used for high voltage conversion. The main requirements are as follows:

(a) It must withstand the high inverse peak voltages between cathode and anode when not conducting.

(b) When a small negative voltage is applied to the grid the valve must not conduct with peak positive voltage on the anode. The instant of firing must be accurately controllable by means of the grid.

(c) The arc between anode to cathode must be stable and not subject to self-quenching.

Extensive research has gone on for many years to develop valves for the very high voltages employed in transmission links. Most of this work has been carried out in Sweden but today extensive research is also going on in other countries. A diagram of the basic form of the valve is shown in figure 9.1.

A converter is required at each end of a d.c. line and operates as a rectifier (a.c. to d.c.) or an inverter (power transfer from d.c. to a.c.). The valves at the sending end of the link rectify the alternating current providing direct current which is transmitted to the inverter. Here it is

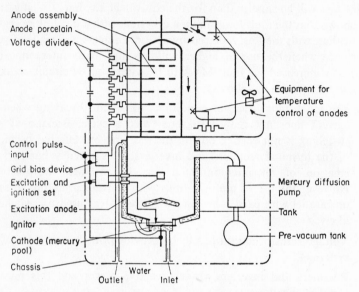

Anode assembly
Anode porcelain
Voltage divider

Equipment for
temperature
control of anodes

Control pulse
input
Grid bias device
Excitation and
ignition set

Excitation anode

Ignitor

Cathode (mercury
pool)

Chassis

Water
Outlet        Inlet

Mercury diffusion
pump

Tank

Pre-vacuum tank

Figure 9.1   Cross-section of a mercury arc valve. (*Permission
Institution of Electrical Engineers.*)

converted back into alternating current which is fed into the connected a.c. system (figure 9.2). If a reversal of power flow is required the inverter and rectifier exchange roles and the direct voltages at each end are reversed (figure 9.2b). This is necessary because the direct current can only flow in one direction (anode to cathode in the valves), so to reverse the direction (or sign) of power the voltage direction must be reversed.

The alternating current waveform injected by the inverter into the receiving-end a.c. system and taken by the rectifier is roughly trapezoidal in shape and thus produces, not only a fundamental sinusoidal wave, but harmonics of an order dependent on the number of valves. For a six-valve bridge the order is $6n \pm 1$, i.e. 5, 7, 11, 13, etc. Filters are incorporated to tune out harmonics up to the twenty-fifth. With the continual increase in the ratings of Thyristors it is likely that they will supplant mercury valves in the future. In this text the term valve will be assumed to embrace solid-state devices.

In the following sections the processes of rectification and inversion will be analysed and the control of the complete link discussed.

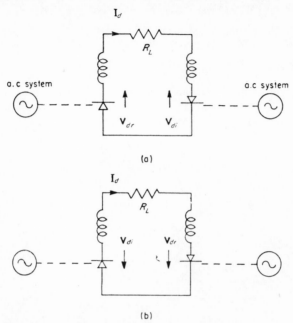

(a)

(b)

Figure 9.2a Symbolic representation of two alternating current systems connected by a direct-current link, $V_{dr}$ = direct voltage across rectifier, $V_{di}$ = direct voltage across inverter.
Figure 9.2b System as in (a) but power flow reversed.

## 9.2 Rectification

Transformer secondary connexions may be arranged to give several phases for supplying the valves. Common arrangements are three, six and twelve phases and high numbers are possible. Six-phase is popular because of the better output voltage characteristics. To begin with a three-phase arrangement will be described but most of the analysis will be for $n$ phases so that results are readily adaptable for any system. In figure 9.3a a three-phase rectifier is shown and in figure 9.3b the current and voltage variation with time in the three phases of the supply transformer. With no grid control, conduction will take place between the cathode and the anode of highest potential. Hence the output voltage wave is the thick line and the current output continuous. In an $n$-phase system the anode change-over occurs at

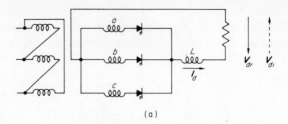

(a)

Figure 9.3a   Three phase rectifier. $V_{di}$ = voltage of inverter.

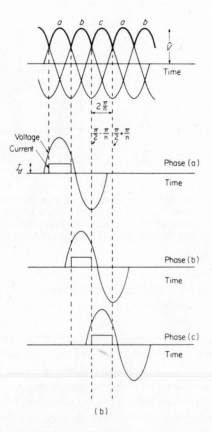

(b)

Figure 9.3b   Waveforms of anode voltage and rectified current in each phase.

$(\pi/2 - \pi/n)$ degrees at a voltage $\hat{V} \sin \pi/n$ and the mean value of the direct output voltage is,

$$V_O = \frac{1}{2\pi/n} \int_{\pi/2-\pi/n}^{\pi/2+\pi/n} \hat{V} \sin \omega t \, d(\omega t)$$

$$= \frac{\hat{V} \sin (\pi/n)}{\pi/n} \tag{9.1}$$

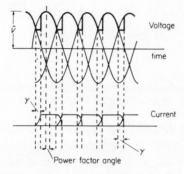

Figure 9.4 Waveforms of voltage and current showing effect of the commutation angle $\gamma$. A lagging power factor is produced.

For 3 phases,

$$V_O = \frac{\hat{V} \sin 60°}{\pi/3}$$

$$= \hat{V}\frac{3\sqrt{3}}{2\pi} = 0\cdot83 \; \hat{V}$$

and for 6 phases,

$$V_O = \frac{3\sqrt{2}}{\pi} V = \frac{3}{\pi} \hat{V} = 0\cdot955 \; \hat{V}$$

where $V =$ R.M.S. voltage.

Owing to the inductance present in the circuit the current cannot change instantaneously from $+I_d$ to 0 in one anode and from 0 to $I_d$ in the next. Hence two anodes conduct simultaneously over a period known as the commutation time or overlap angle ($\gamma$). When valve (*b*) commences to conduct it short-circuits the *a* and *b* phases, the short-circuit current eventually becoming zero in valve (*a*) and $I_d$ in (*b*). This is shown in figure 9.4.

*Grid control*

A positive pulse applied to a grid situated between anode and cathode controls the instant at which conduction commences and once conduction has occurred the grid exercises no further control. In the voltage waveforms shown in figure 9.5 the conduction in the valves has been delayed by

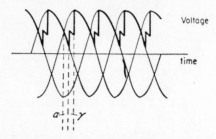

Figure 9.5    Waveforms of rectifier with instant of firing delayed by an angle α by means of grid control.

an angle α by suitably delaying the application of positive voltage to the grids. Considering *n* phases and ignoring the commutation angle γ, the new direct output-voltage with a delay angle of α is,

$$V_0^1 = \frac{1}{2\pi/n} \int_{\pi/2 - \pi/n + \alpha}^{\pi/2 + \pi/n + \alpha} \hat{V} \sin \omega t \, d(\omega t)$$

$$= \frac{n\hat{V}}{2\pi} \int_{-\pi/n + \alpha}^{\pi/n + \alpha} \cos \omega t \, d(\omega t)$$

$$= \frac{n\hat{V}}{2\pi} . 2 . \sin \left(\frac{\pi}{n}\right) \cos \alpha$$

$$\therefore \qquad\qquad V_0^1 = V_0 \cos \alpha \qquad\qquad\qquad (9.2)$$

where $V_0$ is the maximum value of direct output voltage as defined by equation (9.1).

*Bridge connexion*

To avoid undue complexity in describing the basic operations the converter arrangement used so far in this chapter is simple and has disadvantages in practice. Mainly because the d.c. output voltage is doubled the bridge arrangement shown in figure 9.6 is favoured in which there are always two valves conducting in series. The corresponding voltage waveforms are shown in figure 9.7 along with the currents (assuming ideal rectifier operation).

The sequence of events in the bridge connexion is as follows (see figures 9.6 and 9.7). Assume that the transformer voltage $V_A$ is most positive at the beginning of the sequence, then valve (1) conducts and the current flows through (1) and the load then returns through valve (6) as $V_B$ is most negative. After this period $V_C$ becomes the most negative and current

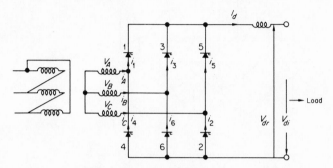

Figure 9.6 Bridge arrangement of valves.

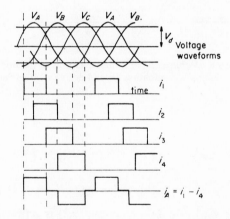

Figure 9.7 Idealized voltage and current waveforms for bridge arrangement.

flows through (1) and (2). Next (3) takes over from (1), the current still returning through (2). The complete sequence of valves conducting is therefore: 1 and 6, 1 and 2, 3 and 2, 3 and 4, 5 and 4, 5 and 6, 1 and 6. Grid control may be obtained in exactly the same manner as previously described and the voltage waveforms with grid delay and commutation time accounted for are shown in figure 9.8.

The direct-voltage output with the bridge may be calculated either by using the line voltage (phase-to-phase) in the formula for six-valve, six-phase rectification or determining the magnitude for three-valve operation and doubling it as both sides of the bridge contribute to the direct voltage.

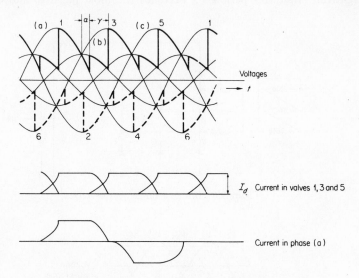

Figure 9.8   Voltage and current waveforms in the bridge connexion including commutation ($\gamma$) and delay ($\alpha$). Rectifier action.

Hence,

$$V_0 = \frac{\hat{V}\,3\sqrt{3}}{2\pi} \times 2 = \sqrt{2}V \times 3\sqrt{3}/\pi$$

$$= \sqrt{2}V_{\rm L} \times 3/\pi \quad (\text{as } V_{\rm L} = \sqrt{3}V)$$

If the analysis used to obtain equation (9.2) is repeated for the bridge it will be shown that $V_{\rm d} = V_0 \cos \alpha$, where $V_0$ is the maximum direct-voltage output for the bridge connexion.

*Current relationships in the bridge circuit*

During the commutation process when two valves are conducting simultaneously the two corresponding secondary phases of the supply transformer are short-circuited and if the arc voltage-drop across the valves is neglected the following analysis applies.

When two phases of the transformer each of inductance $L$ henries are effectively short-circuited, the short-circuit current $(i_s)$ is governed by the equation,

$$2L\frac{di_s}{dt} = \hat{V}_L \sin \omega t = \text{resultant voltage between the two phases.}$$

$$\therefore \qquad i_s = -\frac{\hat{V}_L}{2L}\frac{\cos \omega t}{\omega} + A,$$

where $A$ is a constant of integration and $\hat{V}_L$ the crest value of the line-to-line voltage.

$$\omega t = \alpha \quad \text{when} \quad i_s = 0.$$

$$\therefore \qquad A = \frac{\hat{V}_L}{2\omega L}\cos \alpha$$

Also, when

$$\cdot \omega t = \alpha + \gamma, \quad i_s = I_d$$

$$\therefore \qquad I_d = \frac{\hat{V}_L}{2\omega L}[\cos \alpha - \cos (\alpha + \gamma)]$$

$$= \frac{V_L}{\sqrt{2}\omega L}[\cos \alpha - \cos (\alpha + \gamma)],$$

where $V_L = $ RMS line-to-line voltage

and, as for the bridge circuit

$$V_O = \frac{3\sqrt{2}}{\pi}\hat{V}_L,$$

$$I_d = \frac{\pi V_O}{3\sqrt{2}} \cdot \frac{1}{\sqrt{(2)}X}[\cos \alpha - \cos (\alpha + \gamma)] = \frac{\pi V_O}{6X}[\cos \alpha - \cos (\alpha + \gamma)] \qquad (9.3)$$

The mean direct output voltage, with grid delay angle $\alpha$ only considered, has been shown to be $V_O \cos \alpha$. With both $\alpha$ and the commutation angle $\gamma$, the voltage with $\alpha$ only, will be modified by the subtraction of a voltage equal to the mean of the area under the anode voltage curve lost due to commutation (see figures 9.4 and 9.8).

Referring to figure 9.4 ($\alpha = 0$), the voltage drop due to commutation,

$$\Delta V_0 = \frac{\text{Area lost}}{2\pi/n}$$

$$\Delta V_0 = \frac{n}{2\pi} \int_0^\gamma \hat{V} \sin\frac{\pi}{n} \sin \omega t \, \mathrm{d}(\omega t)$$

$$= \frac{n}{2\pi} \hat{V} \sin\frac{\pi}{n} (1 - \cos\gamma) = \frac{V_0}{2}(1 - \cos\gamma)$$

When $\alpha > 0$, the voltage drop $=$

$$= \int_\alpha^{\alpha+\gamma} \hat{V} \sin \omega t \, \mathrm{d}(\omega t)$$

$$= \frac{V_O}{2}[\cos\alpha - \cos(\alpha+\gamma)]$$

The direct-voltage output,

$$V_d = V_O \cos\alpha - \frac{V_O}{2}[\cos\alpha - \cos(\alpha+\gamma)]$$

$$= \frac{V_O}{2}[\cos\alpha + \cos(\alpha+\gamma)] \tag{9·4}$$

Adding equations (9.3) and (9.4),

$$\frac{V_d/V_O}{2} + \frac{I_d/\pi V_O}{6X}$$

$$\therefore \qquad V_d = V_O \cos\alpha - \frac{3XI_d}{\pi} \tag{9.5}$$

The power factor is given approximately by,

$$\cos\phi = \tfrac{1}{2}[\cos\alpha + \cos(\alpha + \gamma)] \tag{9.6}$$

Equation (9.5) may be represented by the equivalent circuit shown in figure 9.9, the term $(3X/\pi)I_d$ represents the voltage drop due to commutation and not a physical resistance drop. It should be remembered that $V_O$ is the theoretical maximum value of direct output voltage and it is evident

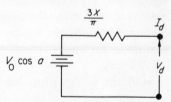

Figure 9.9   Equivalent circuit representing operation of a bridge rectifier. Reactance per phase $X\Omega$.

that $V_d$ can be varied by changing $V_0$ (control of transformer secondary voltage by tap changing) and by changing $\alpha$.

## 9.3 Inversion

With rectifier operation the output current $I_d$ and output voltage $V_d$ are such that power is absorbed by a load. For inverter operation it is required to transfer power from the direct current to the alternating current systems and as current can only flow from anode to cathode (i.e. in the same direction as with rectification) the direction of the associated voltage must be reversed. An alternating-voltage system must exist on the primary side of the transformer and grid control of the converters is essential.

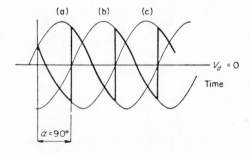

Figure 9.10    Waveforms with operation with $\alpha = 90°$, direct voltage zero. Transition from rectifier to inverter action.

As the bridge connexion is in common use it will be used to explain the inversion process. If the bridge rectifier is given progressively greater delay the output voltage decreases becoming zero when $\alpha$ is 90°. With further delay the average direct voltage becomes negative and the applied direct voltage (from the rectifier) forces current through the valves against this negative or back voltage. The converter thus receives power and inverts. The inverter bridge is shown in figure 9.11a and the voltage and current waveforms in figure 9.11b. Commutation from valve (5) to valve (1) is possible only when phase (a) is positive with respect to (b) and the current changeover must be complete before (F) by a time ($\delta_0$) equal to the dionization time of the valves. From the current waveforms it is seen that the current supplied by the inverter to the a.c. system *leads* the voltage and hence the inverter may be considered as a generator of leading vars or an absorber of lagging vars.

The power factor $\cos \phi \simeq (\cos \delta + \cos \beta)/2$, where $\delta$ and $\beta$ are defined in figure 9.11b. Valve 3 is triggered at time $A$ and as the cathode is held

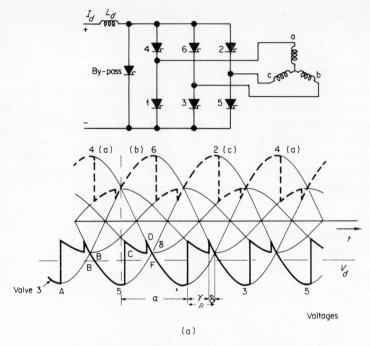

Figure 9.11a    Bridge connexion—inverter operation.

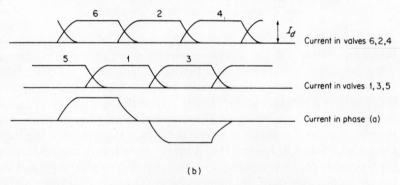

(b)

Figure 9.11b    Bridge connexion—inverter voltage and current
waveforms.

negative to the anode by the applied direct voltage ($V_d$) current flows,
limited only by the circuit impedance. If the voltage drop across the arc
is neglected the cathode and anode are at the same potential in valve (3).
When time $B$ is reached the anode to cathode open-circuit voltage is zero

and the valve endeavours to cease conduction. The large transformer inductance ($L$) however which has previously stored energy now maintains the current constant ($e = -L(\mathrm{d}i/\mathrm{d}t)$ and if $L \rightarrow \infty$, $\mathrm{d}i/\mathrm{d}t \rightarrow 0$). Conduction in (3) continues until time $C$ when valve (5) is triggered. As the anode to cathode voltage for (5) is greater than for (3), (5) will conduct, but for a time (5) and (3) conduct together (commutation time), the current gradually being transferred from (3) to (5) until (3) is non-conducting ($D$). If triggering is delayed to $F$, valve (3) which is not yet extinct would be subject to a positively rising voltage and would continue to conduct into the positive half-cycle with breakdown of the inversion process. Hence triggering must occur to allow valve extinction before time $F$.

The angle ($\delta$) between the extinction of valve (5) and the point ($F$) where the anode voltages are equal is called the extinction angle, i.e. sufficient time must be allowed for the valve to de-ionize and for the grid to regain control. It is usual to replace the delay angle $\alpha$ by $\beta = 180 - \alpha$, hence $\beta$ is equal also to ($\gamma + \delta$). The minimum value of $\delta$ is $\delta_0$.

The action of the inverter is essentially that of the rectifier but with the delay angle $\alpha$ greater than $90°$. With $\alpha < 90°$ the direct voltage output ($V_d$) is in a certain direction, as $\alpha$ increases $V_d$ decreases and when $\alpha = 90°$, $V_d = 0$ volts; with further increase in $\alpha$, $V_d$ reverses and inverter action is obtained. Hence the change from rectifier to inverter action and vice versa is smoothly obtained by control of $\alpha$. This may be seen by consulting figures 9.4 ($\alpha = 0$), 9.10 ($\alpha = 90°$) and 9.11b ($\alpha > 90°$).

Equations (9.3) and (9.4) may be used to describe inverter action. Replacing $\alpha$ by $(180 - \beta)$ and $\gamma$ by $(\beta - \delta)$ the following are obtained;

$$V_d = -[V_O \cos \beta + I_d R_c] \tag{9.7}$$

$$V_d = -[V_O \cos \delta - I_d R_c] \tag{9.8}$$

where

$$R_c = \frac{3X}{\pi}.$$

Therefore two equivalent circuits are obtained for the bridge circuit as shown in figure 9.12a for constant $\beta$ and figure 9.12b for constant $\delta$.

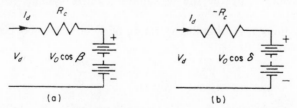

(a)                     (b)

Figure 9.12a   Equivalent circuit of inverter in terms of angle $\beta$.
       b   Equivalent circuit of inverter in terms of angle $\delta$.

### 9.4 Complete Direct Current Link

The complete equivalent circuit for d.c. transmission link under steady-state operation is shown in figure 9.13. If both inverter and rectifier operate at constant delay angles the current transmitted

$$= I_d = \frac{V_{dr} - V_{di}}{R_L}$$

or

$$\frac{V_{Or}\cos \alpha - V_{Oi}\cos \beta}{R_L + R_{c_i} + R_{ci}}$$

where $R_L$ is the loop resistance of the line or cable, $R_{cr}$ and $R_{ci}$ are the effective commutation resistances of the rectifier and inverter, respectively.

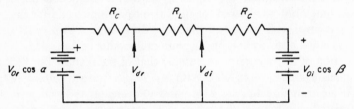

Figure 9.13 Equivalent circuit of complete link with operation with given delay angles, $V_{Or} = V_O$ for rectifier, $V_{Oi} = V_O$ for inverter.

The magnitude of direct current can be controlled by variation of $\alpha$, $\beta$, $V_{Or}$ and $V_{Oi}$ (the last two by tap changing of the supply transformers). Inverter control using constant delay angle has the disadvantage that if $\delta$ and hence $\beta$ are too large excessively high reactive-power demand results (it will be seen from figure 9.11b that the inverter currents are considerably out of phase with the anode voltages and hence a large requirement for reactive power established). Also, a reduction in the direct voltage to the inverter results in an increase in the commutation angle $\gamma$ and if $\beta$ is made large to cover this the reactive-power demand will be excessive again. In view of this it is more usual to operate the inverter with a constant $\delta$ which is achieved by the use of suitable control systems (called *compounding*).

The equations governing the operation of the inverter may be summarized as follows:

$$V_d = \frac{3\sqrt{2}V_L}{\pi} \cos \beta + \frac{3\omega L}{\pi} I_d$$

$$= \frac{3\sqrt{2}V_L}{\pi} \cos \delta - \frac{3\omega L}{\pi} I_d$$

$$= \frac{3\sqrt{2}V_L}{2\pi} (\cos \beta + \cos \delta)$$

and the power factor, $\cos \phi = (\cos \beta + \cos \delta)/2$ leading.

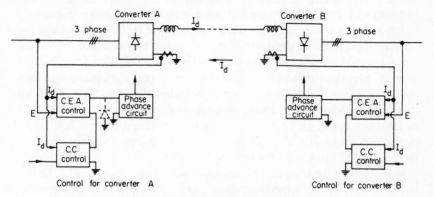

Figure 9.14   Schematic diagram of control of an H.V.D.C. system
CEA = Constant extinction angle
CC = Constant current.

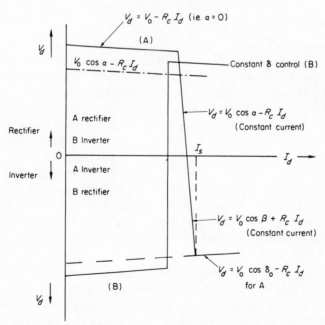

Figure 9.15   Voltage-current characteristics of converters with compounding.

The advantages of operating with $\delta$ fixed and as small as possible have been discussed, it is also advisable to incorporate a facility for constant-current operation. The rectifier and inverter change roles as the required direction of power flow dictates and it is necessary for each device to have dual-control systems. A schematic diagram of the control systems is shown in figure 9.14. In figure 9.15 the full characteristics of the two converters of a link are shown with each converter operating as rectifier and inverter in turn. In the top half of the diagram converter (A) acts as a rectifier and the optimum characteristic with an $\alpha$ of zero is shown. With constant-current control, $\alpha$ is increased and the output voltage-current characteristic crosses the $V_d = 0$ axis, below which (A) acts as an inverter and can be operated on constant-$\delta$ control. A similar characteristic is shown for converter (B) which commences as an inverter and with constant-current control ($\beta$ increasing) eventually changes to inverter operation. It is seen that the current setting for A is larger than that for B.

In figure 9.16 the inverter characteristic for a power flow from A to B is shown drawn in the upper half of the graph and this will facilitate the discussion of the operational procedure.

### Methods of control

To operate an inverter at a constant $\delta$ the instant of valve firing is controlled by a computer which takes into account variations in the instantaneous values of voltage and current. The computer then controls the firing times such that the extinction angle $\delta$ is slightly larger than the deionization angle of the valve. As the current rating of the valves should not be exceeded some measure of current control is desirable the ideal being constant-current operation. This may be achieved in the inverter by increasing $\beta$ beyond the constant-$\delta$ value thus decreasing the back-voltage developed. Similarly constant-current compounding can be incorporated in the rectifier. In figure 9.16 the voltage-current characteristics with both inverter and rectifier current-compounding are shown and it is evident that the transmitted current cannot increase beyond the prescribed value.

The methods of control may be summarized as follows, making reference to figure 9.16.

(a) In the rectifier the transmitted current is regulated by varying the delay angle and hence $V_{dr}$. This is described by ABC. The rectifier tap-changing transformer is used to keep the delay angle within reasonable limits and also to give an excess in voltage over that of the inverter to cover sudden falls in rectifier output voltage which can be recovered in time by the tap changer. It should be remembered that the greater the delay angle the greater the reactive power consumed. Control is normally performed by the rectifier with inverter control used when necessary as follows.

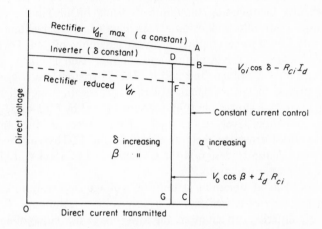

Figure 9.16 Inverter and rectifier operation characteristics with constant-current compounding. Point of operation is where the two characteristics intersect.

(b) The inverter transformer gives the required back direct-voltage by tap-changing. Again this is slow acting and constant-current control is required giving a characteristic indicated by DG. The reference value of current is lower than that for the rectifier (a margin of 10–20 per cent of the rated current).

(c) Point B shows the operating condition with normal rectifier control. Should the rectifier voltage fall below the allowed margin to avoid $I_d$ becoming zero the inverter back direct-voltage (i.e. the mean voltage of the negative anode voltages) is decreased and operation takes place along DFG and the current maintained at value OG. The new operating point is F and the power transmitted is smaller than before. Eventually the rectifier tap changer restores the original conditions. The value of the voltage margin is chosen to avoid frequent operations in the inverter control region.

Summarizing: with normal operation the rectifier operates at constant current and the inverter at constant $\delta$; under emergency conditions the rectifier at zero firing-delay and the inverter at constant current.

## 9.5  Solid-state Converters

Experience is being gained on prototype installations employing stacks of *Thyristors* in series instead of the conventional mercury-arc devices. Two major problems, which now appear to have been overcome are (a) the difficulty in obtaining simultaneous firing of all Thyristors and (b) ensuring correct voltage division across them (due to capacitance effects).

A typical prototype employs a spiral assembly of 144 Thyristors in series employing oil for insulation and cooling. Each Thyristor is rated at 2 kV peak-inverse voltage and 250 A. The Thyristors are triggered by a hybrid optical/magnetic signal isolating system in which light pulses generated by high-power photo-diodes are carried by fibre-optic guides to 12 group-potential levels in the stack. These pulses are then distributed by pulse transformers to the gate circuit of the 12-Thyristor modules in each group. A future design is for a stack rated at 150 kV, 1250 A, d.c. using Thyristors rated at 4 kV.

The problem of obtaining a uniform voltage distribution with many Thyristors in series is well known in low-voltage techniques. Because of the widely differing capacitances, the low-voltage technique of connecting *R–C* circuits in parallel with each Thyristor does not suffice and additional capacitance must be incorporated. Also it is necessary to retain uniform transient voltage distributions with time. To achieve this inductors with a non-linear characteristic may be used. On a sudden rise in voltage the inductor first absorbs the voltage, the capacitors are charged in the opposite direction with the inductor magnetizing current only and the voltage rise across the Thyristor is delayed. Similarly, the rate of change of current may be controlled on switching off.

The relative merits of the mercury-arc valve and Thyristor are summarized as follows:

(a) The serious disadvantage in mercury-arc valves is the possibility of backfire. This does not occur in Thyristors.

(b) High vacuum equipment is needed for mercury-arc values and it is also necessary to bring them up to working temperature before operation can commence. Thyristors require neither.

(c) The failure of an individual Thyristor does not impair the working of the overall Thyristor converter as allowance can be made for this in the design and rating of the converter.

(d) The cooling requirements and temperature control for mercury values are very refined. With Thyristors  ʼmersion in ordinary insulating oil is adequate.

(e) Because of their temperature requirements mercury-arc valves are housed inside buildings whereas Thyristor stacks may be designed for outside installation.

(f) Until the Thyristor converter is in more widespread use an accurate comparison of costs is difficult. However, it must be remembered that with semiconductor converters no building with expensive vacuum and temperature-control systems is required and this comprises a large cost reduction.

## 9.6 Factors Affecting Design and Operation

*Possible transmission systems*

The cheapest arrangement is a single conductor with a ground return. This, however, has various disadvantages. The ground-return current results in the corrosion of buried pipes, cable sheaths, etc., due to electrolysis. With submarine cables the magnetic field set up may cause significant errors in ships' compass readings, especially when the cable runs north–south. This system is shown in figure 9.17a. Two variations on two-conductor schemes are shown in figure 9.17b and c. The latter has the advantage that if the ground is used in emergencies, a double-circuit system is formed.

*Converter faults*

The main types of faults are as follows:

(a) Uncontrolled firing of a valve (valve made to conduct), usually caused by the failure of the grid-bias circuit, inadequate de-ionization time and overvoltages.

(b) Failure of a valve to fire, often caused by arc quenching in the valve caused by high rate of rise of current or failure of the excitation circuits.

(c) Backfire, i.e. conduction through a valve in the reverse direction. This is caused by too high a reverse voltage or too rapid a rate-of-rise of voltage with time. It results in a short-circuit between phases and is the only fault which may result in serious damage. In a bridge inverter, backfire results in a short-circuit of two of the transformer phases. The reverse current in the faulty valve reaches its peak when the phase-to-phase voltage is zero (because of the highly inductive current path) and this is normally before the valve would have de-ionized in normal operation. The result is commutation failure and rapid corrective action is required.

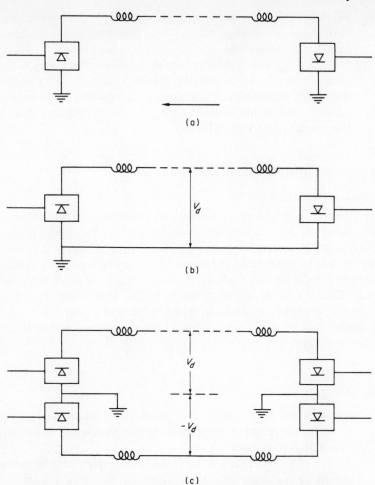

Figure 9.17   Possible conductor arrangements for d.c. transmissions.
(a) Ground return. (b) Two conductors, return earthed at one end.
(c) Double-bridge arrangement.

*Bypass valve*

Due to the various faults which may occur in the d.c. system, it may be
necessary to stop the bridge converter from operating. This may be rapidly
achieved by the cessation of the pulses controlling the grids accompanied
by the firing of a valve connected across the bridge d.c. terminals known
as the bypass valve (see figure 9.11a). For a rectifier unit although blocking
of the grids would cease the operation, normally both converters in the

system incorporate constant-current control. This results in the two valves which will remain conducting when the bridge is blocked trying to provide this specified current and continuing to conduct indefinitely. With an inverter the two valves which conduct at the instant of grid blocking will continue to conduct because of the applied voltage from the rectifier. In both cases, if the bypass valve is made to conduct at the instant of the blocking of the bridges, the d.c. terminals are short-circuited and the bypass valve takes over the current from the main bridge valves.

When recommencing normal operation, it is necessary to unblock the valve grids and open-circuit the bypass valve. The latter, however, will not cease conduction until the anode voltage is negative with respect to the cathode when the grid will regain control. With the rectifier, the bridge valves establish a voltage across the bypass valve making the cathode positive with respect to the anode, and the bridge valves take over the complete current. On the other hand, the required polarity can only be obtained by a reversal of the polarity of the inverter-bridge voltage by advancing the angle $\beta$ to greater than 60°. This angle is only required for about one cycle by when the bypass valve will have ceased conduction.

It should be noted that the inverter analysis given in Section 9.3 only holds for *angle $\beta$ less than 60°*. As indicated above, if $\beta$ becomes greater than 60° the mode of operation becomes very different and a new analysis will be required.

*Harmonics*

A knowledge of the harmonic components of voltage and current in a power system is necessary because of the possibility of resonance and also the enhanced interference with communication circuits. The direct-voltage output of a converter has a waveform containing a harmonic content, which results in current and voltage harmonics along the line. These are normally reduced by a smoothing choke.

The currents produced by the converter currents on the a.c. side contain harmonics. The current waveform in the a.c. system produced by a delta-star transformer bridge converter is shown in figures 9.8 and 9.11. The order of the harmonics produced is $6n \pm 1$, where $n$ is the number of valves. By the use of the Fourier series the equation for the current ($i$) is given by;

$$i = \frac{2\sqrt{3}}{\pi} I_d \left( \cos \omega t + \frac{1}{5} \cos 5\omega t - \frac{1}{7} \cos 7\omega t - \frac{1}{11} \cos 11\omega t + \ldots \right)$$

In figure 9.18 is shown the variation of the 7th harmonic component with both commutation (overlap) angle ($\gamma$) and delay angle ($\alpha$). Generally the harmonics decrease with decrease in $\gamma$ this being more pronounced at higher harmonics. Changes in $\alpha$ for a given $\gamma$ do not cause large decreases in the harmonic components, the largest change being for $\alpha$'s between 0 and 10°. For normal operation $\alpha$ is less than 10° and $\gamma$ is perhaps of the order of 20°, hence the harmonics are small. During faults, however, $\alpha$ may reach nearly 90°, $\gamma$ is small and the harmonics produced are large.

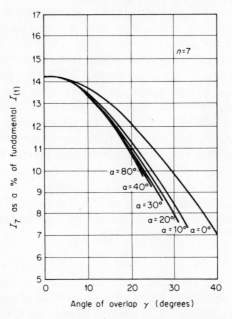

Figure 9.18   Variation of 7th harmonic current with delay angle $\alpha$ and commutation angle. (*Permission of International Journal of Electrical Engineering Education.*)

The harmonic voltages and currents produced in the a.c. system by the converter current waveform may be determined by representing the system components by their reactances at the particular harmonic frequency. Most of the system components have resonance frequencies between the 5th and 11th harmonics.

It is usual to provide filters (L–C shunt resonant circuits) tuned to the harmonic frequencies. A typical installation is that for the England–France link (figure 9.20) and is shown in detail in figure 9.19. At the fundamental frequency the filters are capacitive and help to meet the reactive power requirements of the converters.

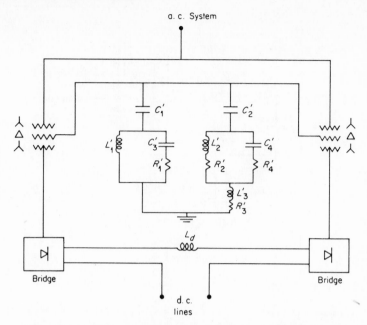

Figure 9.19  Schematic diagram of harmonic filters at Lydd station of U.K.—France Scheme. Component Values per Phase, Based on 33 kV: $C_1' = 36 \cdot 1 \, \mu\text{F}$, $C_2' = 24 \cdot 23 \, \mu\text{F}$, $C_3' = 36 \cdot 1 \, \mu\text{F}$, $C_4' = 192 \cdot 53 \, \mu\text{F}$, $L_1' = 1 \cdot 44 \, \text{mH}$, $L_3' = 13 \cdot 05 \, \text{mH}$, $R_1' = 6 \cdot 3 \, \Omega$, $R_2' = 0 \cdot 1041$, $R_3' = 0 \cdot 2646 \, \Omega$. $L_2' = 1 \cdot 5 \, \text{mH}$. (*Permission International Journal of Electrical Engineering Education.*)

## 9.7  Practical Schemes

The basic six-valve bridge connexion is in general use throughout the world. In order to cater for the high voltages in use several valves are often used in series to form one element of the bridge. The bridges may be interconnected in a number of ways, a common arrangement being shown in figure 9.20 in which the centre connexion at each end of the line is earthed. In the event of a fault on one side of the system the other can continue operating using the earth as the second line.

Several schemes are either in operation or under construction. Brief details of some of these are as follows:

(a) Donbass—Volgograd (U.S.S.R.): 750 MW, eight valve groups per station, 800 kV d.c., 294 miles (overhead line).

(b) Sardinia: 200 MW, two valve groups per station, 200 kV d.c., 65 miles of cable, 193 miles (overhead line).

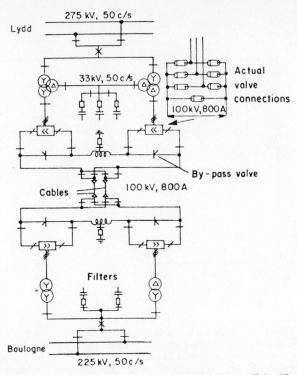

Figure 9.20   Schematic arrangement of Britain–France link. (*Permission of A.S.E.A.*)

(c) Japan: 300 MW, two valve groups per station 250 kV d.c., connexion between 50 Hz and 60 Hz systems.

(d) New Zealand: 600 MW, four valve groups per station, 500 kV (d.c.), 25 miles of cable, 360 miles (overhead line).

(e) Great Britain—France: the main circuit diagram for this link across the English Channel is shown in figure 9.20.

(f) Britain: 51-mile underground cable link between Kingsnorth Power Station and two substations in the London area delivers 640 MW at $\pm 266$ kV and uses 56 valves.

(g) U.S.A.: Pacific coast interties shown in figure 9.21. Six main valves plus bypass valve in each station, each valve rated at 133 kV, 1800 A (6 anodes, 300 A each). Reactive power requirement is 840 MVAr during rectification; of this filters provide 180 MVAr, a.c. system, 200 MVAr and shunt capacitor banks, 460 MVAr.

Figure 9.21 Map of the Pacific Coast interties indicating routes of the 800 kV, 1440 MW d.c. links from the Dalles to Los Angeles and Mead.

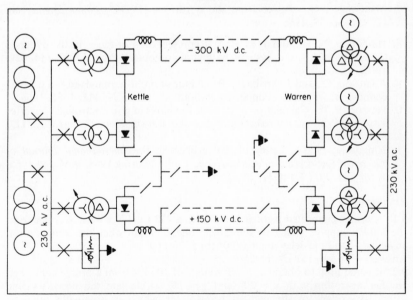

Figure 9.22 Nelson River Scheme (Canada), first stage of operation. 810 MW. Valve groups will be installed for operation at +150 kV, − 300 kV as indicated. (*Permission of Power Engineering.*)

(h) Canada: Kettle Generating Station (Nelson River) to Winnipeg (565 miles), two lines in parallel carry 810 MW. A schematic diagram of the first stage of this connexion is shown in figure 9.22. It is eventually intended to operate the scheme at ±450 kV.

### References

BOOKS

1. Adamson, C., and N. G. Hingorani, *High Voltage Direct Current Power Transmission*, Garraway Ltd., 1960. (*This gives a comprehensive list of references.*)
2. Cory, B. J. (Editor), *High Voltage Direct Current Converters and Systems*, Macdonald, London, 1965.

PAPERS

3. Collection of articles by several authors. *Elec. Rev.*, **178**, No. 8. February 1966.
4. Gorodetzky, S. S., '220–400 kV Direct-Current Cables', (U.S.S.R.), *C.I.G.R.É.*, Paper 206, 1958.
5. 'High Voltage Direct Current Transmission', *Proc. Conf. I.E.E.*, September 1966.
6. Annestrand, S. A. et al., '800 kV d.c. transmission,' *Power Engineering* (*U.S.A.*), p. 46, 1968.
7. MacGregor, G., 'Power from the north via HVDC', *Power Engineering* (*U.S.A.*), p. 38, 1968.
8. Hingorani, N. G., and J. L. Hay, 'Representation of faults in the dynamic simulation of h.v. d.c. systems by digital computer', *Proc. I.E.E.*, **114**, 629, 1967.
9. Adamson, C., and J. Arrillaga, 'Behaviour of multi-terminal a.c.—d.c. interconnexions with series-connected stations', *Proc. I.E.E.*, **115**, 1685, 1968.
10. Persson, E. V., 'Calculation of transfer functions in grid-controlled converter systems. With special reference to h.v. d.c. transmission', *Proc. I.E.E.*, **117**, 989, 1970.
11. Hingorani, N. G., Series of five articles in the *International Journal of Electrical Engineering Education*, Vols. 1 and 2 during 1963, 1964 and 1965, **1**, pp. 273, 393; **2**, pp. 37, 241, 355.

### Problems

9.1. A bridge-connected rectifier is fed from a 230 kV/120 kV transformer from the 230 kV supply. Calculate the direct voltage output when the commutation angle is 15° and the delay angle (a) 0° (b) 30° (c) 60°.
Answer: (176 kV, 141 kV, 65 kV).

9.2. It is required to obtain a direct voltage of 100 kV from a bridge connected rectifier operating with $\alpha = 30°$ and $\gamma = 15°$. Calculate the necessary line secondary voltage of the rectifier transformer which is nominally rated at 345 kV/150 kV; calculate the tap ratio required.
Answer: (164 kV line and 1·085).

9.3. If the rectifier in problem 9.2 delivers 800 A d.c., calculate the effective reactance $X$ ($\Omega$) per phase.
Answer: ($X = 14.5\ \Omega$).
9.4. A d.c. link comprises a line of loop resistance $5\ \Omega$ and is connected to transformers giving secondary voltage of 120 kV at each end. The bridge connected converters operate as follows:

Rectifier:   $\alpha = 10°$            Inverter:   $\delta_0 = 10°$
         $X = 15\ \Omega$                          $\gamma = 15°$
                                        Allow 5° margin
                                        on $\delta_0$ for $\delta$
                                        $X = 15\ \Omega$

Calculate the direct current delivered if the inverter operates on constant $\delta$ control. If all parameters remain constant except $\alpha$, calculate the maximum direct current transmittable.
Answer: (356 A, 637 A).
9.5. The system in problem 9.4 is operated with $\alpha = 15°$ and on constant $\beta$ control. Calculate the direct current.
Answer: (278 A).
9.6. For a bridge arrangement, sketch the current waveforms in the valves and in the transformer windings and relate them in time to the anode voltages. Neglect delay and commutation times. Comment on the waveforms from the viewpoint of harmonics.
9.7. Draw a schematic diagram of an existing or proposed direct current transmission scheme. Give vital parameters e.g. voltage, rating of valves, etc.
9.8. A direct current transmission link connects two a.c. systems via converters, the line voltages at the transformer-converter junctions being 100 kV and 90 kV. At the 100 kV end the converter operates with a delay angle of 10°, and at the 90 kV end the converter operates with a $\delta$ of 15°. The effective reactance per phase of each converter is $15\ \Omega$ and the loop resistance of the link is $10\ \Omega$. Determine the magnitude and direction of the power delivered if the inverter operates on constant-$\delta$ control. Both converters consist of six valves in bridge connexion. Calculate the percentage change required in the voltage of the transformer which was originally at 90 kV to produce a transmitted current of 800 A, other controls being unchanged. Comment on the reactive power requirements of the converters.
Answer: (1.6 kA, 150 MW, 6.85%).
9.9. A Thyristor rectifier supplies direct current to a load. On no-load the voltage across the load is 1.5 kV and the firing or delay angle 35°. The delay is controlled by a feedback system which holds the load voltage constant with changes in current. With this control the transformer reactance causes the ratio (voltage drop/no load voltage) to be 0.08 p.u.

By considering only the fundamental components of the three-phase supply currents plot a curve of volt-amperes reactive as a function of load current. Neglect the commutation angle.

# Overvoltages and Insulation Requirements

## 10.1 Introduction

An area of critical importance in the design of power systems is the consideration of the insulation requirements for lines, cables and stations. At first glance this may appear to be a simple matter once the operating voltage of the system is decided but unfortunately this is far from so. As well as the normal operating voltages, transients causing overvoltages occur in the system due to switching, lightning strokes and other causes; the peak values of these can be much in excess of the operating voltage. Because of this devices must be provided to protect items of plant. The term Extra High Voltage (e.h.v.) has generally been accepted as describing systems of 230 kV up to 765 kV and for voltages above 765 kV the term Ultra High Voltage (u.h.v.) is applied; below 230 kV, High Voltage (h.v.) is in use.

Until recent years lightning has largely determined the insulation requirements, i.e. size of bushings, number of insulators per string and tower clearances of the system and the insulation of equipment tested with voltages of a waveform approximately that of a lightning surge. With the much higher operating voltages now in use and projected the voltage transients or surges due to switching, i.e. the opening and closing of circuit breakers, have become the major consideration. In this context it is of interest to note that voltages of 500 kV and 765 kV are now in use and active discussion and research are taking place to decide the next voltage level which will be in the range 1000 to 1500 kV.

A factor of major importance is the *contamination* of insulator surfaces caused by atmospheric pollution. This considerably modifies the performance of insulation which becomes difficult to assess precisely. The presence of dirt, salt, etc., on the insulator discs or bushing surfaces results in these surfaces becoming slightly conducting and hence flashover occurs.

A few terms frequently used in high-voltage technology need definition. They are as follows.

(i) *Basic Impulse Insulation Level or Basic Insulation Level (B.I.L.)*— Reference levels expressed in impulse crest (peak) voltage with a standard wave not longer than a $1 \cdot 5 \times 50 \, \mu s$ wave. Apparatus insulation as demonstrated by suitable tests shall be equal or greater than the B.I.L. The two standard tests are the power frequency and $1 \cdot 2/50$ impulse wave withstand tests. The *withstand voltage* is the level the equipment will withstand for a given length of time or number of applications without a disruptive discharge occurring, i.e. a failure of insulation resulting in a collapse of voltage and passage of current (sometimes termed sparkover or flashover when the discharge is on the external surface). Normally several tests are performed and the number of flashovers noted. The B.I.L. is usually expressed as a per-unit of the peak (crest) value of the normal operating voltage to earth; e.g. for a maximum operating voltage of 362 kV,

$$1 \text{ p.u.} = \sqrt{2} \times \frac{362}{\sqrt{3}} = 300 \text{ kV},$$

so that a B.I.L. of $2 \cdot 7$ p.u. $= 810$ kV.

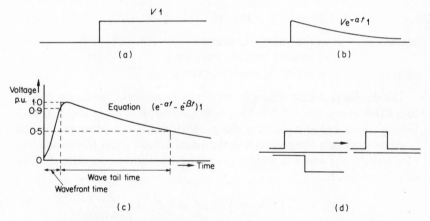

Figure 10.1 Basic Impulse Waveforms. (a) and (b) are frequently used in calculations. (c) Shape of lightning and switching surges; the former have a rise time of say $1 \cdot 2 \, \mu s$ and a fall time of half maximum value of $50 \, \mu s$ (hence $1 \cdot 2/50$ wave). Switching surges are much longer, the duration times varying with situation, a typical wave is $175/3000 \, \mu s$. Equation for $1/50$ wave is, $v = 1 \cdot 036 \, (e^{-0 \cdot 0146t} - e^{-2 \cdot 56t})$. (d) Use of two unit functions to form a wave of finite duration.

(ii) *Critical Flashover Voltage (C.F.O.)*—The peak voltage for a 50 per cent probability of flashover or disruptive discharge (sometimes denoted by $V_{50}$).

(iii) *Impulse Ratio* (for flashover or puncture of insulation)—Impulse peak voltage divided by the crest value of power-frequency voltage to cause flashover or puncture.

Impulse tests are normally performed with a voltage wave which rises in $1\cdot2$ $\mu$s and falls to half the peak value in 50 $\mu$s (see figure 10.1), this is known as a $1\cdot2/50$ $\mu$s wave and typifies the lightning surge. In most impulse generators the shape and duration of the wave may be modified. Basic impulse waves are shown in figure 10.1. Switching surges consist of damped oscillatory waves, the frequency of which is determined by the system configuration and parameters, they are normally of amplitude 2–2·8 p.u. although they can exceed 4 p.u. (per-unit values based on peak line to earth operating voltage as before).

### Corona

Air at normal atmospheric pressure and temperature breaks down at 30 kV/cm (peak or crest value). For smooth cylinders this stress may be determined from the expression,

$$\frac{V}{r \ln (d/r)} V/\text{cm}.$$

where,     $V =$ the voltage to neutral,
           $d =$ spacing between lines (cm),
and,       $r =$ radius of conductor (cm).

The discharge which occurs in the air surrounding a conductor subject to a stress above this value is known as Corona; it is easily detected by a hissing sound and at night by a blue glow around the conductors. Corona is established at a stress $\mathscr{E}_v$ called the visual critical stress (corresponding to a voltage $V_v$) which is greater than the basic breakdown value $\mathscr{E}_0$. For smooth cylinders,

$$\mathscr{E}_v = \mathscr{E}_0 \left(1 + \frac{0\cdot3}{\sqrt{r}}\right) \quad \text{and as} \quad \mathscr{E}_v = \frac{V_v}{r \ln (d/r)}$$

and air breaks down at 21·1 kV (r.m.s.) per cm, it follows that for fair weather conditions,

$$V_v = 21\cdot1 \ r \left(1 + \frac{0\cdot3}{\sqrt{r}}\right) \ln \left(\frac{d}{r}\right) \text{ kV (r.m.s. to earth)}$$

This voltage is dependent on air density and the nature of the conductor surface. The following formula due to Peek[5] incorporates these features,

$$V_v = 21 \cdot 1 \, m\delta r \left(1 + \frac{0 \cdot 3}{\sqrt{\delta r}}\right) \ln \left(\frac{d}{r}\right) \text{ kV (r.m.s.)} \qquad (10.1)$$

where, $\delta$ is the air density factor $= \dfrac{3 \cdot 92b}{273 + T}$

$b$ = barometric pressure in centimetres of mercury
$T$ = temperature in deg. C
$m$ = is typically 0·9 for weathered stranded conductors.

The power loss (kW per mile of conductor) due to corona under fair-weather conditions (Peterson–discussion *A.I.E.E. Transactions*, **52,** 62, 1933) is given by

$$p = \frac{0 \cdot 0000337}{\left[\log_{10}\left(\dfrac{2s}{d}\right)\right]^2} f V^2 F \qquad (10.2)$$

where,  $f$ = frequency (Hz),
$V$ = line to ground voltage,
$s$ = spacing between conductors,
$d$ = conductor diameter,
$F$ = corona factor determined by test.

With fair weather conditions the line loss can approach the same order of magnitude as the insulator leakage losses. These losses have little economic or technical significance for e.h.v. lines. Typical measured values in kW for all three conductors per mile (reference 10.3) are 2–8 for a 500 kV line of conductor diameter 1·465 in. (3·7 cm) and 5–30 for the same conductor at 700 kV (average peak gradient 23·13 kV/cm). The loss for an insulator string (24 discs) at 500 kV is 95 W and at 735 kV (32 discs) 150 W.

Foul-weather corona losses are much more significant. Tests summarized in reference 10.3 indicate that the conductor loss is proportional to the following; logarithm of the rate of rain fall (rain rate), number of conductors, and approximately to the product of the operating voltage and the fifth power of the voltage gradient on the underside of the conductor surface where the loss is greatest due to the accumulation of water. The following formula applies:

Total loss per 3-phase mile in kW

= total fair-weather loss per 3-phase mile (kW)

$$+ \left[ \frac{V}{\sqrt{3}} Jr^2 \ln (1 + KR) \sum_1^n (E^5) \right] \qquad (10.3)$$

where,

$J$ = loss-current constant (approx. $5 \cdot 35 \times 10^{-10}$ at 700 kV and 500 kV and $7 \cdot 04 \times 10^{-10}$ at 400 kV),

$r$ = conductor radius (cm.),

$n$ = number of conductors per bundle × 3,

$E$ = voltage gradient on underside of conductor kV (peak)/cm.,

$K$ = a wetting coefficient = 10 for $R$ in mm/hour,

$R$ = rain rate (mm/hour).

Typical measured values in kW per 3-phase mile (reference 10.3) are as follows: 700 kV, 45·5 ft (14 m) spacing, 190 at 0·1 in./hr, 260 at 0·5 in./hr; 500 kV, 20 ft (6·1 m) spacing, 150 at 0·1 in./hr and 215 at 0·5 in./hr.

### Radio interference (R.I.) or radio noise

Although the presence of corona results in a power loss, a more important effect is that the discharge causes radiations to be propagated in the frequency bands used by radio and television. The most effective way to reduce or avoid corona and radio interference is the use of bundle conductors, i.e. several conductors per phase suspended from common insulators and separated mechanically by spacers of various designs. On some systems four-conductor bundles are in use. The configuration of conductors forming a bundle modifies the surface electric field such that the maximum stress is lower than with a single conductor, also the current rating of the circuit is increased. Troubles with bundle conductor lines have been experienced owing to aerodynamic instability resulting in mechanical oscillations along the line.

The corona discharges occur at discontinuities on the conductor surface and a random generation of pulses occurs. In wet weather water droplets form on the underside of the conductor which deform under the stress. Each droplet elongates and a sharp point is formed providing a strong source of R.I. and loss. Humidity has a marked effect on the corona and R.I. phenomena, increase in humidity causing a marked increase in these quantities. Radio interference is also caused by discharges from insulators and bad contacts.

### Contamination of insulator surfaces

The nature of the contamination varies according to location and often consists of soot, flyash cements, etc. In coastal districts surface films of salt are formed. The surface conductivity ($\sigma_s$) of a conducting film of thickness $t$ cm is defined by $\sigma_s = \sigma t$, where $\sigma$ is the specific (volume) conductivity of the film. For an axi-symmetric insulator surface with a uniform film of contaminating material of surface resistivity $\rho_s$ (i.e. $1/\sigma_s$), the total electrode to electrode resistance,

$$R = \int_0^L \frac{\rho_s dl}{\pi D_1}$$

where,  $L$ = length of leakage path,
$dl$ = element of leakage path,
$D_1$ = diameter of surface at $dl$.

The small leakage current of a dry insulator is determined by the electrostatic field alone. When the pollution layer present becomes moist due to rain, fog or dew it becomes conductive and a surface leakage current of a much higher magnitude than the dry value flows; the electric field becomes very distorted. Where the current density is greatest the heat generated increases the film temperature to boiling point and evaporation of the moisture occurs and the formation of *dry bands*. These bands around the insulation have a high resistance and support almost all the voltage. This results in the breakdown of the air and an arc is formed across the dry band. The arc roots burn upon the moisture film at the dry band boundary and dry out more film under the roots. Hence the arc extends further into the moist area. If the moisture precipitation is low the dry band widens until the arc, when it reaches a certain length, extinguishes. The stress across the dry band is then just below the air breakdown value and should a local increase in precipitation occur breakdown results and the arc

Figure 10.2  Formation of dry bands around polluted insulators.

reappears. If the rate of precipitation balances the loss of moisture due to the arc then a stable condition is reached and the discharge may continue for a considerable time, see figure 10.2.

Various methods are used to prevent flashover due to pollution. The leakage length of the insulator is increased although difficulty is experienced in producing insulators with leakage paths much greater than 1 in. (2·54 cm) per kV. Greasing of the entire insulator surface prevents the formation of a continuous film and is very effective. However, a fresh application of grease is required periodically, e.g. 1 to 2 years. The washing of surfaces with water from hoses when the equipment is live is also effective.

## 10.2 Generation of Overvoltages

### Lightning surges

A thundercloud is bipolar with positive charges at the top and negative at the bottom, usually separated by several kilometres. When the electric field strength exceeds the breakdown value a lightning discharge is initiated. The first discharge proceeds to the earth in steps (stepped leader stroke). When close to the earth a faster and luminous return stroke travels along the initial channel and several such leader and return strokes constitute a flash. The ratio of negative to positive strokes is about 5 to 1 in temperate regions. The magnitude of the return stroke can be as high as 200 kA although an average value is of the order of 20 kA.

Following the initial stroke, after a very short interval, a second stroke to earth occurs usually in the ionized path formed by the original. Again a return stroke follows. Usually several such subsequent strokes (known as dart leaders) occur, the average being between three and four. The complete sequence is known as a multiple stroke lightning flash and a representation of the strokes at different time intervals is shown in figure 10.3. Normally only the heavy current flowing over the first 50 $\mu$s is of importance and the current–time relationship has been shown to be of the form, $i = i_{peak} (e^{-\alpha t} - e^{-\beta t})$.

When a stroke arrives on an overhead conductor equal current surges of the above waveform are propagated in both directions away from the point of impact. The magnitude of each voltage surge set up is therefore $\frac{1}{2}.Z_0.i_{peak} (e^{-\alpha t} - e^{-\beta t})$, where $Z_0$ is the conductor surge impedance. For a current peak of 20 kA and a $Z_0$ of 350 $\Omega$ the voltage surges will have a peak value of $(350/2) \times 20 \times 10^3$, i.e. 3500 kV.

When a ground or earth wire exists over the overhead line a stroke arriving on a tower or on the wire itself sets up surges flowing in both directions along the wire. On reaching neighbouring towers they are partially reflected and transmitted further. This process continues over the length of the line as towers are encountered. If the towers are 300 metres apart the travel time between towers and back to the original tower is $(2 \times 300)/(3 \times 10^8)$, i.e. $2\,\mu s$ where the speed of propagation is $3 \times 10^8$ m/s. The voltage distribution may be obtained by means of the Bewley Lattice Diagram to be described in a later section.

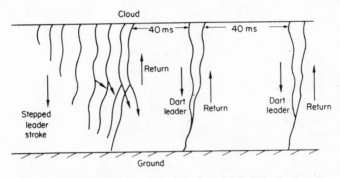

Figure 10.3   Sequence of strokes in a multiple lightning stroke.

An indirect stroke strikes the earth near a line and the induced current which is normally of positive polarity creates a voltage surge of the same waveshape which has an amplitude dependent on the distance from the ground. With a direct stroke the full lightning current flows into the line producing a surge travelling away from the point of impact in all directions. A direct stroke to a tower causes a *back flashover* due to the voltage set up across the tower inductance and footing resistance by the rapidly changing lightning current (typically $10\,kA/\mu s$); this appears as an overvoltage between the top of the tower and the conductors (which are at a lower voltage).

*Switching surges—Interruption of short circuits and switching operations*

When the arc between the circuit breaker contacts breaks, the full system voltage (recovery voltage) suddenly appears across the open gap and hence across the $R$–$L$–$C$ circuit comprising the system. The simplest form of single-phase equivalent circuit is shown in figure 10.4a and b. The resultant voltage appearing across the circuit is shown in figure 10.4c. It consists of a high-frequency component superimposed on the normal system voltage, the total being known as the restriking voltage and constituting a switching surge. The equivalent circuit of figure 10.4 may be analysed by means of the Laplace Transform. When the circuit breaker

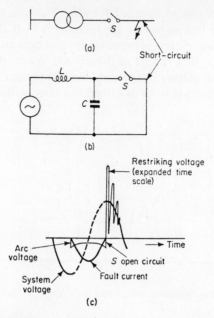

Figure 10.4  Restriking voltage set up on fault interruption. (a) System diagram. (b) Equivalent circuit. (c) Current and voltage waveforms.

opens the cessation of current may be simulated by the injection of an equal and opposite current into the system at time zero. Although this current is sinusoidal in form it may be approximated by a ramp function as shown in figure 10.5a. Let the fault current, $i = \sqrt{2}I \sin \omega t$.

The arc is extinguished at a current pause or zero (see Chapter 11) at which,

$$\frac{\mathrm{d}i}{\mathrm{d}t} = (\sqrt{2}I\omega \cos \omega t)_{t=0}$$

$$= \sqrt{2}I\omega$$

The short-circuit current $= I = V/\omega L$ (r.m.s. value). Hence the equation for the ramp function of injected current is,

$$\left(\frac{\mathrm{d}i}{\mathrm{d}t}\right)_{t=0} = \sqrt{2}\left(\frac{V}{L}\right)t \tag{10.4}$$

The transform of the circuit after breaker opening is shown in figure 10.5b.

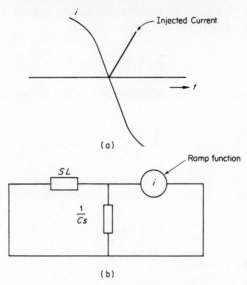

(a)

(b)

Figure 10.5   Application of Laplace Transform to circuit of Figure 4. (a) Ramp function of current. (b) Equivalent circuit.

Transform of the ramp function

$$= i(s) = \frac{\sqrt{2}V}{Ls^2} \tag{10.5}$$

Voltage across the breaker contacts, i.e. across $C$,

$$[v(s)] = \frac{\sqrt{2}V}{Ls^2} \cdot \frac{L/C}{Ls + 1/Cs}$$

$$= \sqrt{2}V \frac{\omega_0^2}{s(s^2 + \omega_0^2)}$$

where $\omega_0^2 = 1/LC$

$$\therefore \quad v(s) = \sqrt{2}V \left( \frac{A}{s} + \frac{Bs + D}{s^2 + \omega_0^2} \right)$$

$$= \sqrt{2}V \left( \frac{1}{s} - \frac{s}{s^2 + \omega_0^2} \right)$$

From a table of transforms,

$$v(t) = \sqrt{2}V \left[ 1 - \cos \left( \sqrt{\frac{1}{LC}} t \right) \right] \tag{10.6}$$

Equation 10.5 assumes that the full system voltage $V$ exists across the open switch, strictly the voltage across $C$ is

$$V\left(\frac{X_c}{X_c - X_L}\right)$$

where $X_c$ and $X_L$ are the capacitive and inductive reactances so that (10.6) becomes,

$$\frac{\sqrt{2}V}{[1 - X_L/X_c)]}(1 - \cos \omega_0 t) \tag{10.7}$$

The difference in practice between these two expressions is small.

When series resistance ($R$ $\Omega$) is significant, equation (10.4) becomes, .

$$\sqrt{2}\left(\frac{V\omega}{R + j\omega L}\right)t$$

and the ramp expression

$$i(s) = \frac{\sqrt{2}\omega V}{(R + j\omega L)s^2} \tag{10.8}$$

and $v(s) = i_s Z(s)$

The analysis of this case yields the expression,

$$v(t) = \omega IL (1 - e^{-\alpha t} \cos \omega_0 t) \tag{10.9}$$

where, $\omega$ is the power frequency angular frequency,

$\omega_0$ is the natural angular frequency of the circuit,

$I =$ short circuit current prior to the breaker opening,

$$\alpha = \frac{R}{2L} \text{ and } (\alpha^2 + \omega_0^2) = \frac{1}{LC}$$

(Note $V \doteqdot I\omega L$)

From (10.9)

$$(dv/dt)_{\text{maximum}} \doteqdot \omega I \sqrt{\left(\frac{L}{C}\right)}e^{-\alpha t} \tag{10.10}$$

The restriking voltage ($v$) can thus rise to a maximum value of $2V$, where $V$ is the peak value of the system recovery voltage. A similar expression is obtained when a line is suddenly energized by the system voltage.

If a resistance $R_s$ be connected across the contacts of the circuit breaker the surge will be critically damped when $R_s = \frac{1}{2}\sqrt{(L/C)}$ and this offers an important method of reducing the severity of the transient. The initial rate of rise of the surge is very important as this determines whether the contact gap, which is highly polluted with arc products, breaks down again after the initial open circuit occurs. In the system shown in figure 10.6a a double-frequency transient (figure 10.6b) is set up, the two frequencies being determined by the circuit on each side of the switch, i.e.

$$\omega_1 = \frac{1}{\sqrt{(L_1 C_1)}} \text{ and } \omega_2 = \frac{1}{\sqrt{(L_2 C_2)}}.$$

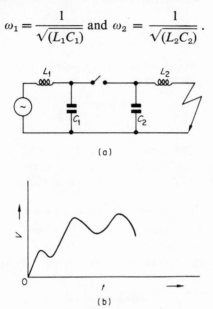

(a)

(b)

Figure 10.6a  System with double frequency restriking transient.
b  Typical waveform of double frequency 'although initial amplitude
is low, rate of rise is high.

*Example 10.1* Determine the relative attenuation occurring in 5 cycles in the overvoltage surge set up on a 66-kV cable fed through an air blast circuit breaker when the breaker opens on a system short circuit. The breaker incorporates resistance switching, i.e. an optimum resistance switched in across contact gap on opening. The network parameters are as follows:

$$R = 7 \cdot 8 \ \Omega, L = 6 \cdot 4 \text{ mH}, C = 0 \cdot 0495 \ \mu\text{F}.$$

*Solution*  Switching resistance $= \frac{1}{2} \sqrt{(L/C)} = 180 \ \Omega$

$$\alpha = \frac{1}{2} \cdot \frac{R}{L} = 610$$

$$\omega = \frac{1}{\sqrt{(LC)}} = 5 \cdot 61 \times 10^4 \text{ rad/s}$$

i.e. transient frequency $= 8 \cdot 93$ kHz

Hence, 5 cycles $= \dfrac{1}{8 \cdot 93 \times 10^3} \times 5$ seconds

and $e^{-\alpha t} = \exp\left(-610 \times \dfrac{5}{8 \cdot 93 \times 10^3}\right) = 0 \cdot 712$

The maximum theoretical voltage set up with the circuit breaker opening on a short-circuit fault would be,

$$2 \times \frac{66000}{\sqrt{3}} \times \sqrt{2}$$

i.e. $107 \cdot 5$ kV and in 560 $\mu$s this becomes $107 \cdot 5 \times (1 - 0 \cdot 712)$, i.e. 31 kV.

Note that resistance switching lowers the current to be broken and raises the power factor so that the voltage is not at peak value at opening.

*Switching surges—Interruption of capacitive circuits*

The interruption of capacitive circuits is shown in figure 10.7. The capacitance is left charged at the instant of arc interruption to value $v_m$ but half a cycle later the system voltage is $-v_m$ giving a gap voltage of

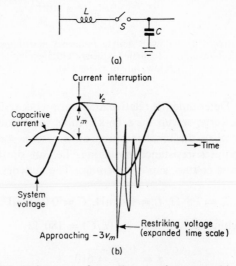

Figure 10.7 Voltage waveform when opening a capacitive circuit.

$2v_m$. If the gap breaks down an oscillatory transient is set up (as previously discussed) which can increase the gap voltage still further as shown in the figure.

*Current chopping* Current chopping arises with air-blast circuit breakers which operate on the same air pressure and velocity for all values of interrupted current. Hence on low-current interruption the breaker tends to open the circuit before the current natural-zero and the electromagnetic energy present is rapidly converted to electrostatic energy, i.e.

$$\tfrac{1}{2}Li_0^2 = \tfrac{1}{2}Cv^2, \text{ and } v = i_0 \sqrt{\frac{L}{C}} \tag{10.11}$$

The voltage waveform is shown in figure 10.8.

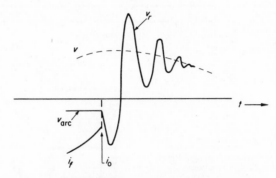

Figure 10.8   Voltage transient due to current chopping.
$i_f$ = fault current, $v$ = system voltage,
$i_o$ = current magnitude at chop.

An extension of 10.11 to include resistance and time yields

$$v = i_0 \sqrt{\frac{L}{C}}\, e^{-\alpha t} \sin \omega_0 t \tag{10.12}$$

where, $\omega_0 = 1\sqrt{LC}$ and $i_0$ is the value of the current at the instant of chopping. High transient voltages may be set up on opening a highly inductive circuit such as a transformer on no load.

*Faults* Over-voltages may be produced by certain types of asymmetrical fault mainly on systems with ungrounded neutrals. The voltages set up are of normal operating frequency. Consider the circuit with a three-phase earth fault as shown in figure 10.9. If the circuit is not grounded the voltage across the first gap to open is $1 \cdot 5\ V_{\text{phase}}$. With the system grounded the gap voltage is limited to the phase voltage. This is discussed more fully in Chapter 6.

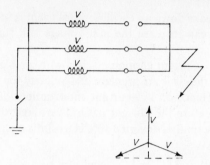

**Figure 10.9   Three-phase system with neutral earthing.**

Table 10.1.    Summary of the more important switching operations

| Switching Operation | System | Voltage across contacts |
|---|---|---|
| 1. Terminal short circuit | | |
| 2. Short line fault | | |
| 3. Two out-of-phase systems – voltage depends on grounding conditions in systems | | |
| 4. Small inductive currents, current chopped (unloaded transformer) | Transformer | |
| 5. Interrupting capacitive currents –capacitor banks, lines and cables on no–load | | |
| 6. Evolving fault – e.g., flashover across transformer plus arc across contacts when interrupting transformer on no–load | Transformer | |
| 7. Switching–in unloaded E.H.V./U.H.V. line (trapped charge) | | |

*Permission of Brown Boveri Review, December 1970*

Table 10.2.    Overvoltages in long e.h.v. and u.h.v. lines

| Means of reducing switching overvoltages | Basic diagram |
|---|---|
| 1. High voltage shunt reactors connected to the line to reduce power-frequency overvoltage | |
| 2. Eliminating or reducing trapped charge by: | |
|    2.1   Line shunting after interruption | |
|    2.2   Line discharge by magnetic potential transformers | |
|    2.3   Low voltage side disconnection of the line | |
|    2.4   Opening resistors | |
|    2.5   Single-phase reclosing | |
|    2.6   Damping of line voltage oscillation after disconnecting a line equipped with h.v. reactors | |
| 3. Damping the transient oscillation of the switching overvoltages | |
|    3.1   Single-stage closing resistor insertion | |
|    3.2   Multi-stage closing resistor insertion | |
|    3.3   Closing resistor in-line between circuit-breaker and shunt reactor | |
|    3.4   Closing resistor in-line on the line side of the shunt reactor | |
|    3.5   Resonance circuit (surge absorber) connected to the line | |
| 4. Switching at favourable switching moments: | |
|    4.1   Synchronized closing | |
|    4.2   Reclosing at voltage minimum of a beat across the breaker | |
| 5. Simultaneous closing at both ends of the line | |
| 6. Limitation by surge arresters when energizing line at no-load (a) disconnecting reactor loaded transformers (b) disconnecting high-voltage reactors (c) | |

1, 2, 3 = Switching sequence    ▌Reactor

*Resonance* It is well known that in a series resonant circuit severe over-voltages occur dependent on the resistance present; the voltage at resonance across the capacitance is $1/(\omega CR)$. Although it is unlikely that resonance in a supply network be obtained at normal supply frequencies it is possible to have this condition at harmonic frequencies. Resonance is normally associated with the capacitance to earth of items of plant and often brought about by an opened phase due to a broken conductor or a fuse operating.

In circuits containing windings such as reactors with iron cores a condition due to the shape of the magnetization curve known as *ferro-resonance* is possible. This can produce resonance with overvoltages and also sudden changes from one condition to another.

A summary of important switching operations is given in Table 10.1 and means of reducing switching overvoltages summarized in Table 10.2.

### 10.3  Protection Against Overvoltages

*Modification of transients*

When considering the protection of a power system against overvoltages the transients may either be modified or even eliminated before reaching the substations, or if this is not possible to protect by various means the lines and substation equipment from flashover or insulation damage. By the use of overhead earth (ground) wires, phase conductors may be shielded from direct lightning strokes and the effects of induced surges from indirect strokes lessened. The shielding is not complete except perhaps for a phase conductor immediately below the earth wire. The effective amount of shielding is often described by an angle $\alpha$ as shown in figure 10.10, a value of 35° appears to agree with practical experience. Obviously two earth wires horizontally separated provide much better shielding. Often for reasons of economy earth wires are installed over the last kilometre or so of line immediately before it enters a substation. It has already been seen that the switching-in of resistance across circuit breaker contacts reduces the high overvoltages produced on opening, especially on capacitive or low-current inductive circuits.

An aspect of vital importance quite apart from the prevention of damage is the maintenance of supply especially as most flashovers cause no permanent damage and therefore a complete and lasting removal of the circuit from operation is not required. This may be achieved by the use of *auto-reclosing* circuit breakers.

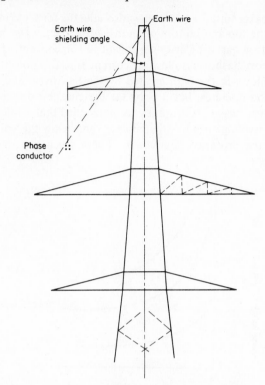

Figure 10.10   Single earth wire protection; shielding angle α normally 35°.

A *surge modifier* may be produced by the connexion of a shunt capacitor between the line and earth or an inductor in series with the line; oscillatory effects may be reduced by the inclusion of damping resistors. It is uneconomical to attempt to modify or eliminate most overvoltages and means are required to protect the various items of power systems. Surge diverters are connected across the equipment and divert to earth the transient; surge modifiers are connected in series with the line entering the substation and attempt to reduce the steepness of the wave front—hence its severity.

*Surge diverters*

The basic requirements for diverters are that they should pass no current at normal voltage, interrupt the power frequency follow-on current after a flashover, and break down as quickly as possible after the abnormal voltage arrives. The simplest form is the *rod gap* as shown in figure 11.2

for a circuit breaker bushing. This may also take the form of rings (arcing rings) around the top and bottom of an insulator string. The breakdown voltage for a given gap is polarity dependent to some extent. It does not interrupt the post flashover follow-on current (i.e. the power-frequency current which flows in the path created by the flashover) and hence the circuit protection operates, but it is by far the cheapest device for plant protection against surges. It is usually recommended that a rod gap be set to a breakdown voltage not less than 30 per cent below the voltage with-stand level of the protected equipment. For a given gap the time for

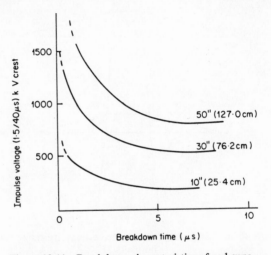

Figure 10.11    Breakdown characteristics of rod gaps.

breakdown varies roughly inversely with the applied voltage; there exists, however, some dispersion of values. The times for positive voltages are lower than those for negative.

Typical curves relating the critical flashover voltage and time to break-down for rod gaps of different spacings are shown in figure 10.11. In figure 10.12 the flashover voltage–rod spacing relationship is shown for a rod-to-rod gap. The flashover voltage is to some extent dependent on the length of the lower (grounded) rod. For low values of this length there is a reasonable difference between positive (lower values) and negative flash-over voltages. Usually a length of 1·5 to 2·0 times the gap spacing is adequate to diminish this effect to a reasonable amount.

*Expulsion gaps or tubes* (*protector tubes*) A disadvantage of the plain rod gap is the power-frequency current (follow-on current) that flows after breakdown which can only be extinguished by circuit breaker operation.

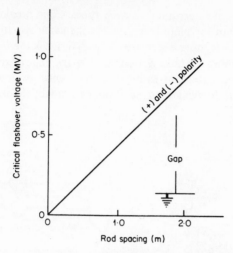

Figure 10.12   Effect of gap spacing on flashover of rod gaps—lower rod grounded.

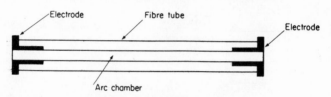

Figure 10.13   Expulsion tube.

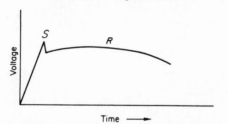

Figure 10.14   Characteristic of lightning arrester.

An improvement is the expulsion tube which consists of a spark gap in a fibre tube as shown in figure 10.13. When a sparkover occurs between the electrodes the follow-on current arc is contained in the relatively small fibre tube. The high temperature of the arc vaporizes some of the organic material of the tube wall causing a high gas pressure to build up in the tube. This gas possesses considerable turbulence and it extinguishes the arc. The hot gas rapidly leaves the tube which is open at the ends. Very high currents have been interrupted in such tubes. The breakdown voltage is slightly lower than for plain rod-gaps for the same spacing.

An improved but more expensive surge diverter is the *lightning arrester*. A porcelain bushing contains a number of spark gaps in series with silicon carbide discs, the latter possessing low resistance to high currents and high resistance to low currents, i.e. it obeys a law of the form $V = aI^{(0\cdot2)}$,

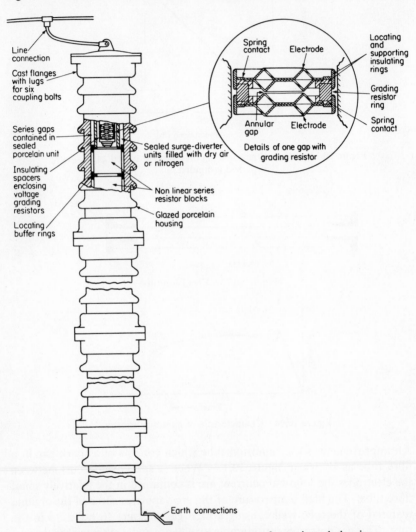

Figure 10.15　Diagram of lightning arrester—four unit stack showing part section of one unit. (*Permission of the Electricity Council.*)

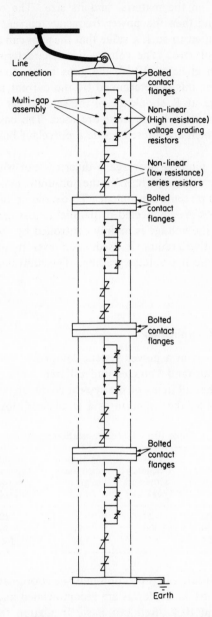

Figure 10.16 Equivalent circuit of components comprising a single-phase four unit surge diverter stack. (*Permission of the Electricity Council.*)

where *a* depends on the material and its size. The overvoltage breaks down the gaps and then the power frequency current is determined by the discs and limited to such a value that the gaps can quickly interrupt it at the first current zero. The voltage–time characteristic of a lightning arrester is shown in figure 10.14, the gaps break down at 'S' and the characteristic after this is determined by the current and the discs; the maximum voltage at R should be in the same order as at S. For high voltages a stack of several such units are used. The basic arrangement of discs and gaps which are housed inside a porcelain bushing is shown in figure 10.15.

Although with multiple spark-gaps diverters can withstand high rates of rise of recovery voltage (R.R.R.V.) the non-uniform voltage distribution between the gaps presents a problem. To overcome this, capacitors and non-linear resistors are connected in parallel across each gap. With the high-speed surge the voltage is mainly controlled by the gap capacitance and hence capacitive grading is used. A power frequencies non-linear resistor provides effective voltage grading. The equivalent circuit is shown in figure 10.16.

### 10.4   Insulation Coordination

The equipment used in a power system comprises items having different breakdown or withstand voltages and different voltage–time character-istics. In order that all items of the system be adequately protected there is a need to consider the situation as a whole and not items of plant in

Table 10.3   British substation practice

| Nominal voltage (kV) | Impulse (peak kV) | Power frequency withstand (peak kV) | Minimum clearance to ground (in.) | Minimum clearance between phases (in.) | Co-ord. gap setting (in.) |
|---|---|---|---|---|---|
| 11 | 100 | 29 | 8 | 10 | $2 \times 1\frac{1}{4}$ |
| 132 | 550 | 300 | 44 | 50 | 26 |
| 400 | 1425 | 675 | 120 | 140 | 60–70 |

isolation, i.e. the insulation protection must be coordinated. To assist this process standard insulation levels are recommended and are summarized in Tables 10.3 and 10.4. Reduced Basic Insulation Impulse Levels are used when considering switching surges and these are summarized in Table 10.4.

Table 10.4 Recommended B.I.L.'s at various operating voltages
(United States Practice)

| Voltage class (kV) | 15 | 23 | 34·5 | 46 | 69 | 92 | 115 |
|---|---|---|---|---|---|---|---|
| B.I.L. (kV) | · 110 | 150 | 200 | 250 | 350 | 450 | 550 |
| Reduced B.I.L. (kV) | | | 125 | | | | 450⎞<br>350⎠ |

| Voltage class (kV) | 138 | 161 | 196 | 230 | 287 | 345 | 500 |
|---|---|---|---|---|---|---|---|
| B.I.L. (kV) | 650 | 750 | 900 | 1050 | 1300 | 1550 | 1800 |
| Reduced B.I.L. (kV) | 550⎞<br>450⎠ | 650⎞<br>550⎠ | | 900⎞<br>825⎬<br>750⎠ | 1175⎞<br>1050⎬<br>900⎠ | 1425⎞<br>1300⎬<br>1050⎠ | 1675⎞<br>1550⎬<br>1300⎠ |

Coordination is rendered difficult by the different voltage–time characteristics of plant and protective devices, for example a gap may have an impulse ratio of 2 for a 20-$\mu$s front wave and 3 for a 5-$\mu$s wave. At the higher frequencies (shorter wavefronts) corona cannot form in time to relieve the stress concentration on the gap electrodes. With a lightning surge a higher voltage can be withstood because a discharge requires a certain discreet amount of energy as well as a minimum voltage and the applied voltage increases until the energy reaches this value. In figure 10.17

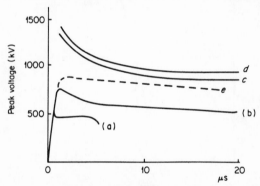

Figure 10.17 Insulation coordination in a H.V. substation. Voltage-time characteristics of plant for a 1·5/40 $\mu$s wave. (a) Characteristic of lightning arrester. (b) Transformer. (c) Line insulator string. (d) Busbar insulation. (e) Maximum surge applied waveform.

the voltage–time characteristics for the system elements comprising a substation are shown and the protection of the weakest items by the arrester is illustrated.

Up to an operating voltage of 345 kV the insulation level is determined by lightning and the standard impulse tests suffice along with normal frequency tests. Above this value, however, the overvoltages resulting from switching are higher in magnitude and therefore decide the insulation. The characteristics of air gaps and some solid insulations are different for switching surges than for the standard impulse waves and closer co-ordination of insulation is required because of the lower attenuation of switching surges, although their amplification by reflexion is less than with lightning. Recent work indicates that for transformers the switching impulse strength is of the order of 0·95 of the standard value while for oil-filled cables it is 0·7 to 0·8.

The design withstand level is selected by specifying the risk of flashover, e.g. for 550 kV towers a 0·13 per cent probability has been used. At 345 kV, design is carried out by accepting a switching impulse level of 2·7 p.u. which corresponds to the lightning level. At 500 kV, however, a 2·7 p.u. switching impulse would require 40 per cent more tower insulation than that governed by lightning. The tendency is therefore for the design switching impulse level to be forced lower with increasing system operating

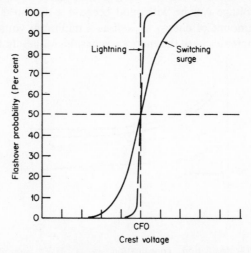

Figure 10.18   Flashover Probability—peak surge-voltage. Lightning flashover is close-grouped near the critical flashover voltage (CFO), whereas switching surge probability is more widely dispersed and follows normal Gaussian distribution. (*Permission of the Westing-house Electrical Corporation, East Pittsburgh, Pennyslvania, U.S.A.*)

voltage and controlling the surges by the more widespread use of resistance switching in the circuit breakers. For example, for the 500 kV network the level is 2 p.u. and at 765 kV it is reduced to 1·7 p.u.; with further increases in system voltage it is hoped to decrease the level to 1·5 p.u.

The problem of switching surges is illustrated in figure 10.18 in which flashover probability is plotted against peak (crest) voltage; *critical flashover voltage* (C.F.O.) is the peak voltage for a particular tower design for which there is a 50 per cent probability of flashover. For lightning the probability of flashover below the C.F.O. is slight but with switching surges having much longer fronts the probability is higher, the curve following the normal Gaussian distribution, and the tower must be designed for a C.F.O. much higher than the maximum transient expected.

An example of the application of lightning arrester characteristics to system requirements is illustrated in figure 10.19. In this the variation of

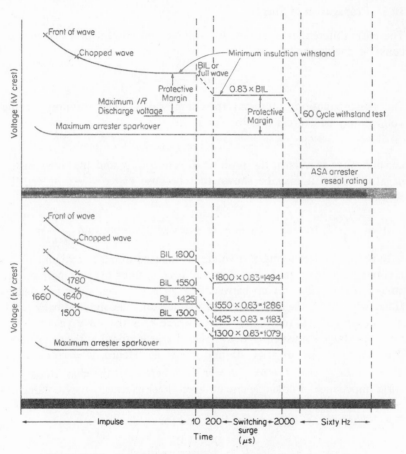

Figure 10.19 Coordination diagrams relating lightning arrester characteristics to system requirements. (*Permission of the Ohio Brass Co.*)

crest voltage for lightning and switching surges with time is shown for an assumed Basic Insulation Level of the equipment to be protected. Over the impulse range (up to 10 $\mu$s) the front of wave and chopped wave peaks are indicated. Over the switching surge region the insulation withstand strength is assumed to be 0·83 of the B.I.L. (based on transformer requirements). The arrester maximum sparkover voltage is shown along with the maximum voltage setup in the arrester by the follow-on current (10 or 20 per cent). A protective margin between the equipment withstand strength and the maximum arrester sparkover of 15 per cent is assumed.

## 10.5  Propagation of Surges

The basic differential equations for voltage and current in a distributed-constant line are as follows:

$$\frac{\partial^2 v}{\partial x^2} = LC \frac{\partial^2 v}{\partial t^2} \text{ and } \frac{\partial^2 i}{\partial x^2} = LC \frac{\partial^2 i}{\partial t^2}$$

and these equations represent travelling waves. The solution for the voltage may be expressed in the form,

$$v = F_1(t - x\sqrt{LC}) + F_2(t + x\sqrt{LC})$$

i.e. one wave travels in the positive direction of $x$ and the other in the negative. Also it may be shown that because $\partial v/\partial x = -L(\partial i/\partial t)$, the solution for current,

$$i = \sqrt{\frac{C}{L}}[F_1(t - x\sqrt{LC}) - F_2(t + x\sqrt{LC})] \text{ noting that } \sqrt{\frac{C}{L}} = \frac{1}{Z_0}$$

In more physical terms, if a voltage is injected into a line (figure 10.20) a corresponding current $i$ will flow and if conditions over a length d$x$ are considered, the flux set up between the go and return wires is equal to $iL$d$x$, where $L$ is the inductance per unit length. The induced back e.m.f. is $-$d$\Phi/$d$t$, i.e. $-Li(\text{d}x/\text{d}t)$ or $-iLU$, where $U$ is the wave velocity. The applied voltage $v$ must equal $iLU$. Also charge is stored in the capacitance over d$x$, i.e. $Q = i \text{ d}t = vC \text{ d}x$ and $i = vCU$. Hence, $vi = viLCU^2$ and $U = 1/\sqrt{LC}$. Also, $i = v\sqrt{C/L} = v/Z_0$, where $Z_0$ is the characteristic or surge impedance. For single-circuit three-phase overhead lines (conductors not bundled) $Z_0$ lies in the range 400 to 600 $\Omega$. $U$ for overhead lines is $3 \times 10^8$ m/sec, i.e. the speed of light and for cables

$$U = \frac{3 \times 10^8}{\sqrt{\epsilon_r \mu_r}} \text{ m/s}$$

where $\epsilon_r$ is usually from 3 to 3·5, and $\mu_r = 1$.

From the above relations,

$$\tfrac{1}{2} Li^2 = \tfrac{1}{2} (iLU) \left(\frac{i}{U}\right)$$

$$= \tfrac{1}{2} \left(\frac{vi}{U}\right) = \tfrac{1}{2} Cv^2.$$

The incident travelling waves of $v_i$ and $i_i$, when they arrive at a junction or discontinuity, produce a reflected current $i_r$ and a reflected voltage $v_r$

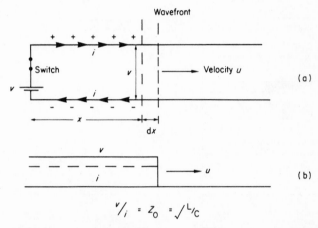

Figure 10.20   Distribution of charge and current as wave progresses along previously unenergized line. (a) Physical arrangement. (b) Symbolic representation.

which travel back along the line. The incident and reflected components of voltage and current are governed by the surge impedance $Z_0$, so that

$$v_i = Z_0 i_i \text{ and } v_r = -Z_0 i_r$$

In the general case of a line of surge impedance $Z_0$ terminated in $Z$ (figure 10.21) the total voltage at $Z$ is $v = v_i + v_r$ and the total current is $i = i_i + i_r$.

Also,        $(v_i + v_r) = Z(i_r + i_i),$

$$Z_0(i_i - i_r) = Z(i_r + i_i)$$

and        $i_r = \left(\frac{Z_0 - Z}{Z_0 + Z}\right) i_i$        (10.13)

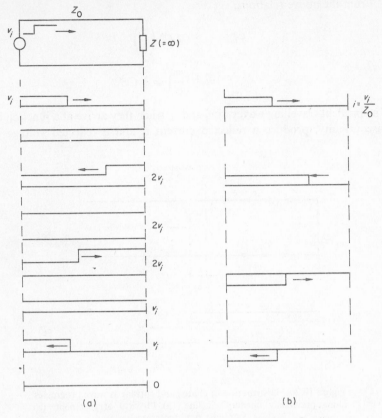

Figure 10.21  Application of voltage to unenergized loss-free line on
open circuit. (a) Distribution of voltage. (b) Distribution of current.
Voltage source is effective short circuit.

Again,              $v_i + v_r = Z(i_i + i_r)$

$$= Z\left(\frac{v_i - v_r}{Z_0}\right)$$

or,                 $v_r = \left(\frac{Z - Z_0}{Z + Z_0}\right) v_i = \alpha v_i$                (10.14)

where $\alpha$ is the *coefficient of reflexion*

Hence,              $v = \left(\frac{2Z}{Z + Z_0}\right) v_i$                (10.15) and, $$i = \left(\frac{2Z_0}{Z + Z_0}\right) i_1 \qquad (10.16)$$

From the above, if $Z \to \infty$, $v = 2v_1$ and $i = 0$. Also if $Z = Z_0$ (matched line) $\alpha = 0$, i.e. no reflexion. If $Z > Z_0$, then $v_r$ is positive and $i_r$ is negative, but if $Z < Z_0$, $v_r$ is negative and $i_r$ is positive. The reflected waves will travel back and forth the line setting up in turn further reflected waves at the ends, and this process will continue indefinitely unless the waves are attentuated due to resistance and corona.

Summarizing, at an open circuit the reflected voltage is equal to the incident voltage and this wave along with a wave $(-i_1)$ travels back along the line; note that at the open circuit the total current is zero. Conversely, at a short-circuit the reflected voltage wave is $(-v_1)$ in magnitude and the current reflected is $(i_1)$, giving a total voltage at the short circuit of zero and a total current of $2i_1$. For other termination arrangements Thevenin's Theorem may be applied to analyse the circuit. The voltage across the termination when it is open-circuited is seen to be $2v_1$ and the equivalent impedance looking in from the open-circuited termination is $Z_0$; the termination is then connected across the terminals of the Thevenin equivalent circuit (figure 10.22).

(a)  (b)

Figure 10.22  Analysis of travelling waves—use of Thevenin equivalent circuit. (a) System. (b) Equivalent circuit.

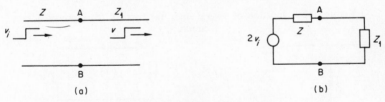

(a)  (b)

Figure 10.23  Analysis of conditions at junction of two lines or cables of different surge impedance.

Consider two lines of different surge impedance in series. It is required to determine the voltage across the junction between them (figure 10.23),

$$v_{AB} = \left(\frac{2v_1}{Z + Z_1}\right) Z_1 = \beta v_1 \qquad (10.17)$$

The wave entering the line $Z_1$ is the refracted wave and $\beta$ is the *Refraction Coefficient*, i.e. the proportion of the incident voltage proceeding along the second line $(Z_1)$.

$$v_r = v_1\alpha = v_1 \left(\frac{Z_1 - Z}{Z_1 + Z}\right)$$

and $i = \dfrac{v_{AB}}{Z_1} = \dfrac{2v_1}{Z_1 + Z} = $ refracted current.

When several lines are joined to the line on which the surge originates (figure 10.24) the treatment is similar, e.g. if the lines have equal surge impedances $(Z_1)$ then,

$$i_A = \frac{2v_1}{\left(Z + \dfrac{Z_1}{3}\right)}, \text{ and } v_{AB} = \left(\frac{2v_1}{Z + \left(\dfrac{Z_1}{3}\right)}\right)\left(\frac{Z_1}{3}\right) \qquad (10.18)$$

An important practical case is that of the clearance of a fault at the junction of two lines and the surges produced. The equivalent circuits are shown in figure 10.25; the fault clearance is simulated by the insertion of

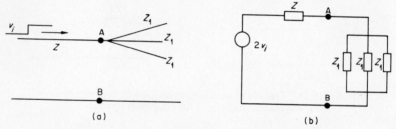

Figure 10.24   Junction of several lines. (a) System. (b) Equivalent circuit.

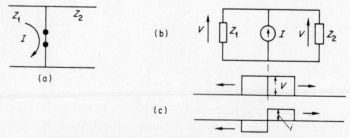

Figure 10.25   Surge set up by fault clearance. (a) Equal and opposite current (I) injected in fault path. (b) Equivalent circuit. (c) Voltage and current waves set up at point of fault with direction of travel.

an equal and opposite current ($I$) at the point of the fault. From the equivalent circuit, the magnitude of the resulting voltage surges ($v$)

$$= I\left(\frac{Z_1 Z_2}{Z_1 + Z_2}\right)$$

and the currents entering the lines are

$$I\left(\frac{Z_1}{Z_1 + Z_2}\right) \text{ and } I\left(\frac{Z_2}{Z_1 + Z_2}\right)$$

The directions are as shown in figure 10.25c.

### Termination in inductance and capacitance

(*a*) *Shunt capacitance* Using the Thevenin equivalent circuit as shown in figure 10.26 the voltage rise across the capacitor C, $v_c = 2v_1(1 - e^{-t/Z_0 C})$,

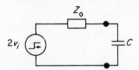

Figure 10.26   Termination of line (surge impedance $Z_0$) in a capacitor (C).

where $t$ is the time commencing with the arrival of the wave at C. The current through C is given by,

$$\left(\frac{2v_1}{Z_0}\right) e^{-t/Z_0 C}$$

The reflected wave,

$$v_r = v_c - v_1 = 2v_1(1 - e^{-t/Z_0 C}) - v_1 = v_1(1 - 2e^{-t/Z_0 C}) \tag{10.19}$$

As to be expected, the capacitor acts initially as a short-circuit and finally as an open circuit.

(*b*) *Shunt inductance* Again from the equivalent circuit the voltage across the inductance,

$$v_L = 2v_1 e^{-\left(\frac{Z_0}{L}\right)t}$$

and

$$v_r = v_L - v_1 = v_1\left(2e^{-\left(\frac{Z_0}{L}\right)t} - 1\right) \tag{10.20}$$

Here the inductance acts initially as an open circuit and finally as a short-circuit.

(*c*) *Capacitance and resistance in parallel*  (figure 10.27a). Open circuit voltage across AB (figure 10.27b),

$$= \left(\frac{2v_1}{Z_0 + R}\right) R$$

Equivalent Thevenin resistance

$$= \frac{R Z_0}{R + Z_0}$$

Voltage across R and C,

$$v = \frac{2v_1}{Z_0 + R} \left(1 - e^{-\frac{t(R + Z_0)}{R Z_0 C}}\right) \qquad (10.21)$$

This is the solution to the practical system shown in figure 10.27c, where C is used to modify the surge. The reflected wave is given by $(v - v_1)$.

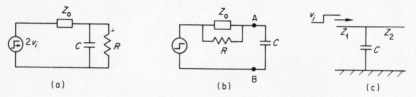

(a)                              (b)                              (c)

Figure 10.27   Two lines surge impedances $Z_1$ and $Z_2$ grounded at their junction through a capacitor C.
(a) and (b) Equivalent circuits. (c) System diagram.

*Example 10.2*   An overhead line of surge impedance 500 Ω is connected to a cable of surge impedance 50 Ω through a series resistor (figure 10.28a). Determine the magnitude of the resistor such that it absorbs maximum energy from a surge originating on the overhead line and travelling into the cable.  Calculate,

(a) the voltage and current transients reflected back into the line, and
(b) those transmitted into the cable, in terms of the incident surge voltage, and
(c) the energies reflected back into the line and absorbed by the resistor.

Let the incident voltage and current be $v_1$ and $i_1$. From the equivalent circuit (figure 10.28b),

$$v_B = \frac{2v_1 \cdot Z_2}{Z_1 + Z_2 + R}$$

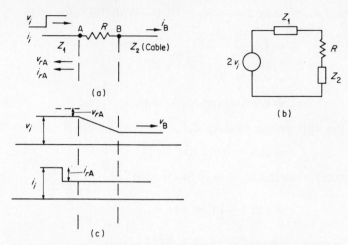

Figure 10.28 (a) System for Example 10.2. (b) Equivalent circuit (c) Voltage and current surges.

and the reflected voltage at A

$$v_{\mathrm{rA}} = \frac{2v_1(Z_2 + R)}{Z_1 + Z_2 + R} - v_1 = \frac{Z_2 + R - Z_1}{Z_1 + Z_2 + R}$$

As $\quad v_1 = Z_1 i_1$

$$i_{\mathrm{B}} = \frac{2v_1}{Z_1 + Z_2 + R} = \frac{2Z_1 i_1}{Z_1 + Z_2 + R}$$

and $\quad i_{\mathrm{rA}} = i_1 \frac{(Z_1 - Z_2 - R)}{Z_1 + Z_2 + R}$

Power absorbed by the resistance

$$= R\left(i_1 \frac{2Z_1}{Z_1 + Z_2 + R}\right)^2 \text{ watt.}$$

This power is a maximum when

$$\frac{\mathrm{d}}{\mathrm{d}R}\left[\frac{R}{(Z_1 + Z_2 + R)^2}\right] = 0$$

if $i_1$ is given and constant; from which, $R = (Z_1 + Z_2)$. With this resistance the maximum energy is absorbed from the surge.

Hence $R$ should be $500 + 50 = 550\,\Omega$. With this value of $R$,

$$i_B = \left(\frac{2 \times 500}{500 + 50 + 550}\right) . i_1 = 0.91\ i_1$$

$$i_{rA} = -0.091\ i_1$$

$$v_B = 0.091\ v_1 \text{ and } v_{RA} = 0.091\ v_1$$

Also the surge energy entering $Z_2$

$$= v_B i_B = 0.082\ v_1 i_1$$

The energy absorbed by $R = (0.91\ i_1)^2 . 550$

$$= 455 \left(\frac{v_1}{Z_1}\right) i_1 = 0.91\ v_1 i_1$$

and the energy reflected $= v_1 i_1\ (1 - 0.082 - 0.91)$

$$= 0.008\ v_1 i_1$$

The waveforms are shown in figure 10.28c.


## 10.6  Determination of System Voltages Produced by Travelling Surges

In the previous section the basic laws of surge behaviour are discussed. The calculation of the voltages set up at any node or busbar in a system at a given instant in time is, however, much more complex than the previous section would suggest. When any surge reaches a discontinuity its reflected waves travel back and are in turn reflected so that each generation of waves sets up further waves which coexist with them in the system.

To describe completely the events at any node involves, therefore, an involved book-keeping exercise. Although many mathematical techniques are available and in fact used, the graphical method due to Bewley[6] indicates clearly the physical changes occurring in time, and this method will be explained in some detail.

### Bewley lattice diagram

This is a graphical method of determining the voltages at any point in a transmission system and is an effective way of illustrating the multiple reflexions which take place. Two axes are established, a horizontal one scaled in distance along the system and a vertical one scaled in time. Lines indicating the passage of surges are drawn such that their slopes give the times corresponding to distances travelled. At each point of change in impedance the reflected and transmitted waves are obtained by multiplying the incidence wave magnitude by the appropriate reflexion and refraction coefficients $\alpha$ and $\beta$.

The method is best illustrated by an example and a loss-free system comprising a long overhead line $(Z_1)$ in series with a cable $(Z_2)$ will be considered. Typically, $Z_1$ is 500 $\Omega$ and $Z_2$ is 50 $\Omega$ and referring to figure 10.29 the following coefficients apply:

Line to cable reflexion coefficient, $\alpha_1 = \dfrac{50 - 500}{50 + 500} = -0{\cdot}818$

Line to cable refraction coefficient, $\beta_1 = \dfrac{2 \times 50}{50 + 500} = 0{\cdot}182$

Cable to line, $\alpha_2 = \dfrac{500 - 50}{500 + 50} = 0{\cdot}818$

Cable to line, $\beta_2 = \dfrac{2 \times 500}{500 + 50} = 1{\cdot}818$

As the line is long, reflexions at its remote end will be neglected. The remote end of the cable is considered to be open-circuited, giving an $\alpha$ of 1 and a $\beta$ of zero.

When the incident wave $v_1$ (see figure 10.29) originating in the line reaches the junction a reflected component travels back along the line $(\alpha_1 v_1)$, the refracted or transmitted wave $(\beta_1 v_1)$ traverses the cable and is reflected from the open-circuited end back to the junction $(1 \times \beta_1 v_1)$. This wave then produces a reflected wave back through the cable $(1 \times \beta_1 \alpha_2 v_1)$ and a transmitted wave $(1 \times \beta_2 \beta_1 v_1)$ through the line. The process continues and the waves multiply as indicated in figure 10.29b. The total voltage at a point P in the cable at a given time $(t)$ will be the sum of the voltages at P up to time $t$; i.e. $v_1 \beta_1 (2 + 2\alpha_2 + 2\alpha_2^2)$ and the voltage at infinite time will be, $2v_1 \beta_1 (1 + \alpha_2 + \alpha_2^2 + \alpha_2^3 + \alpha_2^4 + \ldots)$.

The voltages at other points are similarly obtained. The time scale may be determined from a knowledge of length and surge velocity, for the line the latter is of the order of 1000 ft (300 m) per $\mu$s and for the cable 500 ft/$\mu$s. For a surge 50 $\mu$s in duration and a cable 1000 ft in length there will be 25 cable lengths traversed and the terminal voltage will approach $2v_1$. If the graph of voltage at the cable open-circuited end is plotted against time an exponential rise curve will be obtained similar to that obtained for a capacitor.

The above treatment applies to a rectangular surge waveform but may be modified readily to account for a waveform of the type illustrated in

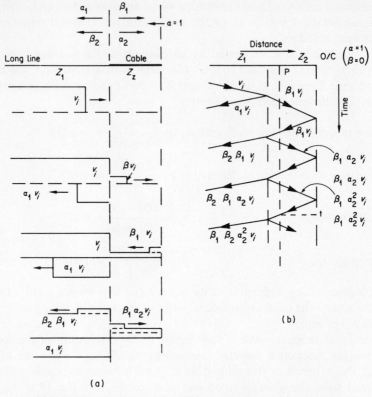

(a)

Figure 10.29   Bewley Lattice Diagram—analysis of long overhead line and cable in series. (a) Position of voltage surges at various instants over first complete cycle of events, i.e. up to second reflected wave travelling back along line. (b) Lattice diagram.

figure 10.1b or c. In this case the voltage change with time must also be allowed for and the process is more complicated.

### Effects of line loss

Attenuation of travelling waves is caused mainly by corona which reduces the steepness of the wavefronts considerably as the waves travel along the line. Attenuation is also caused by series resistance and leakage resistance and these quantities are considerably larger than the power frequency values. The determination of attenuation is usually empirical and use is made of the expression, $v_x = v_1 e^{-\gamma x}$, where $v_x$ is the magnitude of the surge at a distance $x$ from the point of origination. If a value for $\gamma$ is assumed then the wave magnitude of the voltage may be modified to include attenuation for various positions in the lattice diagram. For example, in figure 10.29b, if $e^{-\gamma z}$ is equal to $a_L$ for the length of line traversed and $a_c$ for the cable then the magnitude of the first reflexion from the open circuit is $a_L v_1 a_c \beta_1$ and the voltages at subsequent times will be similarly modified.

Considering the power and losses over a length $dx$ of a line of resistance and shunt conductance per unit length $R$ $\Omega$ and $G$ $\Omega^{-1}$, the power loss,

$$dp = i^2 R dx + v^2 G dx \quad (W)$$

also,

$$p = vi = i^2 Z_0 \text{ and } dp = 2iZ_0 di.$$

As $dp$ is a loss it is considered negative and,

$$-2iZ_0 di = (i^2 R + v^2 G) dx$$

hence,

$$\frac{di}{i} = -\tfrac{1}{2}\left(\frac{R + Z_0^2 G}{Z_0}\right) dx$$

giving

$$i = i_1 e^{-\frac{1}{2}\left(\frac{R}{Z_0} + GZ_0\right)x} \tag{10.22}$$

where

$$i_1 = \text{surge amplitude (A)}.$$

Also it may be shown that

$$v = v_1 e^{-\frac{1}{2}\left(\frac{R}{Z_0} + GZ_0\right)x} \tag{10.23}$$

and the power at $x$

$$v_1 i_1 = vi = v_1 i_1 e^{-\left(\frac{R}{Z_0} + GZ_0\right)x} \tag{10.24}$$

If $R$ and $G$ are realistically assessed (including corona effect) attenuation may be included in the travelling wave analysis.

### Digital methods

The lattice diagram becomes very cumbersome for large systems and normally either analogue or digital methods are applied. Digital methods may use strictly mathematical methods, i.e. the solution of the differential equations or the use of Fourier or Laplace transforms. These methods are capable of high accuracy but require large amounts of data and long computation times. The general principles of the graphical approach described above may also be used to develop a computer program which is very applicable to large systems. Such a method will now be discussed in more detail and an example considered.

A rectangular wave is used, of infinite duration. The theory developed in the previous sections is applicable. The major role of the program is to scan the nodes of the system at each time interval and compute the voltages. In figure 10.30 a particular system (single-phase representation) is shown and will be used to illustrate the method.

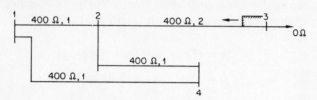

Figure 10.30 Application of digital method—network analysed. Each line labelled with surge impedance and surge travel time (multiples of basic unit), e.g. 400 Ω, 1.

The relevant data describing the system are given in tabular form (see Table 10.5). Branches are listed both ways in ascending order of the first nodal number and are referred to under the name BRANCH. The time taken for a wave to travel along a branch is recorded in terms of a positive integer (referred to as PERIOD) which converts the basic time unit into actual travel time. Reflexion coefficients are stored and referred to as REFLECT and the corresponding refraction coefficients obtained, i.e. $(1 + \alpha_{ij})$. Elements of time as multiples of the basic integer are also shown in Table 10.5.

Table 10.5

| BRANCH $i$ <br> $j$ | 1 <br> 2 | 1 <br> 4 | 2 <br> 1 | 2 <br> 3 | 2 <br> 4 | 3 <br> 2 | 4 <br> 1 | 4 <br> 2 |
|---|---|---|---|---|---|---|---|---|
| PERIOD | 1 | 1 | 1 | 2 | 1 | 2 | 1 | 1 |
| $\alpha_{ij}$ | $-\frac{1}{3}$ | 0 | 0 | $-1$ | 0 | $-\frac{1}{3}$ | 0 | $-\frac{1}{3}$ |
| $1+\alpha_{ij}$ | $\frac{2}{3}$ | 1 | 1 | 0 | 1 | $\frac{2}{3}$ | 1 | $\frac{2}{3}$ |
| TIME 0 | | | | | | 1 | | |
| 1 | | | | | | | | |
| 2 | | | $\frac{2}{3}$ | $-\frac{1}{3}$ | $\frac{2}{3}$ | | | |
| 3 | 0 | $\frac{2}{3}$ | | | | | | |
| 4 | | | | | | | | |

The method is illustrated by examining the system after the arrival of the rectangular wave at node 3 at TIME (0). This voltage (magnitude 1 p.u.) is entered in the BRANCH (3, 2), TIME (0) element of the BRANCH–TIME matrix. On arrival at node 2 at a time equal to zero plus PERIOD (3, 2) two waves are generated, on BRANCH (2, 1) and BRANCH (2, 4) both of magnitude, $1(1 + \alpha_{32})$, i.e. 2/3. A reflected wave is also generated on BRANCH (2, 3) of magnitude $1 \times \alpha_{32}$, i.e. $-1/3$. These voltages are entered in the appropriate BRANCH in the TIME (2) row of Table 10.5. On reaching node 1, TIME (3), a refracted wave of magnitude $\frac{2}{3}(1 + \alpha_{21})$, i.e. 2/3, is generated on BRANCH (1, 4) and a reflected wave $\frac{2}{3} \times \alpha_{21}$, i.e. 0 is generated on BRANCH (1, 2). This process is continued until a specified time is reached. All transmitted waves for a given node are placed in a separate node–time array, a transmitted wave is only considered once even though it could be

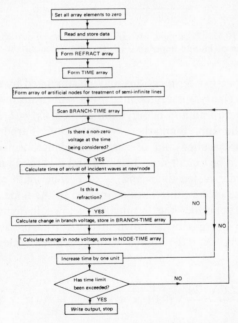

Figure 10.31   Flow diagram of digital method for travelling-wave analysis.

entered into several Branch–Time elements. Current waves are obtained by dividing the voltage by the surge impedance of the particular branch. The flow diagram for the digital solution is shown in figure 10.31.

It is necessary for the programs to cater for semi-infinite lines (i.e. lines so long that waves reflected from the remote end may be neglected) and also inductive/capacitive terminations. *Semi-infinite* lines require the use of artificial nodes labelled (say) 0. For example in figure 10.32 node 2 is open-circuited and this is accounted for by introducing a line of infinite surge impedance between nodes 2 and 0, and hence,

$$\text{if } Z \to \infty, \ \alpha_{12} = \frac{Z - Z_0}{Z + Z_0} \to 1$$

Although the scanning will not find a refraction through node 2 it is necessary for the computer to consider there being one to calculate the voltage at node 2. Lines with short-circuited nodes may be treated in a similar fashion with an artificial line of zero impedance.

$$1 \bullet \!\!\!\!\!\!\!\!\!\!\!\!\!\!\!\!\!\!\!\!\!\!\!\!\!\!\!\!\!\!\!\!\!\!\!\!\!\!\!\!\!\! \underset{\underset{2}{\bullet}}{\overset{Z_0}{\rule{4cm}{0.4pt}}} \overset{Z = \infty}{-\!-\!-\!\bullet} 0$$

Figure 10.32   Treatment of terminations—use of artificial node 0 and line of infinite impedance to represent open circuit at node 2.

Inductive/capacitive terminations may be simulated by *stub lines*. An inductance (L. Henrys) is represented by a stub transmission line short-circuited at the far end and of surge impedance, $Z_L = L/t$, where $t$ is the travel time of the stub. Similarly, a capacitance $C$ (Farads) is represented by a line open-circuited and of surge impedance, $Z_c = t/C$. For the representation to be exact $t$ must be small and it is found necessary for $Z_L$ to be of the order of ten times and $Z_c$ to be one tenth of the combined surge impedance of the other lines connected to the node.

For example, consider the termination shown in figure 10.33a. The equivalent stub line circuit is shown in figure 10.33b. Stub travel times are chosen to be short compared with a quarter cycle of the natural frequency, i.e.

$$\frac{2\pi \sqrt{LC}}{4} = \frac{2\pi}{4} \sqrt{(0{\cdot}01 \times 4 \times 10^{-8})} = 0{\cdot}315 \times 10^{-4} s$$

Let $t = 5 \times 10^{-6} s$ for both $L$ and $C$ stubs (corresponding to the total stub length of 5000 ft or 1524 m). Hence,

$$Z_c = \frac{t}{C} = \frac{2{\cdot}5 \times 10^{-6}}{4 \times 10^{-8}} = 67{\cdot}5 \ \Omega$$

and,
$$Z_L = L/t = \frac{0{\cdot}01}{2{\cdot}5 \times 10^{-6}} = 4000 \ \Omega$$

The configuration in a form acceptable for the computer program is shown in figure 10.33c. Refinements to the program to incorporate attenuation, waveshapes and non-linear resistors may be made without changing its basic form.

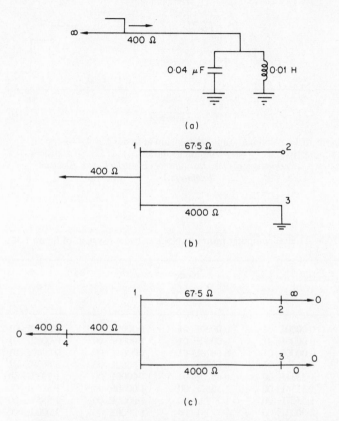

Figure 10.33 Representation of line terminated by L — C circuit by means of stub lines. (a) Original system. (b) Equivalent stub lines. (c) Use of artificial nodes to represent open and short circuited ends of stub lines.

A typical application is the analysis of the nodal voltages for the system shown in figure 10.30. This system has been previously analysed by a similar method by Barthold and Carter[17] and good agreement found. The printout of nodal voltages for the first 20 $\mu$s is shown in Table 10.6 and in figure 10.34 the voltage plot for nodes 1 and 4 is compared with a transient analyser solution obtained by Barthold and Carter[17].

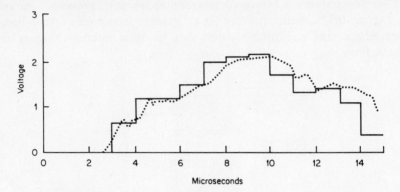

Figure 10.34   Voltage—time relationship at nodes 1 and 4 of system
in Figure 10.30. ——— = computer solution. ......... = transient
analyser. (*Permission of the Institute of Electrical and Electronic
Engineers.*)

Table 10.6   Digital computer printout. Node voltages (System of figure 1.30)

| Time ($\mu$s) | Node 1 | 2 | 3 | 4 |
|---|---|---|---|---|
| 0  | 0·0000E–01 | 0·0000E–01 | 1·0000E 00 | 0·0000E–01 |
| 1  | 0·0000E–01 | 0·0000E–01 | 1·0000E 00 | 0·0000E–01 |
| 2  | 0·0000E–01 | 6·6670E–01 | 1·0000E 00 | 0·0000E–01 |
| 3  | 6·6670E–01 | 6·6670E–01 | 1·0000E 00 | 6·6670E–01 |
| 4  | 1·3334E 00 | 6·6670E–01 | 1·0000E 00 | 1·3334E 00 |
| 5  | 1·3334E 00 | 1·5557E 00 | 1·0000E 00 | 1·3334E 00 |
| 6  | 1·5557E 00 | 1·7779E 00 | 1·0000E 00 | 1·5557E 00 |
| 7  | 2·0002E 00 | 1·7779E 00 | 1·0000E 00 | 2·0002E 00 |
| 8  | 2·2224E 00 | 2·0743E 00 | 1·0000E 00 | 2·2224E 00 |
| 9  | 2·2965E 00 | 1·7779E 00 | 1·0000E 00 | 2·2965E 00 |
| 10 | 1·8520E 00 | 1·8520E 00 | 1·0000E 00 | 1·8520E 00 |
| 11 | 1·4075E 00 | 1·9508E 00 | 1·0000E 00 | 1·4075E 00 |
| 12 | 1·5062E 00 | 1·0617E 00 | 1·0000E 00 | 1·5062E 00 |
| 13 | 1·1604E 00 | 7·6534E–01 | 1·0000E 00 | 1·1604E 00 |
| 14 | 4·1955E–01 | 8·2297E–01 | 1·0000E 00 | 4·1955E–01 |
| 15 | 8·2086E–02 | 2·6307E–01 | 1·0000E 00 | 8·2086E–02 |
| 16 | −7·4396E–02 | 1·6435E–01 | 1·0000E 00 | −7·4396E–02 |
| 17 | 7·8720E–03 | 1·0732E–02 | 1·0000E 00 | 7·8720E–03 |
| 18 | 9·3000E–02 | −2·5556E–01 | 1·0000E 00 | 9·3000E–02 |
| 19 | −1·7043E–01 | 4·1410E–01 | 1·0000E 00 | −1·7043E–01 |
| 20 | 1·5067E–01 | 6·2634E–01 | 1·0000E 00 | 1·5067E–01 |

*Transient analysers*

The main alternative to digital solutions is the equivalent-circuit modelling of systems. These circuits are physically realized and pulses applied by means of pulse generators and voltages measured by cathode ray oscilloscopes. The modelling of large systems is expensive and time-consuming. The exact representation of plant such as generators and transformers is complex because of the distributed capacitance presented by the windings.

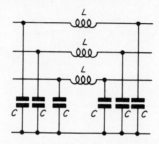

Figure 10.35   Equivalent circuit representation of rotating machines.
C = 1/3 of total capacitance to ground.   L = 0·65 of subtransient inductance.

A typical equivalent circuit representation of a rotating machine is shown in figure 10.35. The difficulty in representing items of plant other than overhead lines holds equally for any digital method.

*Three-phase analysis*

The single-phase analysis of a system as presented in this chapter neglects the mutual effects which exist between the three phases of a line, transformer, etc. The transient voltages due to energization may be further increased by this mutual coupling and also by the three contacts of a circuit breaker not closing at the same instant. The difference resulting between the use of three-phase and single-phase representation has been discussed by Bickford and Doepel[20].

## 10.7   Ultra-High-Voltage Transmission

In this section existing arrangements for Extra-High-Voltage (e.h.v.) lines will be reviewed and factors influencing designs for u.h.v. discussed. At the present time for e.h.v. lines 'V' type insulator strings are used to support bundle conductors in horizontal formation on steel lattice towers with two earth (ground) wires; a typical structure is shown in figure 10.36. In this figure dimensions of interest are labelled and some critical distances are shown. Span lengths (i.e. distances between towers) are of the order of 400–500 m at present and are expected to continue in this range. The number of subconductors in each bundle is determined by radio-interference levels stipulated along the route rather than by current rating.

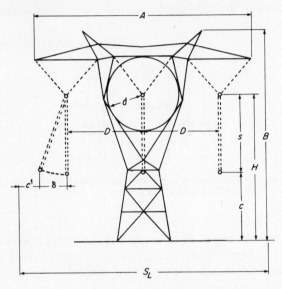

Figure 10.36   U.h.v. tower—critical dimensions. (*Permission of the Institute of Electrical and Electronic Engineers.*)

Tower dimensions are determined by insulation requirements and the importance of the switching surge in this connexion has already been discussed. The tower surge-withstand voltage is set below the critical flashover (i.e. 50 per cent probability) voltage to give a reasonable value of flashover probability. When designing a tower the insulation string length, insulator surface creep distance, tower grounding, and strike distances must be decided upon. Tower strike distance is the distance from the conductor to the tower structure. These dimensions are chosen to make the insulation strength such that the applied surge results in an acceptable surge flashover rate.

An example of laboratory assessment of the strength of tower insulation (V-string insulators) is shown in figure 10.37. It has been shown (figure 10.18) that for a given insulation the flashover voltage follows a Gaussian cumulative distribution curve to at least four standard deviations

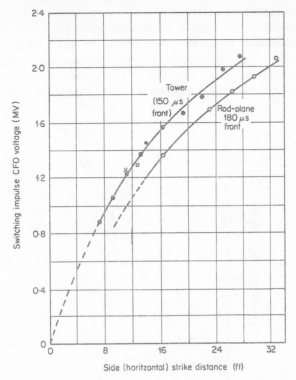

Figure 10.37 Maximum obtainable switching impulse C.F.O. voltage for a specific side strike distance. (*Permission of the Westinghouse Electric Corporation.*)

($\sigma_f$) below the C.F.O.; $\sigma_f$ is about 4·6 per cent to 5 per cent of the C.F.O. The procedure is to equate the switching surge voltage to the withstand strength of the insulation where withstand is defined as that voltage which results in a 0·13 per cent flashover probability, i.e. 3 $\sigma_f$ below the C.F.O. The C.F.O. is determined from a knowledge of $\sigma_f$ and the withstand voltage and from it the strike distances obtained. This process is illustrated in figure 10.38 for C.F.O.s under wet and dry conditions. *Contamination* requirements are expressed in terms of the shortest distance along the insulator disc from cap to pin (creep distance); the creep distance required per kilovolt for flashover-free operation with normal practice at present is between 0·83 in. (2·1 cm) and 1 in. (2·54 cm) per kilovolt. The various factors involving flashover and insulation requirements are summarized in figure 10.39 in which system operating voltage is related to the minimum distance for flashover between the phase conductor and tower sides (strike distance) for a 'V' formation of string insulators. The switching-surge curves refer to various per-unit peak values of switching surge, control over which may be exercised by circuit-breaker resistance switching. It is seen that increases in system voltages require progressively larger increases in the strike distance (conductor-to-tower minimum distance) and the tower dimensions would become intolerable from both

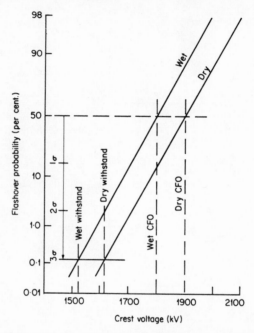

Figure 10.38   Selection of tower surge withstand level. Design withstand level for a tower is established by specifying some acceptable risk of flashover, in this case, three standard deviations below the 50% probability C.F.O. voltage. (*Permission of the Westinghouse Electric Corporation*).

economic and appearance standpoints unless the per-unit value of the surge is reduced.

The method illustrated in this section has been questioned for the u.h.v. region. It does not produce an estimate of the switching surge flashover rate and only matches the withstand voltage to the maximum switching surge. In fact, two probability distributions are involved, one for surge magnitude, the other for insulation strength. The exact form of the surge magnitude probability curves for systems is not as yet known to an acceptable degree of accuracy for the calculation of flashover probability. This

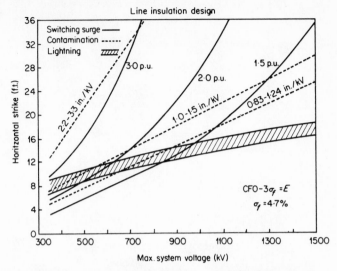

Figure 10.39   Estimates of o.h.v.—u.h.v. tower insulation requirements. (*Permission of the Westinghouse Electric Corporation.*)

Table 10.7   Characteristics of lines at various system voltages

| Highest system voltage ($V_m$), kV: | 420 | 525 | 765 | 1000 | 1300 | 1500 |
|---|---|---|---|---|---|---|
| Overall aluminum section per phase ($S$), mm² | 1240 | 1660 | 2680 | 3780 | 5250 | 6300 |
| Number of subconductors per phase ($n$) | 2 | 3 | 4 | 6 | 8 | 8 |
| Subconductor diameter ($\phi$), mm | 34·5 | 32·4 | 35·8 | 34·7 | 35·5 | 38·8 |
| Conductor-tower clearance ($d$), meters | 3·00 | 3·90 | 5·60 | 7·20 | 8·50 | 9·40 |
| Switching impulse 50% discharge voltage of tower insulation ($V_{50\%}$), per unit | 3·2 | 2·95 | 2·60 | 2·25 | 1·95 | 1·80 |
| Conductor-ground clearance at midspan ($C$), meters | 7·2 | 8·45 | 10·8 | 13·1 | 15·0 | 16·2 |
| Span length ($L$), meters | 400 | 420 | 445 | 475 | 500 | 515 |
| Midspan sag ($s$), meters | 12 | 13·5 | 15 | 17 | 19 | 20 |
| Conductor height at the tower ($H$), meters | 19·2 | 21·7 | 25·8 | 30·1 | 34·0 | 36·2 |
| Interphase distance ($D$), meters | 7·30 | 9·20 | 12·8 | 16·1 | 19·0 | 20·8 |
| Tower width ($A$), meters | 20·0 | 25·4 | 35·6 | 45·2 | 53·3 | 58·4 |
| Tower height ($B$), meters | 24·6 | 28·2 | 35·5 | 42·25 | 47·9 | 51·5 |
| Line-size parameter (right of way) ($S_L$), meters | 35·5 | 42·3 | 52·0 | 62·5 | 72·0 | 76·5 |
| Tower-size parameter ($S_T = 1000AB/L$), m²/km | 1230 | 1700 | 2840 | 4020 | 5110 | 5840 |
| RI limit gradient of lateral phase conductor, kV/cm | 15·8 | 15·7 | 15·35 | 15·5 | 15·25 | 14·85 |
| Voltage gradient at ground ($G$), kV/m | 7·35 | 9·50 | 11·4 | 13·1 | 16·55 | 17·55 |
| Surge impedance ($Z_s$), Ω | 284 | 268 | 264 | 249 | 240 | 245 |
| Surge impedance loading ($P_s$), MW | 560 | 925 | 1970 | 3615 | 6335 | 8265 |

(Permission I.E.E.E.—Reference 21.)

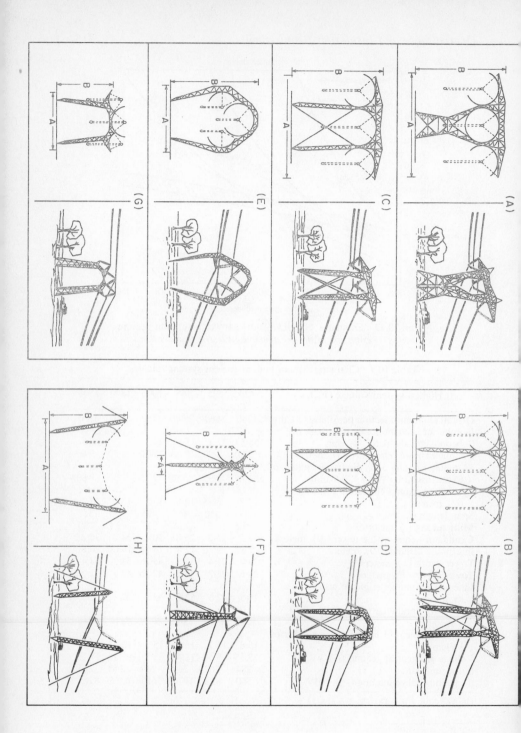

complete probability approach is being studied[23] and revised versions of the curves shown in figure 10.39 obtained.

Paris[21] has suggested various criteria for the design of u.h.v. towers, e.g. dimensions are chosen in such a way as to make the cost of the inactive components a constant ratio (about 0·8) of the active components (i.e. the conductors). The tower size is defined by two parameters: the 'line size'

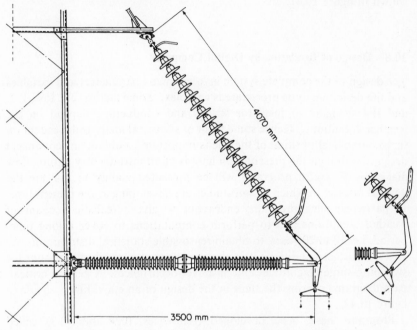

Figure 10.41  Insulating cross-arm for a double-circuit 420 kV line (Italian). (a) Normal conditions. (b) Windy conditions.

giving the width of the 'right of way' (width of the strip of ground required underneath the line) and the 'tower-size' defined as the product of tower width, height and number of towers per kilometre. As the height of the conductors will not increase proportionally with the voltage the magnitude of the electric field at the ground surface becomes a critical factor. Paris's analysis suggests that towers for voltages up to 1500 kV are feasible and some of his suggested towers are shown in figure 10.40 based on line

Figure 10.40  Proposed types of towers for 1500 —kV lines (reference 21). Proposed new tower types, C to H. Traditional tower types, A and B. (*Permission of the Institute of Electrical and Electronic Engineers.*)

characteristics given in Table 10.7 and the criteria mentioned above. The guyed structures already in use lead to the possible use of towers to withstand vertical and transverse forces only with just a small number of special towers to withstand the longitudinal forces as well, say one special tower to five light two-dimensional towers. Other possibilities for reducing size at high voltages include the use of insulated cross arms of the form shown in figure 10.41.

## 10.8  Design of Insulation by Digital Computer

The design of the complete system involves many parameters and variables and the design-analysis procedure is complex. Some factors, e.g. loss, R.I. and the swinging of insulator strings and conductors, depend on the weather conditions. Hence some form of statistical analysis dependent on the geographical position of the line is important to obtain an economical design. Although the presentation here is of an introductory nature, flow diagrams of typical programs will be presented mainly to indicate the general manner in which design studies are developing at the present time. In particular such programs endeavour to give a realistic account of weather conditions and to perform computations to cover a long time-period, i.e. 10 to 20 years, to obtain reasonable statistical distributions. As well as weather conditions it is also necessary statistically to account for switching-surge magnitudes and flashover occurrences. A general critical-path diagram covering the steps in the design of an e.h.v. line is shown in figure 10.42.

Programs have been developed (references 10.9 and 10.3) called METIFOR (METeorologically Integrated FORecasting) and will now be briefly described. METIFOR 1 determines insulation strength and a flow diagram is shown in figure 10.43. Having previously obtained the weather variables the following calculations are made.

(a) Using a set of velocity-force tables previously stored, the wind swings of the conductors are computed and from these all reduced air-gap dimensions obtained.

(b) Flashover strengths of the above gaps (from practical tests) are used and modified to account for precipitation, air density, humidity and fog.

(c) The weakest of the three insulation forms, i.e. of vertical and horizontal air-gaps and insulator strings, is compared with the strength of the configuration under standard dry conditions with zero wind. The ratio of the two is called the relative insulation strength and is stored; it represents the per-unit strength of the tower configurations.

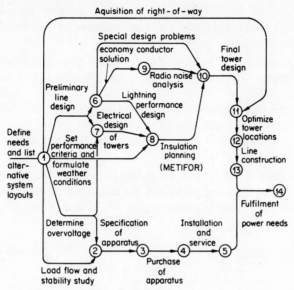

Figure 10.42 Critical path diagram for steps in the design of an e.h.v. line. (*Permission of the Edison Electric Institute.*)

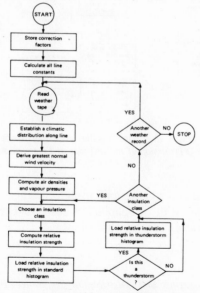

Figure 10.43 Flow chart for digital computation of relative insulation strength—METIFOR 1. (*Permission of the Institute of Electrical and Electronic Engineers.*)

The output of METIFOR 1 is fed into another program called METIFOR 2 which evaluates the performances of alternative designs based on the following,

(a) The probability of experiencing a surge of given magnitude.

(b) The probability that an insulator string will be swung to a certain position.

(c) The probability that weather factors (precipitation, etc.) will produce a certain flashover strength of insulators and gaps.

(d) The probability that a gap or string will flashover above or below its critical strength.

The flow chart of METIFOR 2 is shown in figure 10.44. The output is a tabulation of the flashover paths and the frequency of flashover for each path at specified insulator swing angles. A large number of surges are

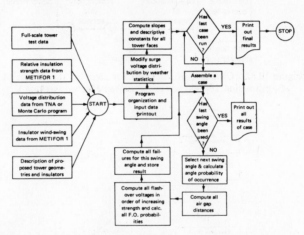

Figure 10.44   Flow chart for insulation performance program— METIFOR 2. (*Permission of the Institute of Electrical and Electronic Engineers.*)

conceptually applied and all gaps and insulator strings examined for flashovers. The final result is the probability of flashover occurring following a circuit breaker operation for the whole line.

### References

**BOOKS**

1. *Electrical Transmission and Distribution Reference Book*, Westinghouse Electric Corp., East Pittsburgh, Pennsylvania.
2. Rüdenburg, R., *Electric Shock Waves in Power Systems*, Harvard University Press, Cambridge, Mass., 1968.

3. *E.H.V. Transmission Line Reference Book*, Edison Electric Institute, New York, 1968.
4. Peterson, H. A., *Power System Transients*, Dover, 1966.
5. Peek, F. W., Jr., *Dielectric Phenomena in High Voltage Engineering*, McGraw-Hill, New York, 1929.
6. Bewley, L. V., *Travelling Waves on Transmission Systems*, Dover, 1961.

PAPERS

7. Paris, L., 'Influence of air-gap characteristics on line-to-ground switching surge strength', *I.E.E.E. Trans., P.A. & S.*, **PAS-86**, 936, 1967.
8. Johnson, I. B. et al., 'Some fundamentals on capacitance switching', *Trans. A.I.E.E.*, **74**, 727, 1955.
9. Anderson, J. G., and L. O. Barthold, 'METIFOR—a statistical method for insulation design of E.H.V. lines', *I.E.E.E. Trans., P.A. & S.*, **83**, 271, 1964.
10. Anderson, J. G., 'Monte Carlo calculation of transmission-line lightning performance', *A.I.E.E. Trans., P.A. & S.*, **80**, 414, 1961.
11. Young, F. S., J. M. Clayton, and A. R. Hilemann, 'Shielding of transmission lines', *I.E.E.E. Trans., P.A. & S.*, **83**, 132, 1963.
12. Armstrong, H. R., and E. R. Whitehead, 'Field and analytical studies of transmission line shielding', *I.E.E.E. Trans., P.A. & S.*, **PAS-87**, 270, 1968.
13. Billings, M. J., and J. T. Storey, 'Consideration of the effect of pollution on the potential distribution of insulator systems', *Proc. I.E.E.*, **115**, 1661, 1968.
14. Ely, C. M. A., and W. J. Roberts, 'Switching-impulse flashover of air gaps and insulators in an artificially polluted atmosphere', *Proc. I.E.E.*, **115**, 1667, 1968.
15. Garrard, C. J. O., 'High voltage switchgear—a review of progress', *Proc. I.E.E.*, **113**, 1523, 1966.
16. Bellaschi, P. L., 'Rationalization of electrical clearances at E.H.V.s 230 kV to 460/500 kV', *A.I.E.E. Trans., P.A. & S.*, **78**, 736, 1959.
17. Barthold, L. O., and G. K. Carter, 'Digital travelling wave solutions, 1—single phase equivalents', *Trans. A.I.E.E.*, **80**, 1961.
18. Uram, R., and R. W. Miller, 'Mathematical analysis of transmission line transients: Part 1. Theory', *I.E.E.E. Trans., P.A. & S.*, **Vol. 83**, 1964.
19. McElroy, A. J., and R. M. Porter, 'Digital computer calculation of transients in electric networks', *I.E.E.E. Trans., P.A. & S.*, **82**, 1963.
20. Bickford, J. P., and P. S. Doepel, 'Calculation of switching transients with particular reference to line energisation', *Proc. I.E.E.*, **114**, 1967.
21. Paris, L., 'The future of U.H.V. transmission lines', *I.E.E.E. Spectrum*, **44**, 1969.
22. I.E.E.E. Committee reports on *Switching Surges, Power Apparatus and Systems*, p. 912, 1948; p. 240, 1961; p. 1091, 1966; p. 173, 1970.
23. Hileman, A. R., P. R. Leblanc, and G. W. Brown, 'Estimating the Switching-Surge Performance of Transmission Lines', *I.E.E.E. Paper 70 TP 38-PWR*, Winter Power meeting, 1970.

**Problems**

10.1. A transmission line consists of three conductors of radius 0·23 in. (0·585 cm) equilaterally spaced 10 ft (306 cm) apart. The line voltage is 139 kV at 60 Hz and the line conductor surfaces are smooth giving a value for *m* of 0·96. The barometric pressure is 72·2 cm of mercury, the temperature 20°c and the weather fair. Calculate the corona loss per mile.

(Answer: 3·6 kW/mile.)

10.2 A 3-phase overhead line consists of three conductors in equilateral formation spaced 244 cm apart. The conductor diameter is 1·04 cm and the conductor surface factor ($m$) is 0·85. The air temperature and pressure are 21·1°c and 74 cm of mercury. Calculate the critical visual voltage for corona.

(Answer: 81 kV.)

10.3 A 10 kV, 64·5 mm² cable has a fault 6 miles from a circuit breaker on the supply side of it. Calculate the frequency of the restriking voltage and the maximum voltage of the surge after 2 cycles of the transient. The cable parameters are (per mile), capacitance per phase = 1·83 $\mu$F, resistance = 8·6 $\Omega$, inductance per phase = 2·75 mH. The fault resistance is 6 $\Omega$.

(Answer: 1·18 kHz and 15·5 kV.)

10.4. The effective inductance and capacitance of a faulted system as viewed by the contacts of a circuit breaker are 2 mH and 500 $\mu\mu$F, respectively. The circuit breaker chops the fault current when it has an instantaneous value of 100 A. Calculate the restriking voltage set up across the circuit breaker. Neglect resistance.

(Answer: 200 kV.)

10.5. A 132 kV circuit breaker interrupts the fault current flowing into a symmetrical three-phase to earth fault at current zero. The fault infeed is 2500 MVA and the shunt capacitance, $C$, on the source side is 0·03 $\mu$F. The system frequency is 50 Hz. Calculate the maximum voltage across the circuit breaker and the restriking-voltage frequency.

If the fault current is prematurely chopped at 50 A, estimate the maximum voltage across the circuit breaker on the first current chop.

(Answer: 215·5 kV, 6·17 kHz, 115·8 kV.)

10.6. Repeat Example 10.2 but with the surge travelling from the cable into the overhead line.

(Current into line = 0·091 × incident surge current; current reflected back into cable = 0·91 × incident current, reflected energy = 0·83 × incident surge energy.)

10.7. Repeat Example 10.2 but with zero resistance between the line and cable.

(Energy reflected back to line = 0·67 × incident surge energy.)

10.8. A cable of inductance 0·3 mH per phase and capacitance per phase 0·4 $\mu$F is connected to a line of inductance 1·5 mH per phase and capacitance 0·012 $\mu$F per phase. All quantities are per mile (1·6 km). A surge of 1 p.u. magnitude travels along the cable towards the line. Determine the voltage set up at the junction of the line and cable.

(Answer: 1·85 p.u.)

10.9. A long overhead line has a surge impedance of 500 $\Omega$ and an effective resistance at the frequency of the surge of 7 $\Omega$ per kilometre. If a surge of magnitude 500 kV enters the line at a certain point, calculate the magnitude of this surge after it has traversed 100 km and calculate the resistive power loss of the wave over this distance. The wave velocity is 3 × 10⁵ km/s.

(Answer: 250 kV, 375 MW.)

10.10. A rectangular surge of 2 $\mu$s duration and magnitude 2 p.u. travels along a line of surge impedance 350 $\Omega$. The latter is connected to another line of equal impedance through an inductor of 800 $\mu$H. Calculate the value of the surge transmitted to the second line.

(Answer: $v = v_1 \left(1 - e^{-\frac{2Z_0}{L}t}\right)$ (i.e. 1·67 p.u.).)

10.11. A lightning arrester employs a thyrite material possessing a resistance characteristic described by $R = (72 \times 10^3)/(I^{0.75})$. An overhead line of surge impedance 500 $\Omega$ is terminated by the arrester. Determine the voltage across the end of the line when a rectangular travelling surge of magnitude 500 kV travels along the line and arrives at the termination. (A graphical method using the voltage-current characteristics is useful.)
(Answer: 500 kV.)

10.12. A rectangular surge of 1 p.u. magnitude strikes an earth (ground) wire at the centre of its span between two towers of effective resistance to ground of 200 $\Omega$ and 50 $\Omega$. The ground wire has a surge impedance of 500 $\Omega$. Determine the voltages transmitted beyond the towers to the earth wires outside the span. (Answer: $0.44v_1$ from 200 $\Omega$ tower and $0.18v_1$ from 50 $\Omega$ tower.)

10.13. A system consists of the following elements in series; a long line of surge impedance 500 $\Omega$, a cable ($Z_0$ of 50 $\Omega$), a short line ($Z_0$ of 500 $\Omega$), a cable ($Z_0$ of 50 $\Omega$), a long line ($Z_0$ of 500 $\Omega$). A surge takes 1 $\mu$s to traverse each cable (they

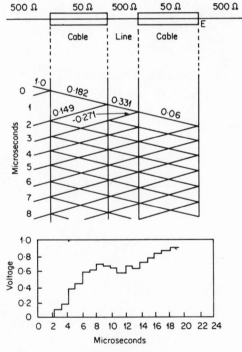

Figure 10.45   Solution of Problem 10.13.

are of equal length) and 0·5 μs to traverse the short line connecting the cables. The short line is half the length of each cable. Determine by means of a lattice diagram the p.u. voltage of the junction of the cable and long line if the surge originates in the remote long line.
(Answer: see figure 10.45.)

# I I

# Protection

## 11.1 Introduction

In Chapter 6 attention is confined to the analysis of various types of faults which may occur in a power system. Although the design of electrical plant is influenced by a knowledge of fault conditions, the major use of fault analysis is in the specification of switchgear and protective gear. Circuit breaker ratings are determined by the fault MVAs at their particular locations. At the time of writing the maximum circuit breaker rating is of the order of 50,000 MVA and this is achieved by the use of several interrupter heads in series per phase in an air-blast system. Not only has the circuit breaker to extinguish the fault current arc, it has also to withstand the considerable forces set up by short-circuit currents which, as indicated in Chapter 1, can be very high.

A knowledge of the currents resulting from various types of fault at a location is essential for the effective operation of what is known as system protection. If faults occur on the system the control engineers, sensing the presence of the fault, can operate the appropriate circuit breakers to remove the faulty line or plant from the network. This, however, takes considerable time and experience. Faults on a power system resulting in high currents and also possible loss of synchronism must be removed in the minimum of time. Automatic means, therefore, are required to detect abnormal currents and voltages and, when detected, to open the appropriate circuit breakers. It is the object of protection to accomplish this. In a large interconnected network considerable knowledge and skill is required to remove the faulty part from the network and leave the healthy remainder working intact.

There are many varieties of automatic protective systems ranging from simple overcurrent electromechanical relays to sophisticated electronic systems transmitting high-frequency signals along the power lines. The simplest but extremely effective form of protection is the electromechanical

relay which closes contacts and hence energizes the circuit breaker opening mechanisms when currents larger than specified pass through the
equipment.

The protection used in a network can be looked upon as a form of
insurance in which a percentage of the total capital cost (about 5 per cent)
is used to safeguard apparatus and ensure continued operation when
faults occur. In a highly industrialized community the maintenance of an
uninterrupted supply to consumers is of paramount importance and the
adequate provision of protection is essential.

Summarizing, protection and the automatic tripping (opening) of
associated circuit breakers has two main functions: (a) to isolate faulty
equipment so that the remainder of the system can continue to operate
successfully, and (b) to limit damage to equipment due to overheating, and
mechanical forces, etc.

## 11.2   Switchgear

Some of the functions of the switches or circuit breakers, as they are
usually called, are obvious and apply to any type of circuit, others are
peculiar to high-voltage equipment. For maintenance to be carried out
on plant, it must be isolated from the rest of the network and hence
switches must be provided on each side. If these switches are not required
to open under working conditions, i.e. with fault or load current and normal
voltage, a cheaper form of switch known as an *isolator* can be used;
this can close a live circuit but not open one. Owing to the high cost of
circuit breakers much thought is given in practice to obtaining the largest
degree of flexibility in connecting circuits with the minimum number of
switches. Popular arrangements of switches are shown in figure 11.1.

High voltage circuit breakers take three basic forms, oil immersed, small
oil volume and air-blast. These will be briefly described below.

(a)  The bulk oil circuit breaker. A cross-section of an oil circuit breaker
with all three phases in one tank is shown in figure 11.2. There are two sets
of contacts per phase. The lower and moving contacts are usually cylindrical
copper rods and make contact with the upper fixed contacts. The fixed
contacts consist of spring loaded copper segments which exert pressure on
the lower contact rod when closed to form a good electrical contact.
On opening, the lower contacts move rapidly downwards and draw an arc.
When the circuit breaker opens under fault conditions many thousands
of amperes pass through the contacts and the extinction of the arc and
hence the effective open circuiting of the switch are major engineering
problems. Effective opening is only possible because the instantaneous

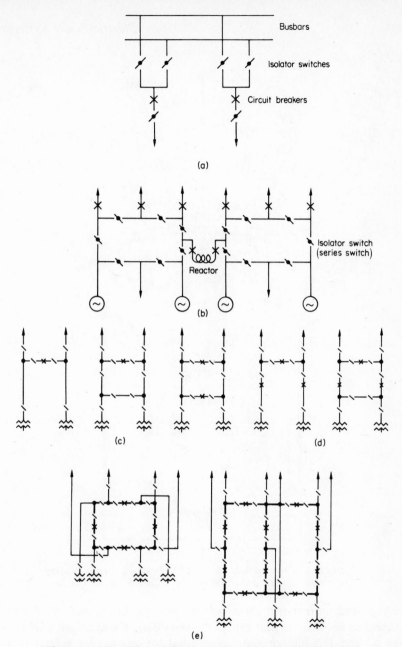

Figure 11.1 Possible switchgear arrangements. (a) Double busbar selection arrangements. (b) Double ring busbars with connecting reactor busbars can be isolated for maintenance but circuits cannot be transferred from one side of the reactor to the other. (c) Open mesh switching stations, transformers not switched. (d) Open mesh switching stations, transformers switched. (e) Closed mesh switching stations. (Isolators are sometimes called series switches.)

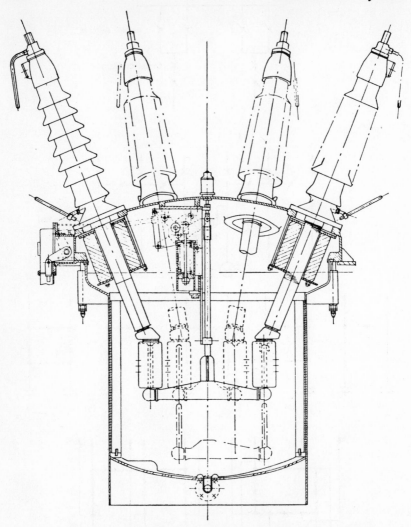

Figure 11.2  Cross-section of a 66 kV bulk oil circuit breaker.
Three phases in one tank. (*Permission of English Electric Co. Ltd.*)

voltage and current per phase reduces to zero during each alternating
current cycle. The arc heat causes the evolution of a hydrogen bubble in
the oil and this high-pressure gas pushes the arc against special vents
in a device surrounding the contacts called a turbulator (figure 11.3).
As the lowest contact moves downwards the arc stretches and is cooled
and distorted by the gas and so eventually breaks. The gas also sweeps

the arc products from the gap so that the arc does not reignite when the voltage rises to its full open-circuit value.

(b) The air blast circuit breaker. For voltages above 120 kV the air blast breaker is popular because of the feasibility of having several contact gaps in series per phase. Schematic diagrams of two types of air-blast head are shown in figure 11.4. Air normally stored at 200 lb/in² is released and directed at the arc at high velocities, thus extinguishing it. The air

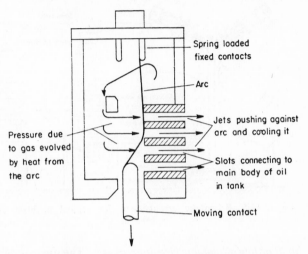

Figure 11.3   Cross jet explosion pot for arc extinction in bulk oil circuit breakers.

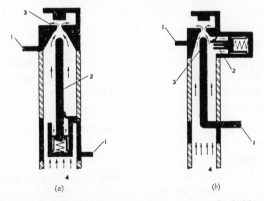

Figure 11.4   Schematic diagrams of two types of air-blast head. (a) Axial flow with axially moving contact. (b) Axial flow with side moving contact. 1. Terminal, 2. Moving contact, 3. Fixed contact and 4. Blast pipe. (*Permission of A. Reyrolle & Company Ltd.*)

also actuates the mechanism of the movable contact. Figure 11.5 shows a 132 kV air blast breaker with two breaks (interrupters) per phase and its associated isolator or series switch. A 245 kV circuit breaker is shown in figure 1.26 in Chapter 1. A similar circuit breaker has been developed using sulphur hexafluoride ($SF_6$) instead of air. When the main contacts have opened, an isolator (or series) switch, in series with the main contacts, opens and then the main contacts reclose when the air pressure is cut off.

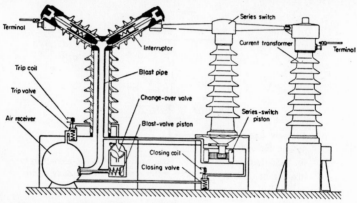

Figure 11.5   Schematic arrangement for typical air-blast circuit breaker.

(c) *Small or low oil-volume circuit breakers*   The quenching mechanisms are enclosed in vertical porcelain insulation compartments and the arc is extinguished by a jet of oil issuing from the moving (lower contact) as it opens. The volume of oil used is much smaller than in the bulk-oil type.

A further arc quenching medium is *vacuum*. An arc is only possible if contact material is vaporized, thus forming a conducting path. The maximum acceptable switching voltage at the present time is about 40 kV (r.m.s.) per vacuum chamber. This type of circuit breaker is under consideration for possible use in high-voltage direct-current schemes.

A circuit breaker must fulfil the following conditions:

(a) open and close in the shortest possible time under any network condition,

(b) conduct rated current,

(c) withstand thermally and mechanically any short-circuit currents,

(d) maintain its voltage to earth and across the open contacts under both clean and polluted conditions,

(e) not create any large overvoltage during opening and closing,

(f) be easily maintained.

## 11.3 Qualities Required of Protection

A few terms often used to describe the effectiveness of protective gear will now be described.

*Selectivity or discrimination*—its effectiveness in isolating only the faulty part of the system.

*Stability*—the property of remaining inoperative with faults occurring outside the protected zone (called external faults).

*Speed of operation*—this property is more obvious. The longer the fault current continues to flow the greater the damage to equipment. Of great importance is the necessity to open faulty sections before the connected synchronous generators lose synchronism with the rest of the system. This aspect is dealt with in detail in Chapter 7. In Britain a typical fault clearance time in h.v. systems is 140 ms. The future requirement is 80 ms, and this encroaches on the basic assessment time of the relays.

*Sensitivity*—this is the level or magnitude of fault current at which operation occurs, which may be expressed in current in the actual network (primary current) or as a percentage of the current transformer secondary current.

*Economic consideration.* In distribution systems the economic aspect almost overrides the technical one owing to the large number of feeders, transformers, etc., provided that basic safety requirements are met. In transmission systems the technical aspects are more important. The protection is relatively expensive but so is the system or equipment protected and security of supply is vital. Two separate protective systems are used, one main (or primary) and one back-up.

*Reliability.* This property is self-evident. A major cause of circuit 'outages' is mal-operation of the protection itself. On average, in the British system (not including faults on generators), nearly 15 per cent of outages are due to this cause.

### Back-up protection

Back-up protection as the name implies is a completely separate arrangement which operates to remove the faulty part should the main protection fail to operate. The back-up system should be as independent of the main protection as possible possessing its own current transformers and relays. Often only the circuit-breaker tripping battery and voltage transformers are common.

Each main protective scheme protects a defined area or *zone* of the power system. It is possible that between adjacent zones a small region, e.g. between the current transformers and circuit breakers, may be unprotected, in which case the back-up scheme (known as remote back-up), will afford protection because it overlaps the main zones, as shown in figure 11.6. In distribution the application of back-up is not as widespread

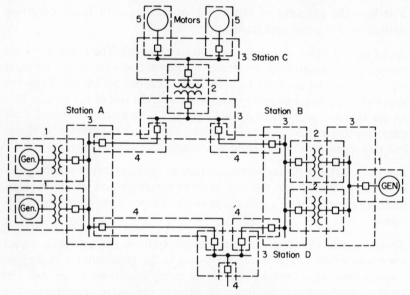

Figure 11.6   Line diagram of a typical system and the overlapping zones of protection. (*Permission of Westinghouse Electrical Corporation.*)

as in transmission systems, it often being sufficient to apply it at strategic points only. Remote back-up is slow and usually disconnects more of the supply system than is necessary to remove the faulty part.

## 11.4   Components of Protective Schemes

*Current transformers.* (*C.T.s*)

In order to obtain currents which are proportional to the system (primary) currents and which can be used in control circuits, current transformers are used. Often the primary conductor itself, e.g. an overhead line, forms a single primary turn (bar primary). Whereas instrument current transformers have to remain accurate only up to slight overcurrents, protection current transformers must retain proportionality up to, say, twenty times normal full load.

A major problem exists when two current transformers are used which should retain identical characteristics up to the highest fault current, e.g. in pilot wire schemes. Because of saturation in the silicon steel used and the possible existence of a direct component in the fault current the exact matching of such current transformers is difficult. The nominal secondary current rating of current transformers is now usually 1 A but 5 A has been used in the past.

*Linear couplers* The problems associated with current transformers have resulted in the development of devices called linear couplers, which serve the same purpose but, having air cores, remain linear at the highest currents.

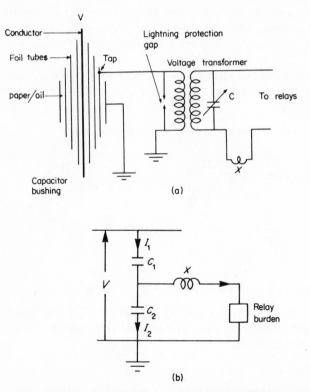

Figure 11.7 Capacitor voltage transformer. (a) Circuit arrangement. (b) Equivalent circuit—burden = impedance of transformer and load referred to primary winding.

*Voltage (or potential) transformers. (V.T.s or P.T.s)*

These provide a voltage which is much lower than the system voltage, the nominal secondary voltage being 110 V. There are two basic types, the wound (electromagnetic), virtually a small power transformer, and the capacitor type. In the latter a tapping is made on a capacitor bushing (usually of the order of 12 kV) and the voltage from this tapping stepped down by a small voltage transformer. The arrangement is shown in figure 11.7, the reactor (X) and the capacitor (C) constitute a tuned circuit which corrects the phase angle error of the secondary voltage.

*Relays*

A relay is a device which when energized by appropriate system quantities indicates an abnormal condition. When the relay contacts close the associated circuit breaker trip-circuits are energized and the breaker contacts open, isolating the faulty part from the system. There are two main forms of relay, electromagnetic and semiconductor. For some purposes (e.g. overload protection) a bimetallic-strip thermal action is used.

The basic forms of the electromagnetic type comprise induction disc, induction cup, hinged armature and plunger action. The hinged armature

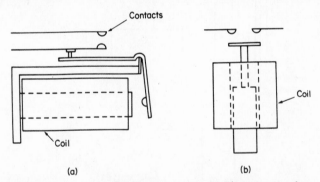

(a)                                         (b)

Figure 11.8   (a) Hinged armature relay. (b) Plunger type relay.

and plunger type devices (figure 11.8) are the simplest and rely on the attraction of an armature or plunger due to an electromagnet which may be energized by a.c. or d.c.

*Induction disc relay*

A copper disc is free to rotate between the poles of an electromagnet which produces two alternating magnetic fields displaced in phase and space. The eddy currents due to one flux and the remaining flux interact to produce a torque on the disc. In early relays the flux displacement was produced by a copper band around part of the magnet pole (shading-ring) which displaced the flux contained by it. Modern relays employ a wattmetric principle in which two electromagnets are employed, as shown in figure 11.9. The current in the lower electromagnet is induced by transformer

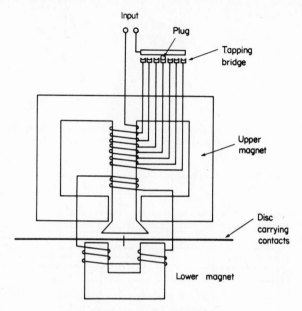

Figure 11.9 Induction disc relay.

action from the upper winding and sufficient displacement between the two fluxes results. This, however, may be adjusted by means of a reactor in parallel with the secondary winding.

The basic mode of operation of the induction disc is indicated in the phasor diagram of figure 11.10. The torques produced are proportional to $\Phi_2 i_1 \sin \alpha$ and $\Phi_1 i_2 \sin \alpha$, so that the total torque is proportional to $\Phi_1 \Phi_2 \sin \alpha$ as $\Phi_1$ is proportional to $i_1$ and $\Phi_2$ to $i_2$.

This type of relay is fed from a C.T. and the sensitivity may be varied by the plug arrangement shown in figure 11.9. The operating characteristics are shown in figure 11.11. To enable a single characteristic curve to be used for all the relay sensitivities (plug settings) a quantity known as the current (or plug) setting multiplier is used as the abscissa instead of current magnitude, as shown in figure 11.11. To illustrate the use of this curve (usually shown on the relay casing) the following example is given.

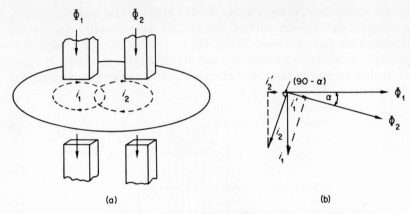

Figure 11.10   Operation of disc type electromagnetic relay. (a) Fluxes.
(b) Phasor diagram. $-i$ and $i_2$ induced currents in disc.

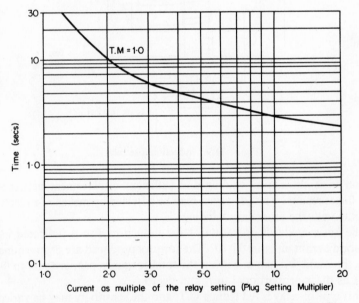

Figure 11.11   Time-current characteristics of a typical induction disc
in terms of the plug setting multiplier. T.M. = Time multiplier.

*Example 11.1*   Determine the time of operation of a 1-A, 3-s over-current
relay having a plug setting of 125 per cent and a time multiplier of 0·6.
The supplying C.T. is rated 400:1 A and the fault current is 4000 A.

*Solution* The relay coil current = (4000/400) × 1 = 10 A. The normal relay coil current is 1 × (125/100) = 1·25 A. Therefore the relay fault current as a multiple of the plug setting = (10/1·25) = 8. From the relay curve (figure 11.11) the time of operation is 3·3 s for a time setting of 1. The Time Multiplier (T.M.) controls the time of operation by changing the angle through which the disc moves to close the contacts. The actual operating time = 3·3 × 0·6 = 2·0 s.

Induction disc relays may be made responsive to power flow by feeding the upper magnet winding in figure 11.9 from a voltage via a potential transformer and the lower winding from the corresponding current. As the upper coil will consist of a large number of turns the current in it lags the applied voltage by 90°, whereas in the lower (small number of turns) coil they are almost in phase. Hence, $\Phi_1$ is proportional to V and $\Phi_2$ is proportional to I, and torque is proportional to $\Phi_1\Phi_2 \sin \alpha$ i.e. to $\Phi_1\Phi_2 \sin (90 - \phi)$, or $VI \cos \phi$ (where $\phi$ is the angle between V and I).

The direction of the torque depends on the power direction and hence the relay is directional. A power relay may be used in conjunction with a current-operated relay to provide a directional property.

*Induction cup relay* The operation is similar to the induction disc, here two fluxes at right-angles induce eddy currents in a bell-shaped cup which

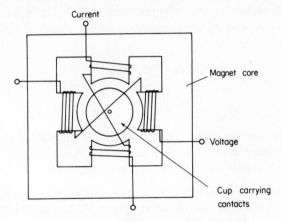

Figure 11.12 Four-pole induction cup relay.

rotates and carries the moving contacts. A four-pole relay is shown in figure 11.12.

*Permanent-magnet moving coil* The action in one type is similar to a moving-coil indicating instrument with the moving-coil assembly carrying the contacts. In a second type the action is basically that of the loud-speaker in which the coil moves axially in the gap of a permanent magnet. The time–current characteristic is inverse with a definite minimum time.

### Balanced beam

The basic form of this relay is shown in figure 11.13. The armatures at the ends of the beam are attracted by electromagnets which are operated by the appropriate parameters, usually voltage and current. A slight mechanical bias is incorporated to keep the contacts open except when operation is required.

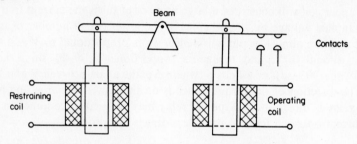

Figure 11.13   Schematic diagram of balanced beam relay.

The pulls on the armatures by the electromagnets are equal to $K_1V^2$ and $K_2I^2$, where $K_1$ and $K_2$ are constants, and for operation (i.e. contacts to close), i.e.

$$K_1V^2 < K_2I^2,$$

$$\frac{V}{I} < \sqrt{\frac{K_2}{K_1}} \text{ or } Z < \sqrt{\frac{K_2}{K_1}}$$

This shows that the relay operates when the impedance it 'sees' is less than a predetermined value. The characteristic of this relay when drawn on $R$ and $jX$ axes is a circle as shown in figure 11.14.

*Distance relays*    The balanced-beam relay, because it measures the impedance of the protected line, effectively measures distance. Two other relay forms also may be used for this purpose:

  (a)  The reactance relay which operates when $\dfrac{V}{I}\sin\phi \leqslant$ constant, having the characteristic shown in figure 11.14.

  (b)  The mho or admittance relay the characteristic of which is also shown in figure 11.14.

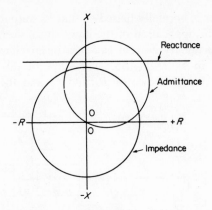

Figure 11.14  Characteristics of impedance, reactance and admittance (mho) relays shown on the R—X diagram.

The above relays operate with any impedance phasor lying inside the characteristic circle or below the reactance line. The mho characteristic may be obtained by devices balancing two torque-producing elements and an induction cup could be used. A rectifier bridge supplying a moving-coil relay is often used.

*Negative-phase sequence.* This is used in generator protection and is sensitive to the presence of negative-sequence currents. The protection comprises a bridge circuit supplying a current-operated relay.

### Semiconductor devices

These relays are extremely fast in operation, having no moving parts and are very reliable. Detection involving phase angles and current and voltage magnitudes are made with appropriate circuits. Most required current–time characteristics may be readily obtained and solid-state devices are now firmly established. Inverse-characteristic, overcurrent and earth-fault relays have a minimum time lag and the operating time is inversely related to some power of the input (e.g. current). In practical static relays it is advantageous to choose a circuit which can accommodate a wide range of alternative inverse time characteristics, precise minimum operating levels, and definite minimum times.

The exponential charging circuit which can be used for definite time functions is not particularly suited to inverse timing functions, and static (non-electromagnetic) inverse relays generally use the constant-current charging of a capacitor. Circuits based on the well-known Miller integrator are particularly suitable, a typical relay being shown in figure 11.15a. The linear amplifier is normally biased so that its output stage is approaching saturation. The application of the input signal derived from the current-transformer secondary current causes a linear change of output voltage $V_0$, as shown in figure 11.15b. The slope of this output waveform is related to the initial choice of $R_1$ and $C$, and the operating time $T$ is

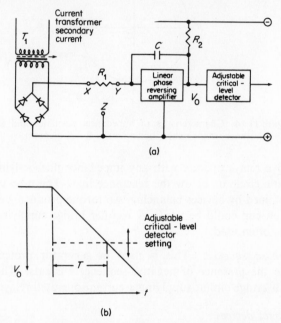

(a)

(b)

Figure 11.15   (a) Basic circuit, inverse timing circuit. (b) Output-voltage waveform—inverse timing circuit. (*Permission of Institution of Electrical Engineers.*)

inversely proportional to the input current. Provided that the gain of the linear amplifier is sufficiently high, the output-voltage waveform is negligibly dependent on the amplifier parameters. The output voltage $V_0$ forms the input to a critical-level detector circuit, the critical level of which can be made adjustable to give a traditional time-multiplier control. The output of the level detector can be arranged to initiate tripping of the circuit breaker through a variety of non-critical devices, using Thyristors or even simple electromagnetic relays.

For an amplitude comparator (for comparing signals from the extremities of the protected zone), a simple development is the use of the well-tried rectifier-bridge circuit, with a static integrating circuit replacing the moving-coil relay. The diodes can be replaced, with advantage, by an arrangement of transistors as shown in figure 11.16. The performance of this type of comparator depends on matching of the gain of the transistors, which should thus incorporate a high degree of negative feedback. Practical arrangements have therefore used transistors in the common-base mode.

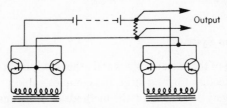

Figure 11.16 Basic amplitude comparator—Transistor-bridge circuit.
(*Permission of Institution of Electrical Engineers.*)

The waveforms generated in this type of comparator are similar to those derived from a phase comparator, and this enables the same form of integrator to be used for the latter.

*Summation transformer*  In some relaying schemes it is necessary to transmit the secondary currents of the current transformers considerable distances in order to compare them with currents elsewhere. To avoid the use of wires from each of the three C.T.s in a three-phase system, a summation transformer is used which gives a single-phase output, the magnitude of which depends on the nature of the fault. The arrangement is shown in figure 11.17 in which the ratios of the turns are indicated. On balanced through-faults there is no current in the winding between c and n. The phase (a) current energizes the 1 p.u. turns between a and b and the phasor sum of $I_a$ and $I_b$ flows in the 1 p.u. turns between b and c.

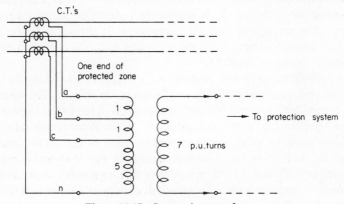

Figure 11.17  Summation transformer.

The arrangement gives much greater sensitivity to earth faults than to phase faults. When used in phase-comparison systems, however, the actual value of output current is not important and the transformer usually saturates on high-fault currents so protecting the secondary circuits against high voltages.

## 11.5 Protection Systems

The application of the various relays and other equipment to form adequate schemes of protection forms a large and complex subject. Also the various schemes are largely dependent on the methods of individual manufacturers. The main intention here is to present a survey of general practice and outline the principles of the methods used. Some schemes are discrimina-tive to fault location and involve several parameters, e.g. time, direction, current, distance, current balance, phase comparison. Others discriminate according to the type of fault, e.g. negative-sequence relays and some use a combination of location and type of fault.

A convenient classification is the division of the systems into *unit* and *non-unit* types. Unit protection signifies that an item of equipment or zone is being uniquely protected independently of the adjoining parts of the system. Non-unit schemes are those in which several relays and associated equipment are used to provide protection covering more than one zone. Examples of both types are classified in figure 11.18. Non-unit schemes represent the most widely used and cheapest forms of protection and these will be discussed first.

*Overcurrent protection*

This basic method is widely used in distribution networks and as a back-up in transmission systems. It is applied to generators, transformers and feeders. The arrangement of the components is shown in figure 11.19. The relay normally employed is the induction disc type with two electro-magnets as shown in figure 11.9.

The application to feeders is illustrated in figure 11.20. Along the radial feeder the relaying points and circuit breakers are shown. The operating times are graded to ensure that only that portion of the feeder remote from the infeed side of a fault is disconnected. When determining selectivity, allowance must be made for the operating time of the circuit breakers. Assume figure 11.20 to be a distribution network with slow-acting breakers operating in 0·3 s and that the relays have true inverse-law characteristics. Selectivity is obtained with a through-fault current of 200 per cent full-load with the fault between D and E as illustrated because the time difference

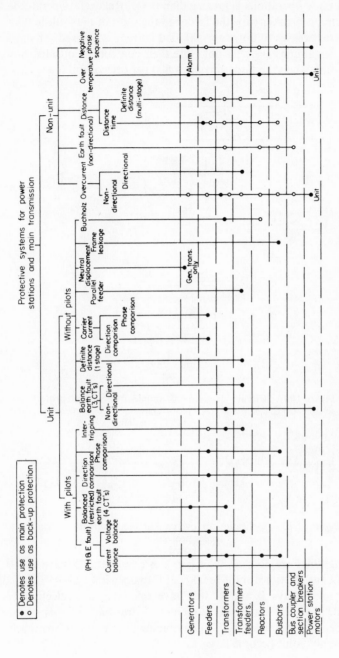

Figure 11.18  Classification of protection schemes.

between relay operations is greater than 0·3 s. Relay (D) operates in 0·5 s and its circuit breaker trips the feeder to the right of the fault in 0·8 s. The fault current ceases to flow (normal load current is ignored for simplicity) and the remaining relays do not close their contacts. Consider, however, the situation when the fault current is 800 per cent of full load. The relay

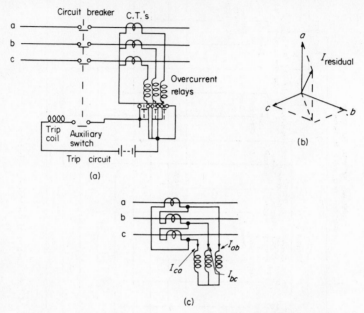

Figure 11.19   Circuit diagram of simple overcurrent protection scheme. (a) C.T.'s in star. (b) Phasor diagram of relay currents, star connexion. (c) C.T.'s in delta.

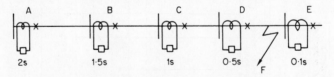

Figure 11.20   Application of overcurrent induction disc relays to feeder protection.

operating times are now: A 0·5 s [i.e. 2 × (200/800)], B 0·375 s, C 0·25 s, D 0·125 s, and the time for the breaker at D to open is 0·125 + 0·3 = 0·325 s. By this time relays B and C will have operated and selectivity is not obtained. This illustrates the fundamental drawback of this system, i.e. that for correct discrimination to be obtained the times of operation close to the supply point become large.

### Overcurrent and directional

In a loop system, to obtain discrimination, relays with an added directional property are required. For the system shown in figure 11.21 directional and non-directional overcurrent relays have time lags for a given fault current as shown. Current feeds into fault at the location indicated from

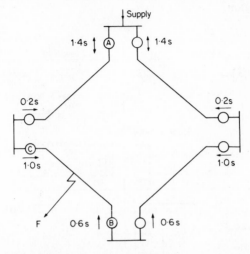

Figure 11.21 Application of directional overcurrent relays to a loop network. ↔ Relay responsive to current flow in both directions. → Relay responsive to current flow in direction of arrow.

both directions and the first relay to operate is at B (0·6 s). The fault is now fed only along route ACB and next the relay at C (1 s) operates and completely isolates the fault from the system. Assuming a circuit breaker clearance time of 0·3 s complete selectivity is obtained at any fault position.

### 11.6 Distance Protection

(*Non-unit protection for feeders*). The shortcomings of graded overcurrent relays have led to the widespread use of distance protection. The distance between any point in the feeder and the fault is proportional to the ratio (voltage/current) at that point and relays responsive to impedance, admittance (mho) or reactance may be used. Although a variety of time–distance characteristics are available for providing correct selectivity the most popular one is the stepped characteristic shown in figure 11.22. A,B,C and D are distance relays with directional properties and A and C only measure distance when the fault current flows in the indicated direction. Relay A trips its associated breaker if a fault occurs within the first 80 per cent of the length of feeder (1). For faults in the remaining 20 per cent of feeder (1) and the initial 30 per cent of (2) (called the stage 2 zone) relay A initiates tripping after a short time delay. A further delay in (A) is introduced for faults further along feeder (2) (stage 3 zone). Relays B and D have similar characteristics when the fault current flows in the opposite direction.

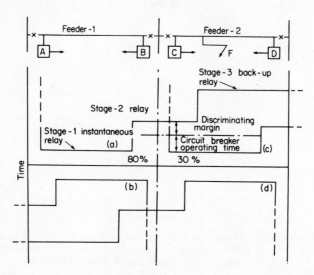

Figure 11.22 Characteristic of three-stage distance protection.

The selective properties of this scheme can be understood by considering a fault such as at F in feeder 2 when fault current flows from A to the fault. For this fault relay A starts to operate, but before the tripping circuit can be completed relay C trips its circuit breaker and the fault is cleared. Relay A then resets and feeder 1 remains in service. The margin of selectivity provided is indicated by the vertical intercept between the two characteristics for relays A and C at the position F less the circuit-breaker operating time.

It will be noted that the stage 1 zones are arranged to extend over only 80 per cent of a feeder from each end. The main reason for this is because practical distance relays and their associated equipment have errors and a margin of safety has to be allowed if incorrect tripping for faults which occur just inside the next feeder is to be avoided. Similarly the stage 2 zone is extended well into the next feeder to ensure definite protection for that part of the feeder not covered by stage 1. The object of the stage 3 zone is to provide general back-up protection for the rest of the adjacent feeders.

The characteristics shown in figure 11.22 require three basic features, viz., response to direction, response to impedance, and timing. These features need not necessarily be provided by three separate relay elements but they are fundamental to all distance protective systems. As far as the directional and measuring relays are concerned, the number required in any scheme is governed by the consideration that three-phase, phase-to-phase, phase-to-earth, and two-phase to-earth faults must be catered for. For the relays to measure the same distance for all types of faults the applied voltages and currents must be different. It is common practice, therefore, to provide two separate sets of relays, one set for phase faults and the other for earth faults, and either of these caters for three-phase faults and double-earth faults. Each set of relays is in practice usually further divided into three, since phase faults may concern any pair of phases, and similarly any phase can be faulted to earth.

## 11.7  Unit Protection Schemes

With the ever-increasing complexity of modern power systems the methods of protection so far described may not be adequate to afford proper discrimination especially when the fault current flows in parallel paths. In unit schemes protection is limited to one distinct part or element of the system which is disconnected if an internal fault occurs. On the other hand the protected part should remain connected with the passage of current flowing into an external fault.

*Differential relaying*  At the extremities of the zone to be protected the currents are continuously compared and balanced by suitable relays. Provided the currents are equal in magnitude and phase no relay operation will occur. If, however, an internal fault (inside the protected zone) occurs this balance will be disturbed (see figure 11.23) and the relay will operate. The current transformers at the ends of each phase should have identical characteristics to ensure perfect balance on through-faults. Unfortunately this is difficult to achieve and a restraining or bias coil is connected (see figure 11.24a) which carries a current proportional to the full system current and restrains the relay operation on large through-fault currents. The corresponding characteristic is shown in figure 11.24b. This principle (circulating current) may be applied to generators, feeders, transformers and busbars and provides excellent selectivity.

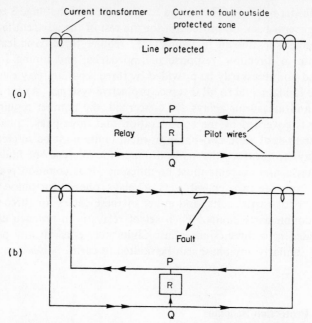

Figure 11.23  Circulating current, differential protection (one phase only shown), (a) Current distribution with through fault—no current in relay. (b) Fault on line, unequal currents from current transformers and current flows in relay coil. Relay contacts close and make operating circuits of circuit breakers at each end of the line.

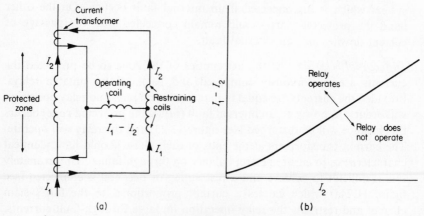

Figure  11.24a   Differential  protection—circuit  connexions  (one phase only)—relay with bias coil.
Figure 11.24b    Characteristic of bias relay in differential-protection. Operating current plotted against circulating or restraint current.

## 11.8 Generator Protection

Large generators are invariably connected to their own step-up transformer and the protective scheme usually covers both items. A typical
scheme is shown in figure 11.25 in which separate differential circulating-
current protections are used to cover the generator alone and the generator
plus transformer. When differential protection is applied to a transformer

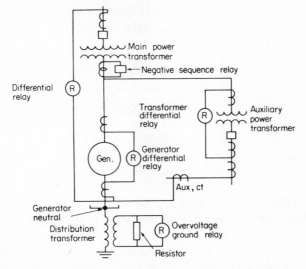

Figure 11.25   Protection scheme for a generator and unit transformer.

the current transformer on each side of a winding must have ratios which
give identical secondary currents. In the U.S.A. the generator neutral is
often grounded through a distribution transformer. This energizes a relay
which operates the generator main and field breakers when a ground fault
occurs in the generator or transformer. The ground fault is usually limited
to about 10 A by the distribution transformer. In Britain the neutral is
usually grounded through a resistor. The field circuit of the generator
must be opened when the differential protection operates to avoid the
machine feeding the fault.

The relays of the differential protection on the stator windings (see
figure 11.26) are set to operate at about 10 to 15 per cent of the circulating
current produced by full-load current to avoid current transformer errors.
If the phase e.m.f. generated by the winding is $E$, the minimum current for
a ground fault at the star-point end and hence with the whole winding in

circuit is $E/R$, where $R$ is the neutral effective resistance. For a fault at a fraction $x$ along the winding from the neutral, the fault current is $xE/R$ and 10–15 per cent of the winding is unprotected. With the neutral grounded via the transformer, $R$ is high and earth faults are detected by a sensitive relay across the transformer secondary. With an interturn fault (turn-to-turn short circuit) on a phase of the stator winding current

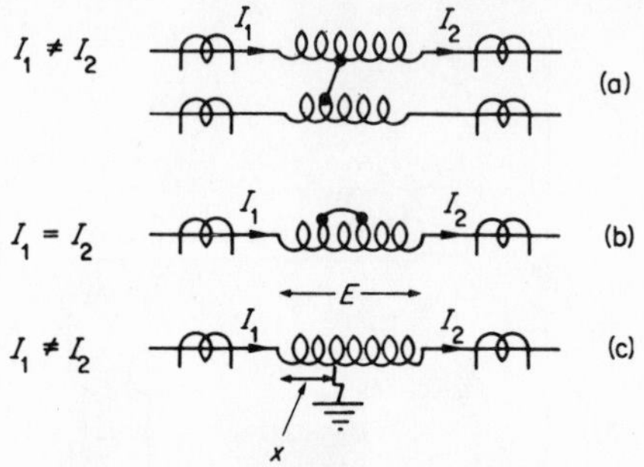

Figure 11.26   Generator winding faults and differential protection. (a) Phase to phase fault. (b) Interturn fault. (c) Phase to earth fault.

balance at the ends is retained and no operation of the differential relay takes place. The relays only operate with phase-to-phase and ground faults.

On unbalanced loads or faults the negative-sequence currents in the generator produce excessive heating on the rotor surface and generally $(I_2^2 t)$ must be limited to a certain value for a given machine (between 3 and 4 for 500 MW machines), where $t$ is the duration of the fault in seconds. To ensure this a relay is installed which detects negative-sequence current and trips the generator main breakers after a prescribed time. When loss of excitation occurs, reactive power $(Q)$ flows into the machine and if the system is able to supply this the machine will operate as an induction generator still supplying power to the network. The generator output will oscillate slightly as it attempts to lock into synchronism. Relays are connected to isolate the machine when a loss of field occurs.

## 11.9   Transformer Protection

A typical protection scheme is shown in figure 11.27a in which the differential circulating current arrangement is used. The specification and arrangement of the current transformers is complicated by the main transformer connexions and ratio. Current magnitude differences are

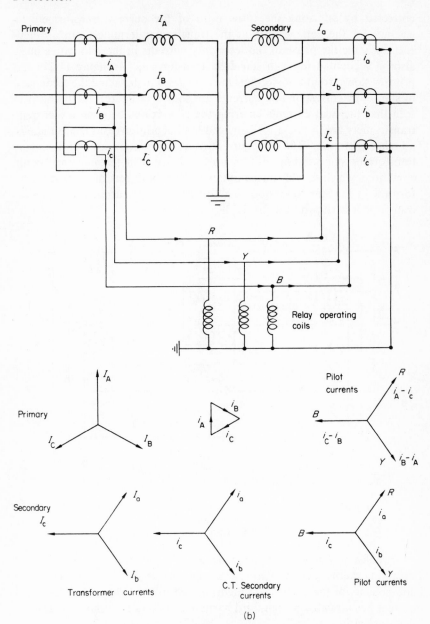

Figure 11.27 Differential protection applied to transformers. (a) Current transformer connexions. (b) Phasor diagrams of currents in current transformers.

corrected by adjusting the turns ratio of the current transformers to account for the voltage ratio at the transformer terminals. In a differential scheme the phase of the secondary currents in the pilot wires must also be accounted for with star-delta transformers. In figure 11.27a the primary side current transformers are connected in delta and the secondary in star. The corresponding currents are shown in figure 11.27b and it is seen that the final currents entering the connexions between the current transformers are in phase for balanced-load conditions and hence there is no relay operation. The delta current-transformer connexion on the main transformer star winding also ensures stability with through earth-fault conditions which would not be obtained with both sets of current transformers in the star connexion. The distribution of currents in a $\lambda - \Delta$ transformer is shown in figure 11.28.

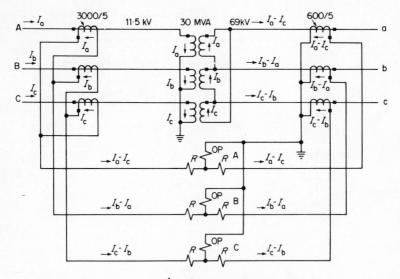

Figure 11.28 Currents in $\lambda - \Delta$ transformer differential protection.
OP = Operating coil.
R = Restraining coil.

Troubles may arise due to the magnetizing current inrush on switching operating the relays and often restraining coils sensitive to third harmonic components of the current are incorporated in the relays. As the inrush current has a relatively high third harmonic content the relay is restrained from operating.

Faults occurring inside the transformer tank due to various causes give rise to the generation of gas from the insulating oil. This may be used as a means of fault detection by the installation in the pipe between the tank and conservator of a gas/oil-operated relay. The relay normally comprises hinged floats and is known as the *Buchholz* relay (see figure 11.29). With a small fault bubbles rising to flow into the conservator are trapped in the relay chamber disturbing the float which closes contacts and operates an

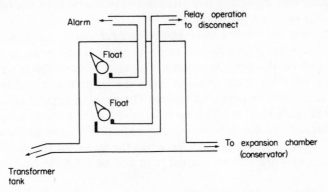

Figure 11.29 Schematic diagram of Buchholz relay arrangement.

alarm. On the other hand, a serious fault causes a violent movement of oil which moves the floats making other contacts which trip the main circuit breakers.

## 11.10 Feeder Protection

### *Differential pilot wire*

The differential system already described can be applied to feeder protection. The current transformers situated at the ends of the feeder are connected by insulated wires known as pilot wires. In figure 11.23, P and Q must be at the electrical mid-points of the pilots and often resistors are added to obtain a geographically convenient mid-point. By reversing the current transformer connexions (figure 11.30) the current transformer

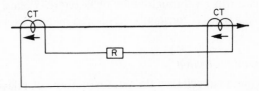

Figure 11.30 Pilot wire differential feeder protection—opposed voltage connexions.

e.m.f.s oppose and no current flows in the wires on normal or through-fault conditions. This is known as the *opposed voltage* method. As under these conditions there are no back ampere-turns in the current transformer secondaries, on heavy through-faults the flux is high and saturation occurs. Also the voltages across the pilots may be high under this condition and unbalance may occur due to capacitance currents between the pilots. To avoid this sheathed pilots are used.

Pilots may be installed underground or strung on towers. In the latter method care must be taken to cater for the induced voltages from the power-line conductors. Sometimes it is more economical to rent wires from the telephone companies, although special precautions to limit pilot voltages are then required. A typical scheme using circulating current is shown in figure 11.31 in which a mixing or summation device is used. With an internal fault the current entering end A will be in phase with the current entering end B, as in h.v. networks the feeder will inevitably be part of a loop network and an internal fault will be fed from both ends. $V_A$

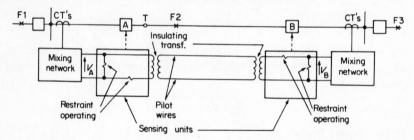

Figure 11.31   Differential pilot wire practical scheme using mixing network (or summation transformer) and biased relays.
$V_A = V_B$ for external faults, e.g. at F1 and F3.
$V_A \neq V_B$ for internal faults, e.g. at F2.

and $V_B$ become additive, causing a circulating current to flow causing relay operation. Thus, this scheme could be looked on as a phase-comparison method. If the pilot wires become short-circuited, current will flow and the relays can trip on through-faults or heavy overloads. In view of this the state of the wires is constantly monitored by the passage of a small d.c. current.

### Carrier-current protection

Because of pilot capacitance the pilot-wire form of relaying is limited to line lengths below 30 miles (48 km). Above this, distance protection may be used although for discrimination of the same order as that obtained with pilots, carrier-current equipment may be used. In carrier-current schemes a high-frequency signal in the band 80 to 500 kHz and of low power level (1 or 2 W) is transmitted via the power line conductors from each end of the line to the other. It is not convenient to superimpose signals proportioned to the magnitude of the line primary current and usually the phases of the currents entering and leaving the protected zone are compared. Alternatively, directional and distance relays are used to start the transmission of a carrier signal to prevent the tripping of circuit breakers at the line ends on through or external faults. On internal faults other directional and distance relays stop the transmission of the carrier signal, the protection operates and the breakers trip.

A further application known as transferred tripping uses the carrier signal to transmit tripping commands from one end of the line to the other. The tripping command signal may take account of, say, the operation or non-operation of a relay at the other end (permissive intertripping) or the signal may give a direct positive instruction to trip alone (intertripping).

Carrier-current equipment is complex and expensive. The high-frequency signal is injected on to the power line by coupling capacitors and may either be coupled to one phase conductor (phase-to-earth) or between two conductors (phase-to-phase), the latter being technically better but more expensive. A schematic diagram of a phase-comparison carrier-current system is shown in figure 11.32. The wave or line trap is tuned to the carrier frequency and presents a high impedance to it but a low impedance to power frequency currents; it thus confines the carrier to the protected line. Information regarding the phase angles of the currents

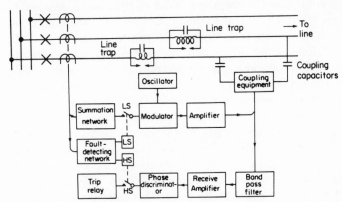

Figure 11.32　Block diagram measuring and control equipment—
for carrier-current phase comparison scheme.
LS = Low set relay
HS = High set relay.

entering and leaving the line is transmitted from the ends by modulation of the carrier by the power current, i.e. by blocks of carrier signal corresponding to half-cycles of power current (figure 11.33). With through or external faults the currents at the line ends are equal in magnitude but 180° phase displaced (i.e. relative to the busbars, it leaves one bus and

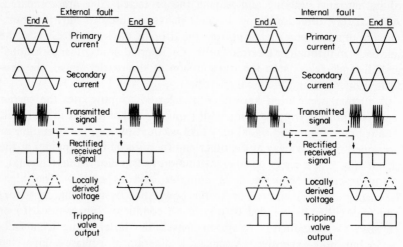

Figure 11.33  Waveforms of transmitted signals in carrier-type line protection.

enters the other). The blocks of carrier occur on alternate half-cycles of power current and hence add to form a continuous signal which is the condition for no relay operation. With internal faults the blocks occur in the same half-cycles and the signal comprises non-continuous blocks; this is processed to cause relay operation (figure 11.32).

The currents from the current transformers are fed into a summation device which produces a single-phase output fed into a modulator (figure 11.32). This combines the power frequency with the carrier to form a chopped 100 per cent modulated carrier signal which is then amplified and passed to the line-coupling capacitors. The carrier signal is received via the coupling equipment, passed through a narrow-bandpass filter to remove any other carrier signals, amplified and then fed to the phase discriminator which determines the relative phase between the local and remote signals and operates relays accordingly. The equipment is controlled by low-set and high-set relays that start the transmission of the carrier only when a relevant fault occurs. These relays are controlled from a starting network. Although expensive, this form of protection is very popular on overhead transmission lines.

## 11.11 Busbar Protection

The protection of the various connexions in a substation is of vital importance. It is essential to obtain correct operation and good discrimination
on external faults, as to isolate busbar faults all source connexions to the
buses must be cut off. Usually differential relaying is used because of its

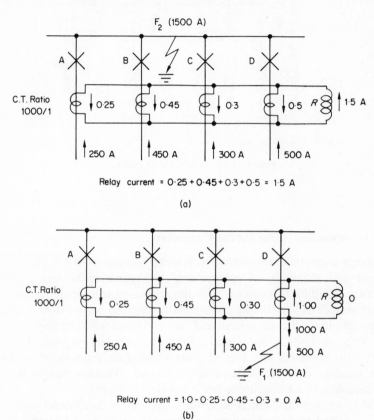

Relay current = 0·25 + 0·45 + 0·3 + 0·5 = 1·5 A

(a)

Relay current = 1·0 − 0·25 − 0·45 − 0·3 = 0 A

(b)

Figure 11.34 Busbar protection—circulating-current differential
scheme. (a) Currents on internal fault. (b) Currents on external fault.

great selectivity. A simple scheme using circulating current is shown (for
one phase only) in figure 11.34a and b. The distribution of secondary
currents for both internal and external faults is shown.

If the switchgear is phase-segregated, all faults are initially earth faults.
A typical arrangement for earth or ground fault protection is shown in
figure 11.35.

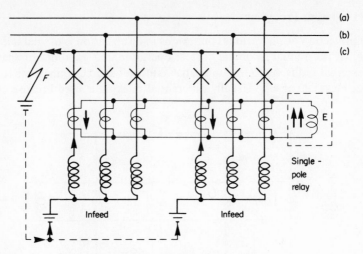

Figure 11.35   Ground fault protection.

## 11.12   Protection using Digital Computers

There is a constant requirement as systems become more complex for the speed and reliability of conventional protection to be improved, an increasingly difficult process. Also the protection is in operation for very brief periods at very infrequent intervals, which to some extent decreases the reliability of the equipment and the confidence which can be placed in it. This could be avoided if local substation computers are used. To avoid instability occurring with a severe short-circuit close to generator terminals, the machine must be isolated very quickly. Whereas this in itself is attainable, correct discrimination is a major problem.

Most measurement and comparison techniques at the moment performed by relays and associated gear could be carried out by digital computers fed by inputs in digital form derived from the system. The instantaneous system quantities would be obtained via current transformers and voltage transformers and converted to digital form using analog to digital conversion techniques. The system parameters are sampled at intervals and then processed according to the particular method in use. A proposed 4 ms tripping of severe faults requires a fault current sampling interval of 0·5 ms and a minimum of six samples is needed to trip thus requiring a total of 3 ms.

## Data storage

The number of quantities to be stored is substantial when the protection of a large complex substation is required. Ten samples of all phase currents and phase-to-ground voltages in the various substation connexions are stored together with two peak magnitudes of these quantities. Also stored are the sample number, i.e. position in time, of the current peaks and other quantities such as differential currents. In all, for an average (500 kV–230 kV–66 kV) substation, storage of the order of 2000 words has been estimated.[13]

Several subroutines are stored and perform the logical and comparison functions previously carried out by relays. These routines include: analog-to-digital conversion, busbar differential, line-current peak determination, current comparison, station overall differential and transformer differential, waveshape analysis and various distance schemes.

## Proposed bus-zone protection[16]

A method for bus-zone protection using current balance has been described. Samples of the currents are simultaneously read in to the digital

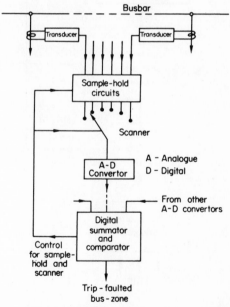

Figure 11.36 Digital protection—basic principle of sample-scan-summate operation. (*Permission of the Institution of Electrical Engineers.*)

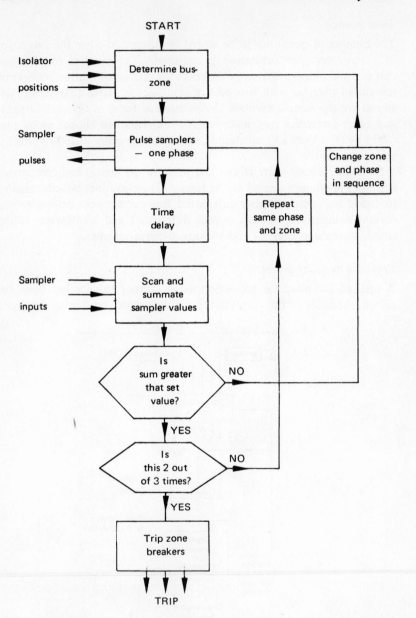

Figure 11.37  Possible flow diagram for direct digital protection—applied to busbars. (*Permission of the Institution of Electrical Engineers.*)

processor and for each set of samples if

$$\sum_{n=1}^{n=f} i_n \geqslant M$$

a trip signal is produced to all the circuit breakers in the protected zone, where $i_n$ is the current in the $n$th feeder and $f$ the number of feeders. If $M$ is not exceeded no internal fault exists. A block diagram for fault detection in one zone of the protection is shown in figure 11.36. The flow diagram of a possible program for this protection is shown in figure 11.37.

Digital protection is at the moment in the initial development stage. It appears to offer several technical advantages including fast operating times. As in all new developments both economic as well as technical factors must be considered.

## References

### BOOKS

1. Lythall, R. T., *The J. & P. Switchgear Book*, Johnson & Phillips, London, 1969.
2. Zajic, V., *High Voltage Circuit Breakers*, Artia, Prague, 1957.
3. Klewe, H. R. J., *Interference between Power Systems and Telecommunication Lines*, Arnold, London, 1958.
4. Willheim, R., and M. Waters, *Neutral Grounding*, Elsevier, New York, 1956.
5. Mason, C. M., *The Art and Science of Protective Relaying*, Wiley, New York, 1956.
6. Warrington, A. R., and C. Van, *Protective Relays*, Vol. 1, 1962, Vol. 2, 1969, Chapman and Hall, London.
7. Westinghouse Electric Corporation, *Applied Protective Relaying*, Newark, New Jersey, U.S.A., 1968.
8. General Electric, *The Art of Protective Relaying*: Series of booklets: *Introduction, Transmission and Subtransmission Lines and E.H.V. Systems*, Switchgear Dept., Philadelphia, U.S.A.
9. Electricity Council (Edited), *Power System Protection*, Vols. 1–3, McDonald and Co., London, 1969.
10. Mathews, P., *Protective Current Transformers and Circuits*, MacMillan, London, 1955.

### PAPERS

Each year many papers are published, most of which are very specialist in nature. The reader is advised to consult the I.E.E.E. committee reports which appear from time to time giving a bibliography of relays and systems of protection. These are published in the *I.E.E.E., P.A. & S. Trans.*

11. Bibliography of Relay Literature for the period 1965–1966. I.E.E.E. Committee Report, *Trans. I.E.E.E., P.A. & S.*, **PAS-88**, 244, 1969.

12. Gross, E. T. B., and E. W. Atherton, 'Application of resonant grounding in power systems in the United States', *A.I.E.E. Trans.*, **70**, 389, 1951.
13. Rockefeller, G. D., 'Fault protection with a digital computer', *I.E.E.E. Trans.*, **PAS-88**, 437, 1969.
14. Hope, G. S., and B. J. Cory, 'Development of digital computer programs for the automatic switching of power networks', *I.E.E.E. Trans.*, **PAS-87**, 1587, 1968.
15. Dy Liacco, T. E., and T. J. Kroynak, 'Processing by logic programming of circuit breaker and protective relaying information', *I.E.E. Trans.*, **PAS-88**, 1969.
16. Cory, B. J., and J. F. Mount, 'Application of digital computers to busbar protection', *Proc. of I.E.E. Conference, The Application of Computers to Power System Protection and Metering*, May 1970.

# Appendix I

**Reactive Power**

In the circuit shown in figure A1.1 let the instantaneous values of voltage and current be

$$e = \sqrt{2}\,E \sin{(\omega t + \phi)}$$

and $$i = \sqrt{2}\,I \sin{(\omega t)}$$

The instantaneous power,

$$p = ei = EI \cos{\phi} - EI \cos{(2\omega t + \phi)}$$

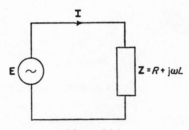

Figure A1.1

Also,

$$-EI \cos{(2\omega t + \phi)} = -EI[\cos{2\omega t} \cos{\phi} - \sin{2\omega t} . \sin{\phi}]$$

$$\therefore \quad p = ei = (EI \cos{\phi} - EI \cos{2\omega t} \cos{\phi}) + (EI \sin{2\omega t} \sin{\phi})$$

$$= (\text{Instantaneous Real Power}) + (\text{Instantaneous Reactive Power})$$

The mean power $= EI \cos{\phi}$

The mean value of $EI \sin{\phi} \sin{2\omega t} = 0$, but the maximum value $= EI \sin{\phi}$.

The voltage source is supplying energy to the load in one direction only. At the same time an interchange of energy is taking place between the source and the load of average value zero, but of peak value $EI \sin{\phi}$. This latter quantity is known as the *reactive power* $(Q)$ and the unit is the VAr (taken from the alternative name, volt-ampere reactive). The interchange of energy between the source and the inductive and capacitive elements (i.e. the magnetic and electric fields) takes place at twice the supply frequency. It is possible to think, therefore, of a power component $P$ (Watts) of magnitude $EI \cos{\phi}$ and a reactive-power component $Q$ (VAr) equal to

$EI \sin \phi$, where $\phi$ is the power factor angle, i.e. the angle between **E** and **I**. It should be stressed, however, that the two quantities $P$ and $Q$ are physically quite different.

The quantity **S** (volt-amperes), known as the complex power, may be found by multiplying **E** by the conjugate of **I** or vice versa. Consider the case when **I** lags **E**, and assume **S** = **E**\***I**. Referring to figure A1.2,

$$\mathbf{S} = \mathbf{E}e^{-j\phi_1} \times \mathbf{I}e^{j\phi_2} = \mathbf{E}\mathbf{I}e^{-(\phi_1-\phi_2)}$$
$$= \mathbf{E}\mathbf{I}e^{-\phi}$$
$$= P - jQ$$

Next assume
$$\mathbf{S} = \mathbf{E}\mathbf{I}^*$$
$$= \mathbf{E}e^{j\phi_1} \times \mathbf{I}e^{-j\phi_2}$$
$$= P + jQ.$$

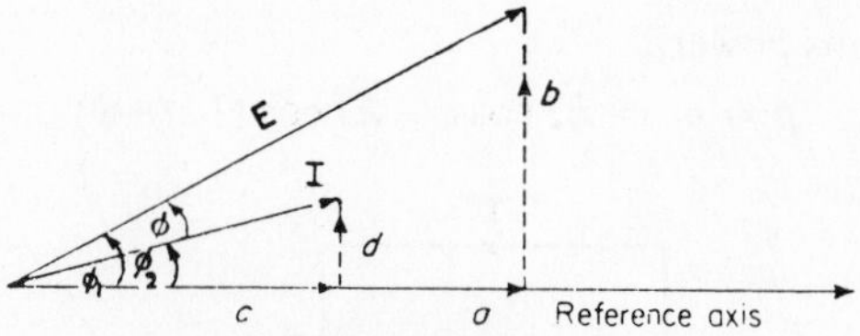

Figure A1.2   Phasor diagram.

Obviously both the above methods give the correct magnitudes for $P$ and $Q$ but the sign of $Q$ is different. The method used is arbitrarily decided and the convention to be adopted in this book is as follows:

the volt-amperes reactive absorbed by an inductive load shall be considered positive, and by a capacitive load negative; hence **S** = **EI**\*. This convention is recommended by the International Electrotechnical Commission.

In a network the nett energy is the sum of the various inductive and capacitive stored energies present. The nett value of reactive power is the sum of the vars absorbed by the various components present, taking due account of the sign. Lagging vars can be considered as either being produced or absorbed in a circuit; a capacitive load can be thought of as generating lagging vars. Assuming that an inductive load is represented by $R + jX$ and that the **EI**\* convention is used then an inductive load *absorbs* positive or lagging vars and a capacitive load *produces* lagging vars. The reactive power flow towards a busbar is positive when the load is lagging; vars exported from a busbar are considered negative for a lagging power factor and positive for leading vars. In this text, reference to the var will imply a lagging power-factor and inductive and capacitive vars distinguished by the sign.

The various elements in a network are characterized by their ability to generate or absorb reactive power. Consider a synchronous generator which can be represented by the simple equivalent circuit shown in figure A1.3. When the generator is over-excited, i.e. its generated e.m.f. is high, it produces lagging vars and a complex power, $P-jQ$. When the machine

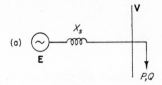

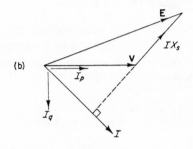

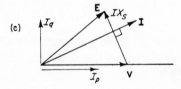

Figure A1.3a  Line diagram of system. (b) Overexcited generator-phasor diagram. (c) Underexcited generator-phasor diagram.

is under-excited the generated current leads $V$ and the generator produces $P+jQ$. It can also be thought of as absorbing lagging vars. The reactive power characteristics of various power-system components are summarized as follows: reactive power is generated by over-excited synchronous machines, capacitors, cables, lightly loaded overhead lines, and absorbed by under-excited synchronous machines, induction motors, inductors, transformers, heavily loaded overhead lines.

# Appendix II

## Summary of Network Theory

*Network reduction*

The analysis of many moderately complicated networks is greatly facilitated by the successive reduction of the networks to much simpler forms. The following useful transformations apply only to linear, passive circuits, i.e. no current or voltage sources are present.

(a) *Delta-star transformation*  The delta network connected between the three terminals A, B and C can be replaced by a star network such that the impedance measured between the terminals is unchanged. The values of the star-connected elements are as follows:

$$Z_{OA} = \frac{Z_{AB}Z_{CA}}{Z_{AB}+Z_{CA}+Z_{BC}}$$

$$Z_{OB} = \frac{Z_{AB}Z_{BC}}{Z_{AB}+Z_{CA}+Z_{BC}}$$

$$Z_{OC} = \frac{Z_{BC}Z_{CA}}{Z_{AB}+Z_{CA}+Z_{BC}}$$

(b) *Star-delta transformation*  A star connected system can be replaced by an equivalent delta connexion if the elements of new network have the following values (figure A2.1):

$$Z_{AB} = \frac{Z_{OA}Z_{OB}+Z_{OB}Z_{OC}+Z_{OC}Z_{OA}}{Z_{OC}}$$

$$Z_{BC} = \frac{Z_{OA}Z_{OB}+Z_{OB}Z_{OC}+Z_{OC}Z_{OA}}{Z_{OA}}$$

$$Z_{CA} = \frac{Z_{OA}Z_{OB}+Z_{OB}Z_{OC}+Z_{OC}Z_{OA}}{Z_{OB}}$$

(c) *General star-polygon transformation*  A star containing $n$ branches of admittances $Y_{10}, Y_{20} \dots Y_{n0}$ (the star point being labelled 0) may be

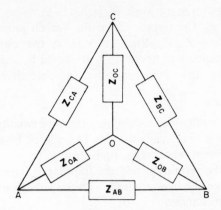

Figure A2.1  Star-delta, delta-star transformation.

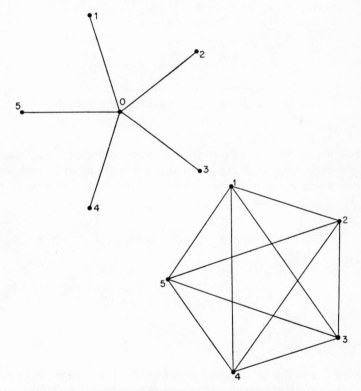

Figure A2.2  Five-element star and its polygon equivalent circuit.
Number of branches of polygon $= \frac{1}{2}n(n-1) = 10$.

replaced by a polygon network of $\frac{1}{2}n(n-1)$ branches joining the outer nodes of the star. The impedance of the branch joining nodes $p$ and $q$ of the star, $\mathbf{Z}_{pq} = \mathbf{Z}_{qo}\,\mathbf{Z}_{po}(\mathbf{Y}_{10} + \ldots \mathbf{Y}_{no})$. A five element star and its replacement are shown in figure A2.2. It should be noted that unlike the star-delta theorem, there is no *converse* to this procedure.

### Thevenin's theorem

This theorem is invaluable for dealing with active circuits. Such a network, when viewed from two terminals, may be replaced by a single voltage in series with an impedance. The equivalent voltage is the open-circuit voltage across the terminals, and the impedance is that of the network measured between the two terminals with the voltage sources replaced by their internal impedances.

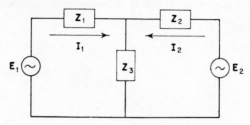

Figure A2.3   Driving point and transfer impedances.

### Driving point and transfer impedances

The Superposition Theorem states that the total current in any branch of a network is the sum of the separate currents produced by each voltage source acting alone with the others short-circuited. Referring to figure A2.3,

$$\mathbf{I}_1 = \frac{\mathbf{E}_1}{\mathbf{Z}_1 + (\mathbf{Z}_2\mathbf{Z}_3/\mathbf{Z}_2 + \mathbf{Z}_3)} - \frac{\mathbf{E}_2}{(\mathbf{Z}_1\mathbf{Z}_2 + \mathbf{Z}_2\mathbf{Z}_3 + \mathbf{Z}_1\mathbf{Z}_3)/\mathbf{Z}_3}$$

or

$$\mathbf{I}_1 = \frac{\mathbf{E}_1}{\mathbf{Z}_{11}} - \frac{\mathbf{E}_2}{\mathbf{Z}_{21}}$$

Similarly,

$$\mathbf{I}_2 = \frac{\mathbf{E}_2}{\mathbf{Z}_{22}} - \frac{\mathbf{E}_1}{\mathbf{Z}_{12}}$$

where $\mathbf{Z}_{22}$ and $\mathbf{Z}_{12}$ are obtained as before. $\mathbf{Z}_{11}$ is the short-circuit driving point impedance of the network as viewed from branch 1. $\mathbf{Z}_{21}$ is the short-circuit transfer impedance between branches 1 and 2.

Generally,

$$Z_{mm} = \frac{\text{driving voltage in branch } m}{\text{current in branch } m}$$

$$Z_{mn} = \frac{\text{driving voltage in branch } n}{\text{current in branch } m}$$

$$Z_{nm} = \frac{\text{driving voltage in branch } m}{\text{current in branch } n}$$

where all other voltages are short-circuited. For linear bilateral circuits $Z_{mn} = Z_{nm}$.

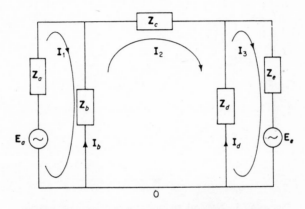

Figure A2.4   Network to illustrate mesh-current method.

### Mesh and nodal analysis

For the solution of large networks a more systematic approach is necessary. There are two methods which enable a large network to be analysed in such a manner.

(a) *Mesh current method*   In this method currents are assumed to circulate in each loop of the network and the usual convention is for these currents to circulate in clockwise directions, as shown in figure A2.4.

The equations governing each mesh are as follows:

$$E_a = (Z_a + Z_b)I_1 - Z_b I_2$$

$$0 = (Z_a + Z_b + Z_c)I_2 - I_1 Z_b - I_3 Z_d$$

$$-E_e = I_3(Z_d + Z_e) - Z_d I_2$$

The above three equations can be expressed in the form:

$\mathbf{E_1} = \mathbf{Z_{11}I_1} - \mathbf{Z_{12}I_2} - \mathbf{Z_{13}I_3}$ (if coupling with loop 3 is present)

$\mathbf{E_2} = -\mathbf{Z_{21}I_1} + \mathbf{Z_{22}I_2} - \mathbf{Z_{23}I_3}$

$\mathbf{E_3} = -\mathbf{Z_{31}I_1} - \mathbf{Z_{32}I_2} + \mathbf{Z_{33}I_3}$

For a solution to be obtained for the currents one equation is required for each loop of the network. In a bilateral network $\mathbf{Z_{12}} = \mathbf{Z_{21}}$, etc.

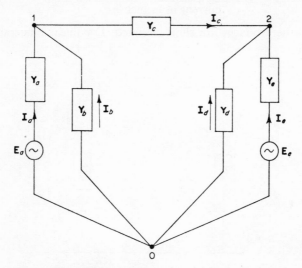

Figure A2.5   Network redrawn for nodal-voltage method.

The mesh currents can be related to the branch currents by Kirchhoff's first law and this can be conveniently achieved by the use of the following connexion matrix.

$$
\begin{array}{cc}
 & \begin{array}{ccc} (2) & (3) & \text{mesh currents} \end{array} \\
\begin{array}{c} \text{Branch (a)} \\ \text{currents (b)} \\ \text{(c)} \\ \text{(d)} \\ \text{(e)} \end{array} &
\begin{bmatrix} 1 & 0 & 0 \\ -1 & 1 & 0 \\ 0 & 1 & 0 \\ 0 & -1 & 1 \\ 0 & 0 & -1 \end{bmatrix}
\end{array}
$$

The branch current $\mathbf{I_a}$ is, from the matrix, equal to $1 \times \mathbf{I_1} + 0 \times \mathbf{I_2} + 0 \times \mathbf{I_3}$, i.e. $\mathbf{I_a} = \mathbf{I_1}$. Similarly, $\mathbf{I_b} = \mathbf{I_2} - \mathbf{I_1}$, $\mathbf{I_c} = \mathbf{I_2}$, $\mathbf{I_d} = \mathbf{I_3} - \mathbf{I_2}$, and $\mathbf{I_e} = -\mathbf{I_3}$. The impedance $\mathbf{Z_{11}}$ is the sum of all the impedances around mesh 1 and is called the mesh self impedance. $\mathbf{Z_{12}}$ is the mesh mutual impedance between meshes 1 and 2. The general mesh equations for an $n$ loop network may be conveniently expressed in matrix form as follows:

$$
\begin{bmatrix} E_1 \\ E_2 \\ \cdot \\ \cdot \\ \cdot \\ E_n \end{bmatrix} = \begin{bmatrix} Z_{11} & -Z_{12} & -Z_{13} \ldots -Z_{1n} \\ -Z_{21} & Z_{22} & -Z_{23} \ldots -Z_{2n} \\ \cdot & \cdot & \cdot \quad\quad \cdot \\ \cdot & \cdot & \cdot \quad\quad \cdot \\ \cdot & \cdot & \cdot \quad\quad \cdot \\ -Z_{n1} & -Z_{n2} & +Z_{n3} \ldots -Z_{nn} \end{bmatrix} \begin{bmatrix} I_1 \\ I_2 \\ \cdot \\ \cdot \\ \cdot \\ I_n \end{bmatrix}
$$

If the network is bilateral (most power networks are) the matrix is symmetrical about the principal diagonal formed by the self impedance terms.

(b) *Nodal voltage method* Taking the same network as in figure A2.4 the equivalent single-phase circuit is redrawn and relabelled as shown in figure A2.5. The emphasis now is the clear identification of nodes rather than loops and the nodes have now been numbered. For each node Kirchhoff's first law is expressed in terms of voltage and admittances.

For node 1, $\qquad 0 = Y_a(V_{10} - E_a) + Y_b V_{10} + Y_c V_{12}$

for node 2, $\qquad 0 = Y_c(V_{21}) + Y_d V_{20} + Y_e(V_{20} - E_e)$

Rearranging, $\qquad Y_a E_a = (Y_a + Y_b + Y_c)V_{10} - Y_c V_{20}$

replacing $V_{12}$ by $V_{10} - V_{20}$

$$Y_e E_e = -Y_c V_{10} + (Y_d + Y_c + Y_e)V_{20}.$$

Rewriting these equations in a general form,

$$
\begin{aligned}
YE_1 &= Y_{11}V_1 + Y_{12}V_2 + Y_{13}V_3 \ldots Y_{1n}V_n \\
YE_2 &= Y_{12}V_1 + Y_{22}V_2 + Y_{23}V_3 \ldots Y_{2n}V_n \\
&\quad \cdot \qquad \cdot \qquad \cdot \qquad\qquad \cdot \\
&\quad \cdot \qquad \cdot \qquad \cdot \qquad\qquad \cdot \\
&\quad \cdot \qquad \cdot \qquad \cdot \qquad\qquad \cdot \\
&\quad \cdot \qquad \cdot \qquad \cdot \qquad\qquad \cdot \\
YE_n &= Y_{n1}V_1 + Y_{n2}V_2 \ldots + \qquad Y_{nn}V_n
\end{aligned}
$$

$Y_{nn}$ is the self admittance of the node $n$ and is equal to the sum of all the admittances connected to that node. $Y_{12}$ is the mutual admittance between nodes 1 and 2 and equals the negative of the sum of the admittances connected between 1 and 2. $YE_1$ can be replaced by a current $I_1$ directed towards the node 1. Similarly, currents can be injected at the other nodes. The solution of a network in matrix form is therefore,

$$\begin{bmatrix} I_1 \\ I_2 \\ I_3 \\ \cdot \\ \cdot \\ \cdot \\ \cdot \\ \cdot \\ I_n \end{bmatrix} = \begin{bmatrix} Y_{11}+Y_{12}+Y_{13} \ldots Y_{1n} \\ Y_{21}+Y_{22}+Y_{23} \ldots Y_{2n} \\ Y_{31}+Y_{32}+Y_{33} \ldots Y_{3n} \\ \cdot \quad \cdot \quad \cdot \quad \cdot \\ \cdot \quad \cdot \quad \cdot \quad \cdot \\ \cdot \quad \cdot \quad \cdot \quad \cdot \\ \cdot \quad \cdot \quad \cdot \quad \cdot \\ \cdot \quad \cdot \quad \cdot \quad \cdot \\ Y_{n1}+Y_{n2}+Y_{n3} \ldots Y_{nn} \end{bmatrix} \begin{bmatrix} V_1 \\ V_2 \\ V_3 \\ \cdot \\ \cdot \\ \cdot \\ \cdot \\ \cdot \\ V_n \end{bmatrix}$$

i.e.                                    $[I] = [Y][V].$

Branch currents can be obtained as before using $I_a = (E_c - V_1)Y_a$, etc., and the appropriate connexion matrix.

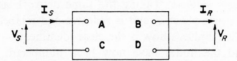

Figure A2.6 Representation of a four-terminal (two-port) network.

There are various advantages and disadvantages in the use of the two methods but for power networks the nodal voltage method is preferable for the following reasons:

(i) the number of equations involved is usually less;
(ii) no initial reduction of parallel branches is required;
(iii) it is easier to deal with branches which cross.

*Example* A power system may be represented by the equivalent single phase current shown in figure A2.7. Calculate the current in branch en. The branches have negligible resistance.

*Solution* Three methods will be used to solve this problem. The first two, Thevenin and Superposition, will require the network to be simplified or reduced. This will be accomplished by replacing the delta network between terminals c, d and e by the equivalent star.

$$Z_{oc} = \frac{j1 \times j0 \cdot 6}{j1 + j0 \cdot 6 + j0 \cdot 4} = j0 \cdot 3 \text{ p.u.}$$

Similarly,

$$Z_{od} = j0 \cdot 2 \text{ p.u.} \quad \text{and} \quad Z_{oe} = j0 \cdot 12 \text{ p.u.}$$

Adding reactances in series the circuit in figure A2.7c is obtained. First this will be solved by superposition.

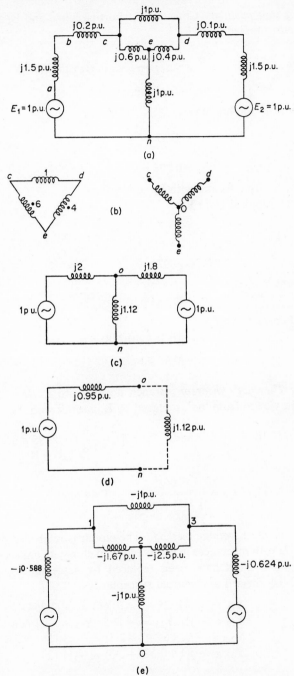

Figure A2.7(a) Network for example. (b) Delta-star transformation. (c) Simplified network. (d) Thevenin equivalent circuit. (e) Network redrawn for nodal-voltage method.

(a) *Using superposition* Consider $E_2$ short-circuited and $E_1$ acting alone, the current supplied from $E_1$

$$= \frac{1}{j[2 + (1 \cdot 12 \times 1 \cdot 8)/(1 \cdot 12 + 1 \cdot 8)]} \text{ p.u.}$$

$$= -j0 \cdot 37 \text{ p.u.}$$

The current in the branch en

$$= -j0 \cdot 37 \times \frac{1 \cdot 8}{2 \cdot 92}$$

$$= -0 \cdot 228 \text{ p.u.}$$

Similarly, shorting $E_1$ and with $E_2$ acting alone,
the current supplied from $E_2$

$$= \frac{1}{j[1 \cdot 8 + (2 \times 1 \cdot 12)/3 \cdot 12]} \text{ p.u.}$$

$$= -j \cdot 397 \text{ p.u.}$$

$\therefore$   current in en

$$= -j0 \cdot 254 \text{ p.u.}$$

$\therefore$   total current in en

$$= -j(0 \cdot 254 + 0 \cdot 228) \text{ p.u.}$$

$$= -j0 \cdot 48 \text{ p.u.}$$

(b) *Using Thevenin's theorem* Branch no in figure A2.7c is opened and the circuit, viewed from 'no', replaced by E in series with $X_e$.

$$E = 1 \text{ p.u.} \quad \text{and} \quad X_e = j\left(\frac{2 \times 1 \cdot 8}{2 + 1 \cdot 8}\right) = j0 \cdot 95 \text{ p.u.}$$

Hence

$$\mathbf{I}_{on} = \mathbf{I}_{en} = \frac{1}{j1 \cdot 12 + j0 \cdot 95} = -j0 \cdot 48 \text{ p.u.}$$

(c) *Using the nodal voltage method* The circuit is relabelled as shown in figure A2.7e with reactances in series added to eliminate unnecessary work. The reactance values have been replaced by susceptances, i.e. $(1/jX)$. First establish the self and mutual susceptances.

$$\mathbf{Y}_{11} = -j(0 \cdot 588 + 1 + 1 \cdot 666) = -j3 \cdot 254$$
$$\mathbf{Y}_{13} = -j1, \; \mathbf{Y}_{23} = -j2 \cdot 5, \; \mathbf{Y}_{12} = -j1 \cdot 666$$
$$\mathbf{Y}_{22} = -j(1 \cdot 666 + 1 + 2 \cdot 5) = -j5 \cdot 166$$
$$\mathbf{Y}_{33} = -j(1 + 0 \cdot 624 + 2 \cdot 5) = -j4 \cdot 124$$

Writing the nodal equations,

$$I_1 = -j0 \cdot 588 \times 1 = -j3 \cdot 254 V_1 + j1 \cdot 666 V_2 + j V_3$$
$$I_2 = 0 = 1 \cdot 666 j V_1 - 5 \cdot 166 j V_2 + j2 \cdot 5 V_3$$
$$I_3 = -j0 \cdot 624 \times 1 = j V_1 + 2 \cdot 5 j V_2 - j4 \cdot 124 V_3$$

Solve for $V_2$ by elimination,

$$-1 \cdot 388 = + 2 \cdot 78 V_2 + 1 \cdot 666 V_3$$
$$0 = -18 \cdot 1\ V_2 + 8 \cdot 75\ V_3$$

$$\overline{-1 \cdot 388 = -15 \cdot 42 V_2 + 10 \cdot 416 V_3}$$

Similarly by eliminating $V_1$ from the $I_2$ and $I_3$ equations;

$$1 \cdot 04 = -9 \cdot 336 V_2 + 9 \cdot 37 V_3$$

Eliminating $V_3$

$$V_2 = 0 \cdot 48 \text{ p.u.}$$

Hence

$$I_2 = \frac{0 \cdot 48}{-j1} = -j0 \cdot 48 \text{ p.u.}$$

If in this example the p.u. quantities were based on 132 kV and 100 MVA base quantities, the base current

$$= \frac{100 \times 10^6}{\sqrt{3} \times 132,000} = 437 \text{ A.}$$

$\therefore$      the current in en $= -j0 \cdot 48 \times 437 = -j210$ A.

### Four terminal networks

A lumped-constant circuit, provided that it is passive, linear and bilateral, can be represented by the four-terminal network shown in the diagram in figure A2.6. The complex parameters **A**, **B**, **C** and **D** describe the network in terms of the sending and receiving end voltages and currents as follows,

$$V_S = A V_R + B I_R$$
$$I_S = C V_R + D I_R$$

and it can be readily shown that $AD - BC = 1$.

**A**, **B**, **C** and **D** may be obtained by measurement and certain physical interpretations made, as follows:

(a) Receiving end short-circuited:

$$V_R = 0, \quad I_S = D I_R \quad \text{and} \quad D = \frac{I_S}{I_R}$$

Also,

$$V_S = BI_R \quad \text{and} \quad B = \frac{V_S}{I_R} = \text{short-circuit transfer impedance.}$$

(b) Receiving end open-circuited:

Here

$$I_R = 0, \quad I_S = CV_R \quad \text{and} \quad C = \left(\frac{I_S}{V_R}\right)$$

$$V_S = AV_R, \quad A = \left(\frac{V_S}{V_R}\right)$$

Expressions for the constants can also be found by measurements carried out solely at the sending end with the receiving end open and short-circuited.

Often it is useful to have a single four terminal network for two or more items in series or parallel, e.g. a line and two transformers in series.

It is shown in most texts on circuit theory that the generalized constants for the combined network, $A_0$, $B_0$, $C_0$ and $D_0$ for the two networks (1) and (2) in series are as follows:

$$A_0 = A_1A_2 + B_1C_2, \quad B_0 = A_1B_2 + B_1D_2, \quad C_0 = A_2C_1 + C_2D_1,$$
$$D_0 = B_2C_1 + D_1D_2.$$

For two four-terminal networks in parallel it can be shown that the parameters of the equivalent single four-terminal network are:

$$A_0 = \frac{A_1B_2 + A_2B_1}{B_1 + B_2}, \quad B_0 = \frac{B_1B_2}{B_1 + B_2}, \quad D_0 = \frac{B_2D_1 + B_1D_2}{B_1 + B_2}$$

$C_0$ can be found from $A_0D_0 - B_0C_0 = 1$.

**Matrix Analysis of Networks**

Although the methods described in Appendix 2 are effective for small networks, for large networks a more systematic building up of the [Y] and [Z] matrices via connexion matrices is useful. This is facilitated by the initial description of a network in terms of its components alone, i.e. when unconnected. For example, consider again the network shown in figure A2.4. The elements are described by

$$[\mathbf{Z}_{br}] = \begin{vmatrix} \mathbf{Z}_a & 0 & 0 & 0 & 0 \\ 0 & \mathbf{Z}_b & 0 & 0 & 0 \\ 0 & 0 & \mathbf{Z}_c & 0 & 0 \\ 0 & 0 & 0 & \mathbf{Z}_d & 0 \\ 0 & 0 & 0 & 0 & \mathbf{Z}_e \end{vmatrix}$$

where $[\mathbf{Z}_{br}]$ is the primitive unconnected impedance matrix or branch impedance matrix. A similar matrix may be obtained in terms of admittance. A matrix connecting the branch and loop currents has been obtained in the previous analysis of figure A2.4, it is called a connexion matrix and is denoted by $[C]$; $[\mathbf{I}_{branch}] = [C] [\mathbf{I}_{loop}]$ or $[\mathbf{I}] = [C] [\mathbf{i}]$. The loop voltages (v) and branch voltages (V) in the network of figure A2.5 are related as follows:

| | Branch a | b | c | d | e | |
|---|---|---|---|---|---|---|
| Loop (1) | 1 | −1 | | | | |
| (2) | | 1 | 1 | −1 | | $= [A]$ |
| (3) | | | | +1 | −1 | |

i.e. $[\mathbf{v}] = [A] [\mathbf{V}]$

where $\mathbf{v}$ = loop voltage

and $\quad \mathbf{V}$ = branch voltage

It is seen that $[A]$ is the transpose of $[C]$, i.e. $[A] = [C_t]$. The reason for this is as follows. In an unconnected primitive network the power, $P = V_a I_a + V_b I_b + V_c I_c + \ldots$ where the elements are resistive, $V_a$ and $I_a$ relate to element $R_a$ and so on.

$$\therefore \qquad P = \boxed{V_a\, V_b\, V_c} \cdot \begin{vmatrix} I_a \\ I_b \\ I_c \end{vmatrix} = [V_t]\,[I]$$

As power is independent of the ırame of reference, loop currents and voltages may be used instead of branch values

and $\qquad\qquad P = [v_t]\,[i] = [V_t]\,[I]$

from which $\qquad\quad [v_t]\,[i] = [V_t]\,[C]\,[i]$

hence, $\qquad\qquad [v_t] = [V_t]\,[C]$

and transposing both sides, $\quad [v] = [C_t]\,[V]$

thus, $\qquad\qquad [A] = [C_t]$

Again, complex power $S = VI^*$

$$[v_t]\,[i^*] = [V_t]\,[I^*] = [V_t]\,[C^*]\,[I^*]$$

$$\therefore \qquad\qquad [v_t] = [V_t]\,[C^*]$$

$$[v] = [C_t^*]\,[V] = [C_t^*]\,[Z_{br}]\,[I] = [C_t^*]\,[Z_{br}]\,[C]\,[i]$$

if $\qquad\qquad\qquad [v] = [Z]\,[i]$

then, $\qquad\qquad [Z] = [C_t^*]\,[Z_{br}]\,[C]$

where $[Z]$ is the mesh impedance matrix which was previously derived by inspection of the network.

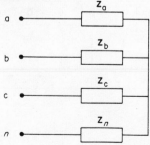

Figure A3.1    Unbalanced three-phase load.

## Application to unbalanced three-phase systems

Consider the arrangement in figure A3.1 in which mutual coupling (due to magnetic or capacitive coupling) exists between all the elements. The unconnected primitive relationships are given by

$$[V] = [Z_{br}] [I], \text{ i.e.}$$

$$
\begin{bmatrix} V'_a \\ V'_b \\ V'_c \\ V'_n \end{bmatrix}
=
\begin{bmatrix}
Z_a & Z_{ab} & Z_{ac} & Z_{an} \\
Z_{ba} & Z_b & Z_{bc} & Z_{bn} \\
Z_{ca} & Z_{cb} & Z_c & Z_{cn} \\
Z_{na} & Z_{nb} & Z_{nc} & Z_n
\end{bmatrix}
\begin{bmatrix} I'_a \\ I'_b \\ I'_c \\ I'_n \end{bmatrix}
$$

where $V'_a$, etc., are the voltage drops across the impedances. With no mutual coupling the matrix reduces to a more familar result.

$$
\begin{bmatrix}
Z_a & 0 & 0 & 0 \\
0 & Z_b & 0 & 0 \\
0 & 0 & Z_c & 0 \\
0 & 0 & 0 & Z_n
\end{bmatrix}
$$

The connexion between these primitive unconnected currents and the actual currents, $I_a$, $I_b$ and $I_c$ is given by the matrix $[C]$ i.e. $[I'] = [C] [I]$

and

$$
\begin{bmatrix} I'_a \\ I'_b \\ I'_c \\ I'_n \end{bmatrix}
=
\begin{bmatrix}
1 & 0 & 0 \\
0 & 1 & 0 \\
0 & 0 & 1 \\
-1 & -1 & -1
\end{bmatrix}
\begin{bmatrix} I_a \\ I_b \\ I_c \end{bmatrix}
$$

as $I'_n = -(I_a + I_b + \dot{I}_c)$ in a three-phase system, i.e.

$$
\begin{bmatrix} V'_a \\ V'_b \\ V'_c \\ V'_n \end{bmatrix}
=
\begin{bmatrix}
Z_a & Z_{ab} & Z_{ac} & Z_{an} \\
Z_{ba} & Z_b & Z_{bc} & Z_{bn} \\
Z_{ca} & Z_{cb} & Z_c & Z_{cn} \\
Z_{na} & Z_{nb} & Z_{nc} & Z_n
\end{bmatrix}
[C]
\begin{bmatrix} I_a \\ I_b \\ I_c \end{bmatrix}
$$

$$= \begin{vmatrix} Z_a \ - \ Z_{an} & Z_{ab} \ - \ Z_{an} & Z_{ac} \ - \ Z_{an} \\ Z_{ba} \ - \ Z_{bn} & Z_b \ - \ Z_{bn} & Z_{bc} \ - \ Z_{bn} \\ Z_{ca} \ - \ Z_{cn} & Z_{cb} \ - \ Z_{cn} & Z_c \ - \ Z_{cn} \\ Z_{na} \ - \ Z_n & Z_{nb} \ - \ Z_n & Z_{nc} \ - \ Z_n \end{vmatrix} \begin{vmatrix} I_a \\ I_b \\ I_c \end{vmatrix}$$

The loop voltage drops are related to the unconnected voltage drops $V'_a$, etc., by the matrix

Branches →

Loops ↓ $\begin{vmatrix} 1 & 0 & 0 & -1 \\ 0 & 1 & 0 & -1 \\ 0 & 0 & 1 & -1 \end{vmatrix}$ , i.e. $[v] = [A] [V']$

as $V'_n = -(V_a + V_b + V_c)$

Hence, total impedance voltage per loop $[v] = [Z] [I]$

$$= \begin{vmatrix} Z_a - Z_{an} - Z_{na} - Z_n & Z_{ab} + Z_{an} - Z_{nb} + Z_n & Z_{ac} - Z_{an} - Z_{nc} + Z_n \\ Z_{ba} - Z_{bn} - Z_{na} + Z_n & Z_b - Z_{bn} - Z_{nb} + Z_n & Z_{bc} - Z_{bn} - Z_{nc} + Z_n \\ Z_{ca} - Z_{cn} - Z_{na} + Z_n & Z_{cb} - Z_{cn} - Z_{nb} + Z_n & Z_c - Z_{cn} - Z_{nc} + Z_n \end{vmatrix} \cdot \begin{vmatrix} I_a \\ I_b \\ I_c \end{vmatrix}$$

In each loop the terminal voltage = source voltage $[E] - [v]$

The above equation may be simplified by the following assumptions which in practical systems are often reasonable.

$$Z_a = Z_b = Z_c,$$
$$Z_{ab} = Z_{bc} = Z_{ca},$$
$$Z_{ba} = Z_{cb} = Z_{ac}$$
$$Z_{an} = Z_{bn} = Z_{cn} = Z_{na} = Z_{nb} = Z_{nc}$$
but $Z_{ab} \neq Z_{ba}$ (in a rotating machine).

and $[Z]$ becomes,

$$\begin{vmatrix} Z_a + (Z_n - 2Z_{an}) & Z_{ab} + (Z_n - 2Z_{an}) & Z_{ba} + (Z_n - 2Z_{an}) \\ Z_{ab} + (Z_n - 2Z_{an}) & Z_a + (Z_n - 2Z_{an}) & Z_{ab} + (Z_n - 2Z_{an}) \\ Z_{ab} + (Z_n - 2Z_{an}) & Z_{ba} + (Z_n - 2Z_{an}) & Z_a + (Z_n - 2Z_{an}) \end{vmatrix}$$

Also if the impedances are balanced $Z_{ab} = Z_{ba}$

[Z] becomes
$$\begin{bmatrix} Z_p & Z_m & Z_m \\ Z_m & Z_p & Z_m \\ Z_m & Z_m & Z_p \end{bmatrix}$$

where,  $\qquad Z_p = Z_a + Z_n - 2Z_{an}$

and,  $\qquad Z_m = Z_{ab} + Z_n - 2Z_{an}$

$\qquad\qquad\quad = Z_{ba} + Z_n - 2Z_{an}$

If the network is unbalanced but free from mutual coupling,

[Z] becomes

| $Z_a + Z_n$ | $Z_n$ | $Z_n$ |
|---|---|---|
| $Z_n$ | $Z_b + Z_n$ | $Z_n$ |
| $Z_n$ | $Z_n$ | $Z_c + Z_n$ |

# Appendix IV

## Optimal System Operation

The limitations resulting from the 'B coefficient' method for economic load dispatching have been referred to in Chapter 3. A more general approach involving the use of non-linear programming has been reported by Carpentier[1] and Sirioux[2].

Non-linear programming classifies the method used to optimize non-linear problems. Normally it is required to minimize a function $f(x)$ subject to certain constraints, e.g.

$$\left. \begin{array}{l} g_i(x) \geqslant 0 \text{ when } i = 1 \ldots, m \\ h_j(x) = 0 \text{ when } j = 1 \ldots, p \end{array} \right\} \tag{1}$$

and

The simplest case resolves to the solution of the set of simultaneous equations

$$\frac{\partial f(x)}{\partial x} = 0$$

to obtain the minimum but the solution is often very impracticable. Several numerical methods are available for minimization without constraints especially gradient methods, e.g. steepest descents[3], parallel tangents, conjugate gradients[4,5] and the generalized Newton–Raphson[3]. With constrained minimization the classical approach is that of the Lagrange Multipliers[6]. $f(x)$ is minimized with $m = 0$ and $p > 0$.

The solution of the constraints, $h_j(x) = 0$ together with

$$\frac{\partial f}{\partial x} + \sum_{j=1} \lambda_j \frac{\partial h_j(x)}{\partial x} = 0$$

is required and this is often difficult. An extension of the Lagrange method by Kuhn and Tucker[7] deals with equation (1) with $p > 0$ and $m > 0$. This method requires the solution of the constraints, $h_j(x) = 0$, together with

$$\frac{\partial f}{\partial x} + \sum_{j=1}^{p} \lambda_j \frac{\partial h_j(x)}{\partial x} - \sum_{i=1}^{m} u_i \frac{\partial g_i(x)}{\partial x} = 0$$

and,

$$u_i g_i(x) = 0; \, i = 1 \ldots, m$$

$$u_i \geqslant 0; \, i = 1 \ldots, m$$

Again it may be difficult to solve these equations but there are non-linear programming numerical methods which can cope with the solution.

The solution of the power-system dispatching problem describes in fact an operating point for the power system. This may be considered the result of a standard load flow given optimal boundary conditions. The following equations describe the nature of the problem to be solved. No attempt to give a solution will be made as this is outside the scope of this text. It is hoped, however, that the reader will gain some appreciation of this important topic and with the references given be able to commence a detailed study if required.

The more general approach involving the use of non-linear programming reported by Carpentier[1] and Sirioux[2] is formulated as follows. Let $P_i$ and $Q_i$ be the active and reactive powers generated at the $i$th node of an $n$ node system, and $C_i$ and $D_i$ be the active and reactive powers consumed at the $i$th node.

The total cost, $F_t = f(P_i, Q_i)$ for $i = 1 \ldots, n$.

As for modern systems $\dfrac{\partial F_i}{\partial P_i} \gg \dfrac{\partial F_i}{\partial Q_i}$, $F_t = f(P_i)$, $i = 1$ to $n$.

The steady-state equations for a transmission link (see section 2.3) between nodes $i$ and $k$ are,

$$P_i - C_i = \Sigma_k \frac{V_i V_k}{Z_{ik}} \sin (\theta_i - \theta_k - \delta_{ik}) + \Sigma_k \frac{V_i^2 \sin \delta_{ik}}{Z_{ik}} \qquad (2)$$

and, $Q_i - D_i = -\Sigma_k \dfrac{V_i V_k}{Z_{ik}} \cos (\theta_i - \theta_k - \delta_{ik}) + \Sigma \dfrac{V_i^2 \cos \delta_{ik}}{Z_{ik}} - V_i V_i^2$

$$\ldots (3)$$

where, $i = 1 \ldots n$, $i \neq k$, $\delta_{ik}$ = angle across link = $\cos^{-1} \dfrac{R_{ik}}{Z_{ik}}$ and

$\theta_i$ and $\theta_k$ are the angles at nodes $i$ and $k$, respectively. Along with these two equations the following inequality constraints apply,

$$P_i^2 + Q_i^2 - {}_M S_i^2 \leqslant 0 \qquad (4)$$

$${}_m P_i - P_i \leqslant 0 \qquad (5)$$

$$Q_i - {}_M Q_i \leqslant 0 \qquad (6)$$

$${}_m Q_i - Q_i \leqslant 0 \qquad (7)$$

$$V_i - {}_M V_i \leqslant 0 \qquad (8)$$

$${}_m V_i - V_i \leqslant 0 \qquad (9)$$

where the prefixes $m$ and $M$ denote lower and upper limits, respectively, determined by MVA ratings, generator limits, transformer tap ranges and boiler limits. Also synchronous stability requires that

$$\theta_i - \theta_k - {}_M\theta_{ik} \leqslant 0 \tag{10}$$

The cost $F_t$ has to be minimized with respect to $P_i$, $Q_i$, $V_i$ and $\theta_i$ while satisfying equations (2) to (10). To obtain the solution the use of non-linear programming is used. When the system contains appreciable hydro-electric generation the problems become more complex.

**References**

1. Carpentier, J., 'Contribution a l'etude du dispatching economique', *Bulletin de la Societe Française des Electriciens*, Ser. 8, **3**, 1962.
2. Carpentier, J., and J. Sirious, 'L'optimization de la production a l'electricite de France', *Bulletin de la Societe Française des Electriciens*, March, 1963.
3. Householder, A. S., *Principles of Numerical Analysis*, McGraw-Hill, New York, 1953.
4. Powell, M. J. D., 'An efficient method for finding the minimum of a function of several variables without calculating derivatives', *Computer Journal*, **7**, 155, 1964.
5. Fletcher, R., and C. M. Reeves, 'Function minimization by conjugate gradients', *Computer Journal*, **7**, 149, 1964.
6. Hancock, H., *Theory of Maxima and Minima*, p. 100, Dover, New York, 1960.
7. Kuhn, H. W., and A. W. Tucker, 'Nonlinear programming', *Proc. 2nd Berkeley Symposium on Mathematical Statistics and Probability*, p. 481, 1951.

Table A5.2   United States. Approximate reactance values of three-phase 60-cycle generating equipment
(Values in per unit on rated kVA base)

| Apparatus | Positive sequence | | | | | | Negative sequence $X_2$ | | Zero sequence $X_0$ | |
| | Synchronous $X_d$ | | Transient $X_d'$ | | Subtransient $X_d''$ | | | | | |
| | Average | Range | Average | Range | Average | Range | Average | Range | Average | Range |
|---|---|---|---|---|---|---|---|---|---|---|
| 2-pole Turbine gen. (45 psig inner-cooled H₂) | 1·65 | 1·22–1·91 | 0·27 | 0·20–0·35 | 0·21 | 0·17–0·25 | 0·21 | 0·17–0·25 | 0·093 | 0·04–0·14 |
| 2-pole Turbine gen. (30 psig H₂ cooled) | 1·72 | 1·61–1·86 | 0·23 | 0·188–0·303 | 0·14 | 0·116–0·17 | 0·14 | 0·116–0·17 | 0·042 | 0·03–0·073 |
| 4-pole Turbine gen. (30 psig H₂ cooled) | 1·49 | 1·36–1·67 | 0·281 | 0·265–0·30 | 0·19 | 0·169–0·208 | 0·19 | 0·169–0·208 | 0·106 | 0·041–0·1825 |
| Salient pole gen. and motors—with dampers | 1·25 | 0·6–1·5 | 0·3 | 0·2–0·5 | 0·2 | 0·13–0·32 | 0·2 | 0·13–0·32 | 0·18 | 0·03–0·23 |
| Salient pole gen.—without dampers | 1·25 | 0·6–1·5 | 0·3 | 0·2–0·5 | 0·3 | 0·2–0·5 | 0·48 | 0·35–0·65 | 0·19 | 0·03–0·24 |
| Synchronous condensers air-cooled | 1·85 | 1·25–2·20 | 0·4 | 0·3–0·5 | 0·27 | 0·19–0·3 | 0·26 | 0·18–0·4 | 0·12 | 0·025–0·15 |
| Syn. cond.—H₂ cooled at ½ psig rating | 2·2 | 1·5–2·65 | 0·48 | 0·36–0·6 | 0·32 | 0·23–0·36 | 0·31 | 0·22–0·48 | 0·14 | 0·03–0·18 |

(Permission *Westinghouse Corp.*)

# Appendix V

Table A5.1  Typical percentage reactances of synchronous machines at 50 Hz—British

| Type and rating of machine | Positive sequence | | | Negative sequence $X_2$ | Zero sequence $X_0$ | Short-circuit ratio |
|---|---|---|---|---|---|---|
| | $X_{st}$ | $X_t$ | $X_s$ | | | |
| 11 kV Salient-pole alternator without dampers | 22·0 | 33·0 | 110 | 22·0 | 6·0 | — |
| 11·8 kV 60 MW 75 MVA Turbo-alternator | 12·5 | 17·5 | 201 | 13·5 | 6·7 | 0·55 |
| 11·8 kV 56 MW 70 MVA Gas-turbine turbo-alternator | 10·0 | 14·0 | 175 | 13·0 | 5·0 | 0·68 |
| 11·8 kV 70 MW 87·5 MVA Gas-turbine turbo-alternator | 14·0 | 19·0 | 195 | 16·0 | 7·5 | 0·55 |
| 13·8 kV 100 MW 125 MVA Turbo-alternator | 20·0 | 28·0 | 206 | 22·4 | 9·4 | 0·58 |
| 16·0 kV 275 MW 324 MVA Turbo-alternator | 16·0 | 21·5 | 260 | 18·0 | 6·0 | 0·40 |
| 18·5 kV 300 MW 353 MVA Turbo-alternator | 19·0 | 25·5 | 265 | 19·0 | 11·0 | 0·40 |
| 22 kV 500 MW 588 MVA Turbo-alternator | 20·5 | 28·0 | 255 | 20·0 | 6·0–12·0 | 0·40 |
| 23 kV 600 MW 776 MVA Turbo-alternator | 23·0 | 28·0 | 207 | 26·0 | 15·0 | 0·50 |

Table A5.3  Principal data of 200–500 MW turbogenerators—Russian

| | Units | Values of parameters of turbogenerators of various types | | | | | |
| --- | --- | --- | --- | --- | --- | --- | --- |
| | | 1 | 2 | 3 | 4 | 5 | 6 |
| Power | MW/MVA | 200/235 | 200/235 | 300/353 | 300/353 | 500/588 | 500/588 |
| Cooling of winding (stator) | | Water | Hydrogen | Water | Hydrogen | Water | Water |
| (rotor) | | Hydrogen[3] | Hydrogen | Hydrogen | Hydrogen | Hydrogen | Water |
| Rotor diameter | m | 1·075 | 1·075 | 1·075 | 1·120 | 1·125 | 1·120 |
| Rotor length | m | 4·35 | 5·10 | 6·1 | 5·80 | 6·35 | 6·20 |
| Total weight | kg/VA (N/kVA) | 0·93 (9·1) | 1·3 (12·7) | 0·98 (9·6) | 1·05 (10·3) | 0·64 (6·26) | 0·63 (6·17) |
| Rotor weight | kg/kVA (N/kVA) | 0·18 (1·76) | 0·205 (2·01) | 0·16 (1·57) | 0·158 (1·55) | 0·11 (1·08) | 0·1045 (1·025) |
| $X_d^1$ | % | 188·0 | 184·0 | 169·8 | 219·5 | 248·8 | 241·3 |
| $X_d'$ | % | 27·5 | 29·5 | 25·8 | 30·0 | 36·8 | 37·3 |
| $X_d''$ | % | 19·1 | 19·0 | 17·3 | 19·5 | 24·3 | 24·3 |
| $X_q$ | % | 188·0 | 184·0 | 169·8 | 219·5 | 248·8 | 241·3 |
| $X_q''$ | % | 28·6 | 28·5 | 26·0 | 29·2 | 36·0 | 36·0 |
| $X_0$ | % | 8·5 | 8·37 | 8·8 | 9·63 | 15·0 | 14·6 |
| $\tau_1^2$ | sec | 2·3 | 3·09 | 2·1 | 2·55 | 1·7 | 1·63 |

[1] For all reactances unsaturated values are given.

[2] Without the turbine.

[3] Hydrogen pressure for columns 1–4 is equal to 3 atm (304 000 N/m²) and for columns 5–4 atm (405 000 N/m²).

(C.I.G.R.E., 1968, Paper 11–07—paper by Glebov, I. A. et al.)

Table A5.4  Standard impedance limits for power transformers above 10,000 kVA (60 Hz)

| Highest Voltage winding: BIL kV | Low voltage winding: BIL kV  For intermediate BIL use value for next higher BIL listed | Self-cooled (OA), self-cooled rating of self-cooled/forced-air cooled (OA/FA), self-cooled rating of self-cooled/forced-air, forced-oil cooled (OA/FOA)  Standard impedance: Percent | | | | Forced-oil cooled (FOA and FOW)  Standard impedance: Percent | | | |
|---|---|---|---|---|---|---|---|---|---|
| | | Ungrounded neutral operation | | Grounded neutral operation | | Ungrounded neutral operation | | Grounded neutral operation | |
| | | Min. | Max. | Min. | Max. | Min. | Max. | Min. | Max. |
| 110 and below | 110 and below | 5·0 | 6·25 | | | 8·25 | 10·5 | | |
| 150 | 110 | 5·0 | 6·25 | | | 8·25 | 10·5 | | |
| 200 | 110 | 5·5 | 7·0 | | | 9·0 | 12·0 | | |
| | 150 | 5·75 | 7·5 | | | 9·75 | 12·75 | | |
| 250 | 150 | 5·75 | 7·5 | | | 9·5 | 12·75 | | |
| | 200 | 6·25 | 8·5 | | | 10·5 | 14·25 | | |
| 350 | 200 | 6·25 | 8·5 | | | 10·25 | 14·25 | | |
| | 250 | 6·75 | 9·5 | | | 11·25 | 15·75 | | |
| 450 | 200 | 6·75 | 9·5 | 6·0 | 8·75 | 11·25 | 15·75 | 10·5 | 14·5 |
| | 250 | 7·25 | 10·75 | 6·75 | 9·5 | 12·0 | 17·25 | 11·25 | 16·0 |
| | 350 | 7·75 | 11·75 | 7·0 | 10·25 | 12·75 | 18·0 | 12·0 | 17·25 |
| 550 | 200 | 7·25 | 10·75 | 6·5 | 9·75 | 12·0 | 18·0 | 10·75 | 16·5 |
| | 350 | 8·25 | 13·0 | 7·25 | 10·75 | 13·25 | 21·0 | 12·0 | 18·0 |
| | 450 | 8·5 | 13·5 | 7·75 | 11·75 | 14·0 | 22·5 | 12·75 | 19·5 |
| 650 | 200 | 7·75 | 11·75 | 7·0 | 10·75 | 12·75 | 19·5 | 11·75 | 18·0 |
| | 350 | 8·5 | 13·5 | 7·75 | 12·0 | 14·0 | 22·5 | 12·75 | 19·5 |
| | 450 | 9·25 | 14·0 | 8·5 | 13·5 | 15·25 | 24·5 | 14·0 | 22·5 |

Table A5.4  Standard impedance limits for power transformers above 10,000 kVA (60 Hz) (contd.)

| Highest Voltage winding: BIL kV | Low voltage winding: BIL kV — For intermediate BIL use value for next higher BIL listed | At kVA base equal to 55C rating of largest capacity winding | | | | | | | |
| --- | --- | --- | --- | --- | --- | --- | --- | --- | --- |
| | | Self-cooled (OA), self-cooled rating of self-cooled/forced-air cooled (OA/FA), self-cooled rating of self-cooled/forced-air, forced-oil cooled (OA/FOA) — Standard impedance: Percent | | | | Forced-oil cooled (FOA and FOW) — Standard impedance: Percent | | | |
| | | Ungrounded neutral operation | | Grounded neutral operation | | Ungrounded neutral operation | | Grounded neutral operation | |
| | | Min. | Max. | Min. | Max. | Min. | Max. | Min. | Max. |
| 750 | 250 | 8·0 | 12·75 | 7·5 | 11·5 | 13·5 | 21·25 | 12·5 | 19·25 |
| | 450 | 9·0 | 13·75 | 8·25 | 13·0 | 15·0 | 24·0 | 13·75 | 21·5 |
| | 650 | 10·25 | 15·0 | 9·25 | 14·0 | 16·5 | 25·0 | 15·0 | 24·0 |
| 825 | 250 | 8·5 | 13·5 | 7·75 | 12·0 | 14·25 | 22·5 | 13·0 | 20·0 |
| | 450 | 9·5 | 14·25 | 8·75 | 13·5 | 15·75 | 24·0 | 14·5 | 22·25 |
| | 650 | 10·75 | 15·75 | 9·75 | 15·0 | 17·25 | 26·25 | 15·75 | 24·0 |
| 900 | 250 | | | 8·25 | 12·5 | | | 13·75 | 21·0 |
| | 450 | | | 9·25 | 14·0 | | | 15·25 | 23·5 |
| | 750 | | | 10·25 | 15·0 | | | 16·5 | 25·5 |
| 1050 | 250 | | | 8·75 | 13·5 | | | 14·75 | 22·0 |
| | 550 | | | 10·0 | 15·0 | | | 16·75 | 25·0 |
| | 825 | | | 11·0 | 16·5 | | | 18·25 | 27·5 |
| 1175 | 250 | | | 9·25 | 14·0 | | | 15·5 | 23·0 |
| | 550 | | | 10·5 | 15·75 | | | 17·5 | 25·5 |
| | 900 | | | 12·0 | 17·5 | | | 19·5 | 29·0 |
| 1300 | 250 | | | 9·75 | 14·5 | | | 16·25 | 24·0 |
| | 550 | | | 11·25 | 17·0 | | | 18·75 | 27·0 |
| | 1050 | | | 12·5 | 18·25 | | | 20·75 | 30·5 |

(Permission *Westinghouse Corp.*)

Table A5.5   Overhead line parameters—50 Hz

| Parameter | 275 kV | | 400 kV | |
|---|---|---|---|---|
| | $2 \times 0.175$ in.$^2$ | $2 \times 0.4$ in.$^2$ | $2 \times 0.4$ in.$^2$ | $4 \times 0.4$ in.$^2$ |
| $Z_1$ $\Omega$/ml | $0.143+j0.508$ | $0.063+j0.51$ | $0.063+j0.527$ | $0.032+j0.446$ |
| $Z_0$ $\Omega$/ml | $0.326+j1.39$ | $0.222+j1.38$ | $0.234+j1.38$ | $0.167+j1.27$ |
| $Z_{mo}$ $\Omega$/ml | $0.183+j0.78$ | $0.174+j0.74$ | $0.173+j0.72$ | $0.136+j0.68$ |
| $Z_p$ $\Omega$/ml | $0.204+j0.801$ | $0.116+j0.801$ | $0.120+j0.811$ | $0.077+j0.721$ |
| $Z_{pp}$ $\Omega$/ml | $0.061+j0.293$ | $0.053+j0.291$ | $0.057+j0.284$ | $0.045+j0.275$ |
| $B_1$ $\mu$mho/ml | 5·77 | 5·85 | 5·65 | 6·56 |
| $B_0$ $\mu$mho/ml | 3·20 | 3·30 | 3·31 | 3·71 |
| $B_{mo}$ $\mu$mho/ml | 9·5 | 11·2 | 12·4 | 13·6 |

$Z_1$ Positive-sequence impedance.
$Z_0$ Zero-sequence impedance of a D.C.1. line.
$Z_{mo}$ Zero-sequence mutual impedance between circuits.
$Z_p$ Self-impedance of one phase with earth return.
$Z_{pp}$ Mutual-impedance between phases with earth return.
$B_1$ Positive-sequence susceptance.
$B_0$ Zero-sequence susceptance of a D.C.1. line.
$B_{mo}$ Zero-sequence mutual susceptance between circuits.

Note: A D.C.1. line refers to one circuit of a double circuit line in which the other circuit is open at both ends.

## Table A5.6 Overhead line data—a.c. (60 Hz) and d.c. lines

| Region | 500–550 kV—a.c. | | 700–750 kV | | HVDC |
|---|---|---|---|---|---|
| **Region** | Pacific | Canada | Canada | Mountain | Pacific |
| **Utility** | So. California Edison Co. | Ontario Hydro | Quebec Hydroelectric Com. | U.S. bureau of reclamation | Los Angeles Dept. of W.& P. |
| Line name or number | Lugo-Eldorado | Pinard-Hanmer | Manics.-Levis | Oregon-Mead | Dalles-Sylmar |
| Voltage (nominal), kV; a.c. or d.c. | 500 a.c. | 500; a.c. | 735.; a.c. | 750 (±375); d.c. | 750 (±375); d.c. |
| Length of line, miles | 176 | 228 | 235;·92 | 560 | 560 |
| Originates at | Lugo sub | Pinard TS | Manicouagan | Oregon border | Oregon border |
| Terminates at | Eldorado sub | Hanmer TS | Levis sub | Mead substa. | near, Los Angeles |
| Year of construction | 1967–69 | 1961–63 | 1964–65 | 1966–70 | 1969 |
| Normal rating/cct, MVA | 1000 | — | 1700 | 1350 mW | 1350 mW |
| **STRUCTURES** | | | | | |
| (S)teel, (W)ood, (A)lum | S | S, A | S | S | S or A |
| Average number/mile | 4;21 | 3·7 | 3·8 | 4·5 | 4·5 |
| Type: (S)q, (H), guy (V) or (Y) | S (S-56) | V (A-51, S-51) | —; (S-71) | S, T (S-72, 73) | S, T (S-72, 73) |
| Min 60-cps flashover, kv | 850 | 1110 | | | 915 |
| Number circuits: Initial; Ultimate | 1; 1 | 1; 1 | 1; 1 | 1; 1 | 1; 1 |
| Crossarms (S)teel, (W)ood, (A)lum | | S, A | S | S | S or A |
| Bracing (S)teel, (W)ood, (A)lum | | | S | S | S or A |
| Average weight/structure, lb. | 19,515 | 10,800; 4730 | 67,3000 | 11,000 | — |
| Insulation in guys (W)ood, (P)orc; kV | No guys | Nil | Nil | None | — |
| **CONDUCTORS** | | | | | |
| Al, ACSR, ACAR, AAAC, 5005 | ACSR 84/19 | ACSR 18/7 | ACSR 42/7 | ACSR 96/19 | ACSR 84/19 |
| Diameter, in.; MCM | 1·762; 2156 | 0·9; 583 | —; 1361 | —; 1857 | 1·82, 2300 |
| Weight 1 cond./ft, lb | 2·511 | 0·615 | 1·468 | 2·957 | 2·68 |
| Number/phase; Bundle spacing, in. | 2; 18 | 4; 18 | 4; 18 | 2; 16 | 2; 18 |
| Spacer (R)igid, (S)pring, (P)reform | S | R | R | S or P | S or A |
| Designed for—amp/phase | 2400 | 2500 | | 1800 | 3380 |
| Ph. config. (V)ertical, (H)orizontal, (T)riangular | H | H | H | H | H |
| Cond. offset from vertical, ft | | | | | |
| Phase separation, ft | 32 | 40 | 50 | 38–41 | 39–40 |
| Span: Normal, ft; Maximum, ft | 1500; 2764 | 1400; 2550 | 1400; 2800 | 1150; 1800 | 1175; — |
| Final sag, ft; at F; Tension, 10³ lb | 46 at 130F; 19·1 | 45·0 at 60F; 3·4 | 226 at 120F; 6·2 | 40 at 120F; 12·4 | 35 at 60F; 13·0 |
| Minimum clearance: Ground, ft; Structure, ft | 40; 12·5 | 33; 10·5 | 45; 18·3 | 35; 7·75 | 35; 7·8 |
| Type armor at clamps | Performed | Nil | None | — | None |
| Type vibration dampers | Stockbridge | Spacer damper | Tuned spacer | — | Stockbridge |
| Line altitude range, ft | 1105–4896 | 800–1300 | 50–2500 | 2000–7000 | 400–8000 |
| Max corona loss, kW/3-ph mile | 5 | — | 74 (½ in. snow/hr) | — | — |

| | 500–550 kV—a.c. | | 700–750 kV | HVDC | |
|---|---|---|---|---|---|
| Region | Pacific | Canada | Canada | Mountain | Pacific |
| Utility | So. California Edison Co. | Ontario Hydro | Quebec Hydroelectric Com. | U.S. bureau of reclamation | Los Angeles Dept. of W.& P. |
| **Insulation** | | | | | |
| Basic impulse level, kV | 2080 | 1800 | — | — | — |
| Tangent: (D)isk or (L)ine-post | D | D | D | D | D |
| Number strings in (V) or (P)arallel | 2V | 1 or 2P | 4V | Single | Single |
| Number units; Size; Strength, $10^3$ lb | 27; $5\frac{3}{4}\times 10$; 30 | 23; $5\frac{3}{4}\times 10$; 25 | 35; $5\times 10$; 15 | —; —; — | —; —; — |
| Angle: (D)isk or (L)ine-post | D | D | D | D | D |
| Number strings in (V) or (P)arallel | 2V | 3P | 4P | Single | — |
| Number units; Size; Strength, $10^3$ lb | 25; $6\frac{1}{4}\times 10\frac{3}{8}$; 40 | 26; $5\frac{3}{4}\times 10$; 25 | 35; $6\frac{1}{4}\times 10$; 36 | —; —; — | —; —; — |
| Insulators for struts | None | None | — | — | None |
| Terminations: (D)isk or (L)ine-post | D | D | — | D | D |
| Number strings in (P)arallel | 2P | 3P | — | — | 4P |
| Number units; Size; Strength, $10^3$ lb | 25; $6\frac{1}{4}\times 10\frac{3}{8}$; 40 | 26; $5\frac{3}{4}\times 10$; 25 | —; —; — | —; —; — | —; —; — |
| BILreduction, kV or steps | 3·5 steps | — | — | — | — |
| BIL of term. apparatus, kV | 1150–1425 | 1675–1900 | 2050–2200 | — | 1300 |
| **Protection** | | | | | |
| Number shield wires; Metal and Size | 2; 7 No. 6 AW | 2; 5/16 in. galv. | 2; 7/16 in. galv. | 1; — | 2; — |
| Shield angle, deg; Span clear, ft | 24·5; 28 | 20; 48 | 20; 50 | 30; — | 20 ;— |
| Ground resistance range, ohms | — | 15 | — | — | 30 max |
| Counterpoise: (L)inear, (C)rowfoot | C | L | L | — | — |
| Neutral gdg: (S)ol, (T)sf; Eq. ohms | S; 0 | S; — | S; — | S; — | S; — |
| Arrester rating, kV; Horn gap, in. | 420; None | 480/432; None | 636; — | —; — | —; — |
| Arc ring diameter, in.: Top; Bottom | None | None | —; — | —; — | —; — |
| Expected outages/100mi/year | 0·5 | 1 | — | — | 0·2 |
| Type relaying | Ph. compar., carr. | Dir'l compar. | — | — | — |
| Breaker time, cyc; Reclos, cyc | 2; None | 3; None | 6; 24 | —; — | —; — |
| **Voltage Regulation** | | | | | |
| Nominal, % | ±10 | — | — | Variable | Grid control |
| Synch. cond. MVA; Spacing, miles | None | None | —; — | —; — | None |
| Shunt capac. MVA; Spacing, miles | None | None | 330; 236 | —; — | On a.c. terminals |
| Series capac. MVA; Compens., % | None | None | None | —; — | —; — |
| Shunt reactor, MVA; Spacing, mi | —; — | — | — | — | — |
| **Communication** | | | | | |
| (T)elephone circuits on structures | None | None | None | None | None |
| (C)arrier; (F)requency, kc | C; — | C; 50–98 | C; 42·0 and 50·0 | None | None |
| (P)ilot wire | None | — | P | None | None |
| (M)icrowave; Frequency, Mc | M; 6700 | —; — | M; 7777·35 | M; — | — |

(Permission *Edison Electric Institute*)

# Appendix VI

**Transmission Study–190**

*(Information reproduced by permission of the United States Department of the Interior)*

This is a reconnaissance study conducted by the United States Bureau of Reclamation, Bonneville Power Administration, and Southwestern Power Administration to investigate e.h.v. interconnexions between the utilities of the North Western and Central areas of the United States of America. In particular these interconnexions would carry *seasonal* peak diversity interchanges between the Northwest and the Central States. Also a limited consideration is made of the magnitude of the time-zone diversity between these regions. In figure A6.1 the seasonal load diversity expected by 1980 is shown for the areas under consideration which are designated 1, N, C and S, respectively.

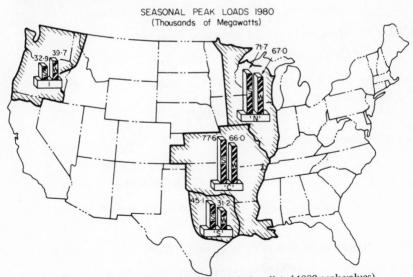

SEASONAL PEAK LOADS 1980
(Thousands of Megawatts)

Figure A6.1 Loads in areas under study (predicted 1980 peak values).
S = summer, W = winter.

*Apportioning non-committed resource additions in two areas*

The supply and need for the two annual peak-load periods in an area are balanced with the lesser seasonal load diversity and also the need of the combined areas at the season of greater load is considered. Consider areas 1 and C for the summer and winter of 1980.

Let
$R_1$ = resource addition required in area 1 (MW),
$R_C$ = resource addition required in area C (MW),
$P_s$ = sending-end value of interchange power,

(applies to '(1) end' for summer and '(C) end' for winter),

$P_r$ = receiving end value of interchange power,

('(1) end' in winter and '(C) end' in summer).

| Summer 1980 | Area 1 | Area C | Total |
|---|---|---|---|
| Resource (MW) | 27,571 | 48,538 | 76,109 |
| Need (MW) | 32,947 | 78,311 | 111,258 |
| Surplus or deficit | −5376 | −29,773 | −35,149 |

Balancing the supply and need for area (1) in the summer of 1980,

$$27{,}571 + R_1 = 32{,}947 + P_s$$

For the same area in the winter,

$$(\text{Resource}) \; 28{,}086 + R_1 + P_R = 39{,}707 \; (\text{Need})$$

Balancing supply and need for combined areas for the summer,

$$R_1 + R_C = 35{,}149 + P_s - P_r$$

Also assuming a 10 per cent loss in interchange power at maximum seasonal exchanges,

$$P_s - P_r = 0 \cdot 1 \, (P_s + P_r)/2$$

Solving these four equations yields,

$$R_1 = 8655 \text{ MW}, \; R_C = 26{,}806 \text{ MW}$$
$$P_s = 3278 \text{ MW}, \; P_r = 2966 \text{ MW}$$

$$\left(\frac{P_s + P_r}{2}\right) = \text{average interchange power} = 3122 \text{ MW}$$

and the transmission loss = $P_s - P_r = 312$ MW.

This indicates a 3000 MW seasonal interchange from east to west in winter and west to east in summer.

Similar analysis indicates the following equal and opposite seasonal interchanges:

Areas (1) and (N),   2500 MW N → 1 in winter
                     and 1 → N in summer

Areas (1) and (C),   3100 MW C → 1 in winter
                     and 1 → C in summer

Areas (1) and (S),   3100 MW S → 1 in winter
                     and 1 → S in summer

The minimum requirement for transmission capacity between areas will result from equal and opposite seasonal exchanges.

*Transmission system (a.c.)*

The above data indicates a firm power transfer of the order of 3000 to 4000 MW with a distance between end terminals of over 1500 miles (2400 km). In view of this a.c. systems operating at 700 kV (nominal), 765 kV (maximum) are proposed. All lines have 70 per cent series capacitor compensation with the capacitors rated to cover an outage of one circuit. The quad-bundle conductors proposed reduce the line inductive reactance by 35 per cent compared with the single conductor line. Connexions are made every 300 to 500 miles and the angle across a single line section is 8 to 12°. The total angle across the whole system is 40° for 3000 MW.

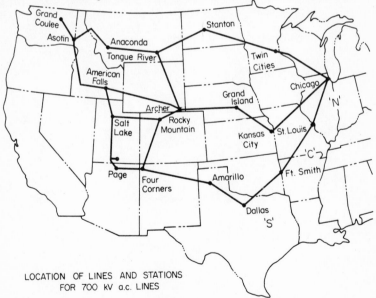

LOCATION OF LINES AND STATIONS
FOR 700 kV a.c. LINES

Figure A6.2   Possible 700 kV a.c. network for interchanges.

The general requirements would be:

> 4300 circuit miles, 20,000 towers,
> 80,000 insulation strings (35 units in each),
> 43 series capacitor stations (22,540 MVAr),
> 45 shunt reactors at 700 kV (13,500 MVAr),
> 80 shunt reactors at 15 kV (4000 MVAr),
> 83,700 kV circuit breakers and finally transformers 700/500
> kV 8, 700/345 kV 5, 700/230 kV 1.

A geographical outline of the a.c. scheme is shown in figure A6.2.

*Transmission system (d.c.)*

A system operating at 1200 kV (± 600 kV) is proposed. Firm capability is ensured by appropriate switching to transfer full power on to the remaining lines in the event of an outage. The general routing of the d.c. system (untapped) is similar to that for a.c. shown in figure A6.2. An economic analysis shows the d.c. scheme with no tappings to be cheaper than a tapped d.c. one, both d.c. schemes are cheaper than the a.c. proposals.

# Subject Index